Broadband Fixed Wireless Networks

ABOUT THE AUTHOR

Neil P. Reid is a Solutions Engineer at Cisco Systems with worldwide responsibilities for systems engineers engaged in IP fixed wireless solutions and for defining fixed wireless product requirements for the service provider market. He has more than 20 years of experience in high technology, is well recognized as an authority in the field, and is a speaker at numerous national and international symposiums on wireless technologies. He lives in Texas and when not fully engaged in technology, enjoys time with his wife and two children and the pursuit of the perfect bar-b-qued steak.

ABOUT THE TECHNICAL EDITORS

Mitch Taylor is a Systems Engineer Manager for Cisco System's Service Provider Line of Business. Prior to this role, he was part of the team that introduced Cisco's first broadband fixed wireless solution. Mr. Taylor has over 18 years of experience in RF communications systems. He has an in-depth knowledge of wireless modulation techniques and network design. Before joining Cisco, he marketed custom microwave test solutions at Hewlett-Packard (now Agilent Technologies), involving advanced signal simulation for the communication markets.

Mike Tibodeau is a Solutions Engineer for Cisco Systems. For the past five years, he has supported enterprise and service providers with core and access network architectures. Currently he focuses on developing products, technologies, and architectures for metropolitan-area and multi-tenant unit networks.

Mike holds a Bachelor's degree in Systems Engineering from the University of Virginia and a Master's degree in Systems Engineering and Management, concentrating on telecommunications.

In his free time he enjoys scuba diving, biking, traveling, and watching movies.

Broadband Fixed Wireless Networks

NEIL P. **REID**

McGraw-Hill/Osborne
New York Chicago San Francisco
Lisbon London Madrid Mexico City Milan
New Delhi San Juan Seoul Singapore Sydney Toronto

McGraw-Hill/Osborne
2600 Tenth Street
Berkeley, California 94710
U.S.A.

To arrange bulk purchase discounts for sales promotions, premiums, or fund-raisers, please contact **McGraw-Hill**/Osborne at the above address. For information on translations or book distributors outside the U.S.A., please see the International Contact Information page immediately following the index of this book.

Broadband Fixed Wireless Networks

1234567890 CUS CUS 01987654321

ISBN 0-07-213366-X

Publisher
Brandon A. Nordin
Vice President & Associate Publisher
Scott Rogers
Editorial Director
Tracy Dunkelberger
Acquisitions Editor
Steven Elliot
Senior Project Editor
Carolyn Welch
Acquisitions Coordinator
Alex Corona
Technical Editors
Mitch Taylor
Mike Tibodeau
Copy Editor
Bill McManus
Proofreader
Nancy McLaughlin
Indexer
Valerie Perry
Computer Designers
Michelle Galicia, Lucie Ericksen,
Melinda Moore Lytle
Illustrators
Michael Mueller, Lyssa Wald
Series Design
Peter F. Hancik

This book was composed with Corel VENTURA™ Publisher.

To my son Martin and daughter Kelly—you will never know how much I love you until you have your own children. This is for you.

AT A GLANCE

Part I	Network and Broadband Fixed Wireless Fundamentals	
▼ 1	Selected Network Fundamentals	3
▼ 2	Types of Wireless	37
▼ 3	Selected RF Fundamentals	55
Part II	**Business and Market Strategies for BBFW Providers**	
▼ 4	BBFW Versus Other Broadband Methodologies	79
▼ 5	Primary Uses for BBFW	91
▼ 6	Customer System Selection	105
Part III	**Deployment and Implementation**	
▼ 7	Deployment: The Make-or-Break Proposition	129

Part IV Wireless Technologies

▼ 8 Modulation Schemes 155
▼ 9 System Performance Metrics 175
▼ 10 Differences Between Headend and CPE Gear . . 191
▼ 11 BBFW Security 205
▼ 12 Comparison of Unlicensed vs. Licensed BBFW 229
▼ 13 802.11 Wireless Local Area Networks 249

Part V Comparison of Technologies

▼ 14 Frequency Selection Issues 287

Part VI Standards

▼ 15 WLAN Standards Comparison 303
▼ 16 Technology Standards 311

Part VII Government and Regulatory Issues

▼ 17 The FCC 331
▼ 18 Licensing 347

Part VIII airBand Networks: An Interview

▼ 19 Deploying a Network: An Interview with airBand Networks of Dallas, TX 363

▼ Appendix 371

▼ Glossary 375

▼ Index 397

CONTENTS

Acknowledgments xvii
Introduction xix

Part I

Network and Broadband Fixed Wireless Fundamentals

▼ 1 Selected Network Fundamentals 3
The Most Commonly Used BBFW Terms and Acronyms 4
Network and Broadband Fixed Wireless Fundamentals 11
- OSI Reference Stack 12
- Transmission Control Protocol/Internet Protocol (TCP/IP) . 14
- Asynchronous Transfer Mode (ATM) 21
- Layer 3 BBFW Solution Versus Layer 2 26
- Operational System Support (OSS) Selected Fundamentals . . 28
- Elements of a Total Network Solution 30

▼ 2 Types of Wireless 37
Wireless Technology Today 38
Residential vs. Commercial Systems 40
Typical BBFW System Block Diagram 45

IP BBFW Services . . . 48
Increase Bandwidth . . . 48
Decrease Monthly Charge . . . 48
Provide Added Services . . . 48
Point-to-Multipoint Architecture . . . 49
Supercell vs. Microcell Networks . . . 52

▼ 3 Selected RF Fundamentals . . . 55

Frequency . . . 58
Power Transmission . . . 59
Frequency Bands . . . 62
Modulation Schemes . . . 63
Multi-Path . . . 67
OFDM . . . 68
WLAN Spread Spectrum . . . 70
Typical BBFW PT-PT RF System Block Diagram . . . 71
Router or Switch . . . 71
Outdoor Unit (ODU) . . . 72
Miscellaneous Cables, Connectors, and Antenna Mast . . . 72
Antenna . . . 73
Earth Curvature Calculation for Line-Of-Sight Systems . . . 74

Part II

Business and Market Strategies for BBFW Providers

▼ 4 BBFW Versus Other Broadband Methodologies . . . 79

Broadband Comparisons . . . 80
Fiber . . . 81
DSL . . . 83
Cable . . . 84
BBFW . . . 84
The Pros and Cons of BBFW . . . 84
BBFW Pros . . . 85
BBFW Cons . . . 86

▼ 5 Primary Uses for BBFW . . . 91

IP BBFW Services . . . 92
VPNs . . . 92
Content Data Networks (CDNs) . . . 96
Differentiated Services . . . 100
Service Delivery Architecture . . . 101
Wireless Local Area Networks (WLANs) . . . 102

▼ 6 Customer System Selection . . . 105
Types of Customers . . . 106
ILECs . . . 106
CLECs . . . 109
IXCs . . . 111
ISPs . . . 111
BLECs . . . 112
Customer Engagement Questions . . . 113
The Service Providers Target Market and User Benefits . . . 114
Five Key Winning Service Provider Strategies . . . 122
Secure Pent-Up Demand by Deploying in Geographies That Lack Broadband Service . . . 123
Know What Service to Deploy First . . . 123
Prioritize BBFW Targets . . . 123
Deploy a Scalable Network That Can Readily Migrate to QoS-Sensitive Services . . . 124
Make an Informed Decision on Unlicensed Radios . . . 124
Value Proposition for BBFW . . . 125
SMB Access Offerings . . . 126

Part III
Deployment and Implementation

▼ 7 Deployment: The Make-or-Break Proposition . . . 129
The Deployment Phase . . . 130
Preliminary Site Evaluation . . . 132
RF Site Survey . . . 132
Cable Routing, Customizing for Lengths and Connectors, and Selection for Minimal Loss . . . 134
Mast or Tower Construction . . . 134
Lightning Protection Systems . . . 134
Mechanical Assembly . . . 135
Construction . . . 136
Antenna Selection, Pointing, and Weatherproofing . . . 136
Final Assembly, Test, and Documentation . . . 137
Selection, Integration, Configuration, and Documentation of Routers and Switches . . . 137
Provision of Emergency Spares at Local Depots . . . 138
Negotiation of Right-of-Way Access (Rooftop or Tower Leases) . . . 139
Selection of Deployment Partner . . . 139
FCC Certification (for Installing Licensed Frequencies) . . . 140

Financial Solvency 140
Local Environmental Experience 141
Depth of Resources 141
Exceptional Logistics and Project Management 141
Established Relationships with Subcontractors 142
Local Permits 142
Finance 142
Systems Integration 143
Forward-Going Risk Profile Assessment 143
WLAN (802.11) Deployment Issues 144
In-Building Design Considerations 144
Sample Installations 148

Part IV

Wireless Technologies

▼ 8 Modulation Schemes 155
Modulation Schemes 156
Multipath and OFDM 160
WLAN Spread Spectrum 168
Duplexing Techniques 168
Error Control Schemes 170

▼ 9 System Performance Metrics 175
RF Equipment Selection 176
Link Margin Management 177
Link Metrics 178
References to Link Performance 185

▼ 10 Differences Between Headend and CPE Gear 191
Point-to-Point Architecture 192
Point-to-Multipoint Architecture 193
Radios 195
Antennas 195
Routers/Switches 197
Slave Status 198
Asymmetric Data Flows 198
Cost 199
Use 200
Detailed CPE Registration Process 201

▼ 11 BBFW Security 205

Threats to Security Ever Increasing 206
BBFW Link Access As Hacking Medium 208
Denial of Service 208
Theft of Information 209
Theft of Service 209
Use of Networks As Stooges 209
Challenges of Penetrating a BBFW Link 210
Extensive RF Hardware and Software 211
Extensive RF Expertise 211
State-of-the-Art Schemes Such As OFDDM, 64 QAM, and so on Are Very Difficult to Crack . . 212
State-of-the-Art BBFW Link Security 212
Wireline Access As a Hacking Medium 216
Unix-Based Hacker Software 217
Viruses 219
The Four Types of Viruses 220
Web Browser Vulnerabilities 222
Security Conclusions 226
Additional Reading 227

▼ 12 Comparison of Unlicensed vs. Licensed BBFW 229

Advantages of Unlicensed Spectrum 230
Sharing of Spectrum 234
Mobility of Wireless Applications 235
Efficient Use of Spectrum Through Intermittent Use 235
Innovation and Research for New Fixed Wireless Devices and Applications 235
High Potential for Interference, and Coexistence with Other U-NII Networks 236
100 Percent Availability? 238
High Pre-Equipment Investment? 240
Guarantee from Interference 240
Guarantee from Link Degradation 240
Restricted Footprint 240
Layer 3 Options 241
Recourse for Co-Channel Interference 241
In-Band Market Competition 242
U-NII As Final Mile for Fiber 242
The Fiber Provider Can't Access Large Buildings Fast Enough 243
The BBFW Service Provider Often Requires a Large Backhaul Pipe 243

The Tenants in the Large Building Wish to Bypass the Local Telco Exchange 243
U-NII for Extension of DSL and Cable 244
Venture Capital Aspects of Unlicensed vs. Licensed Spectrum . . 244
BBFW Is Still an Emerging Broadband Medium 246
They View a Broadcast License As an Asset 247
Resolution of Co-Channel Interference 247
Competition from Numerous U-NII Regional Service Providers 247
Underestimating the Value of Management Execution Over Frequency Selection 247

▼ 13 802.11 Wireless Local Area Networks 249

Wireless Local Area Networks: The Final 400 Feet 250
Broadband Mobility 250
Pervasive Computing 251
WLAN Radios 252
Outdoor WLAN Radios 253
Data versus Distances 255
Indoor WLAN Radios and the Five Elements of WLAN Communication 256
Infrared 802.11 258
Common Uses for a WLAN 258
Indoor versus Outdoor WLAN 261
Most Common Modulation Schemes for 802.11 267
802.11 Site Surveys 272
Typical Site Survey Steps 276
Sample Deployments 280
802.11 WLAN Summary: A Compelling Technology and Standard 284

Part V

Comparison of Technologies

▼ 14 Frequency Selection Issues 287

Laying the Foundation for the Comparison 288
Technology Is Easy—Business Is Hard 288
The Primary Business Issues 289
Decision Criteria for Selecting the Most Appropriate Frequency 292
Decision Table: Which BBFW Technology Has the Advantage? 298
The Final Analysis 299

Part VI

Standards

▼ 15 WLAN Standards Comparison 303
Competing WLAN Standards 304
The 802.11 Standard 306
The HomeRF Standard 308
The Bluetooth Standard 309

▼ 16 Technology Standards 311
Globalization of Standards 312
Early Technology Adopters and Standards 314
Technological Maturity 315
Stability of Basic Design 315
Interoperability 315
How BBFW and Other Standards Are Developed 316
Introduction of New Technology 316
Relatively High Interest by Developers 317
Deployment of Technology to Early Adopters 317
Standards Definition by One or More Technology Providers . . 318
Establishment of the Standard by a Standards Body 318
Ratification of the Standard by Technology Providers 319
BBFW Standards Bodies and Supporting Bodies 319
Institute of Electrical and Electronics Engineers 320
Three Key BBFW Entities That Support IEEE Standards . . . 325

Part VII

Government and Regulatory Issues

▼ 17 The FCC 331
FCC Organization and Bureaus 332
FCC Staff Offices 333
The Wireless Telecommunications Bureau 334
Unlicensed Systems 335
Universal Licensing System (ULS) 336
Auctions 338
Deployment Requirements 345

▼ 18 Licensing 347
U.S. BBFW Licenses 348
Maximum Power Output 351

The Most Common U.S. BBFW Licenses 353
Licensing Summary 358

Part VIII

airBand Networks: An Interview

▼ 19 Deploying a Network: An Interview with airBand Networks of Dallas, TX . . 363

▼ Appendix 371

▼ Glossary 375

▼ Index 397

ACKNOWLEDGMENTS

If it takes a village to raise a child, it surely takes a neighborhood block to write a book. The truth about writing a technology book is that a lot of the content requires research by the author and verification by colleagues who are the true experts.

With this in mind, the following folks were not only key assets in the writing of this book but should be recognized as truly world class professionals. I am also proud and grateful to say they are my friends.

Mitch Taylor—your professionalism, technical prowess in BBFW, and wonderful sense of humor is always a delight and gratefully accepted. Your imprint on this work is indelible. An honorary Texan if ever I met one.

Chris Lewis—a genuinely successful author with both technical and non-technical books and articles on the market. One of the best IP people on the planet and an esteemed colleague. Thanks for the introduction to the great people at McGraw-Hill/Osborne.

Mike Tibodeau—one of the best systems engineers in the industry. If it's network related, he's been there and done that. Many thanks for the friendly but critical eye and more than a few laughs along the way.

Dale Roznowski—one of the rising stars at Cisco Systems and a good friend with a superb and rare technical intellect. Glad you got a new Mule.

Bruce Alexander and Ron Seide—my main go-to guys on 802.11 issues. Thanks so much to you both for the technical edit on Chapter 13. Two of the best 802.11 people to be found anywhere.

Steven Elliot, Tracy Dunkelberger, Alexander Corona, Carolyn Welch, and Bill McManus of McGraw-Hill/Osborne—you folks know your business! Simply great people personally and professionally. You were nothing if not patient, accommodating, skilled, gracious, and a pleasure to work with.

John Nuno—my manager. It's very difficult not to admire someone who at the time of this writing has more than 11 years at Cisco Systems. Legendary within the organization for his leadership and unsurpassed dedication to the company, John is one of the three most important people in my life with regard to my career and professional objectives.

Saving the best for last—my wonderful wife Mary. The best person I know in a storm and even better in good weather. You are and always will be the best friend, mentor, comedian, and navigator I've ever had. None of this was possible without your support. Not even Job's wife had to live with an author.

INTRODUCTION

The world of broadband fixed wireless (BBFW) is a marvelously interesting and relevant technical subject in this day and age. Having stated that, the subject is rife with mystery and lore. I was tempted therefore to title this book *The Secret Life of Broadband Fixed Wireless*, but in the end decided on something more polished.

Even the most experienced people in this industry have a generally limited scope of expertise; in other words, their expertise tends to be very deep in one or two subject areas and not very deep in any of the other areas. One of the key objectives of this book is to shed light on the full scope of key issues that are important but not well known or understood.

Indeed, the scope of issues in BBFW is enormous and include technical, economic, regulatory, deployment, scaling and migration items, among others. Because each of these areas requires a deep level of expertise, there is no single repository for all of this information that I'm aware of, even though each of these areas is mission critical to understand. The rationale for this book became evident when I realized the BBFW information I've been stockpiling for years on a wide array of subjects was something not a lot of people have.

It's my hope that two of the key values of this book will be to provide hard-to-find information and also to provide you with a single reference guide that covers a wide array of issues. The third value I'm attempting to provide you is an array of information in a manner that is easy to understand.

Best of good luck to you in your experience in this industry. I hope you'll take the time to provide feedback on this book—it would be great to hear from you.

Neil P. Reid

Network and Broadband Fixed Wireless Fundamentals

CHAPTER 1

Selected Network Fundamentals

Having taught technology to many hundreds of people over many years, it's my experience that if readers are new to a technology, one of the first things they need is possession of the most basic building blocks, which in this case are the most commonly used terms and acronyms. The next item they need are the subject fundamentals, which then lead up to the new principles being received. In this case, readers must have a basic understanding of how the Internet works.

While the Internet is surprisingly complex and no single person could possibly understand every nuance of the entire system, the most essential parts of its architecture can be readily learned with a bit of time, diligence, effort, and patience. Having a working knowledge of how the Internet works will provide the context for how and where broadband fixed wireless systems fit into this ecosystem of hardware and software. That's where this book will come in.

Be patient; go over the material enough times until it makes sense to you and then press forward to the next level. Remember that at one point even the most legendary Internet guru knew less about the Internet than you do now. Don't be intimidated by the scope of the entire Internet and hang on the old saw about eating an elephant one bite at a time. The Internet is one of the most fascinating inventions that has had some of the biggest impact on humanity since the invention of the wheel. Enjoy!

THE MOST COMMONLY USED BBFW TERMS AND ACRONYMS

The following list of terms was selected relative to their frequency of use when discussing broadband fixed wireless (BBFW) interests. This list should not be considered comprehensive with respect to the fixed wireless industry in general. Further, some of these acronyms remain in a state of flux and should therefore be considered accurate per the date of this document.

Adjacent Channel is a channel or frequency that is directly above or below a specific channel or frequency.

Amplitude is the magnitude or strength of a varying waveform.

Analog Signal is the representation of information with a continuously variable physical quantity such as voltage. Because of this constant changing of the wave shape with regard to its passing a given point in time or space, an analog signal may have an infinite number of states or values. This contrasts with a digital signal, which has a very limited number of discrete states.

Antenna is a device for transmitting or receiving a radio frequency (RF). Antennas are usually designed for specific and relatively tightly defined frequencies and are quite varied in design. As an example, an antenna for a 2.5 GHz (MMDS) system will generally not work for a 28 GHz (LMDS) design.

Antenna Gain is the measure of an antenna *assembly* performance relative to a theoretically perfect antenna called an isotropic radiator ("radiator" is another term for antenna). Certain antenna designs feature higher performance relative to radiating a specific area or with regard to frequencies.

Bandwidth is the frequency range necessary to convey a signal measured in units of hertz (Hz). For example, voice signals typically require approximately 7 KHz of bandwidth, and data traffic typically requires approximately 50 KHz of bandwidth, but this depends greatly on modulation scheme, data rates, and how many channels of a radio spectrum are used.

BTA (Basic Trading Area) is the geographical area frequently used by the FCC for assigning licensed frequencies. BTAs are typically contiguous counties or trading areas and were first described by the Rand McNally mapping company. Rand McNally eventually licensed these area descriptions to the FCC.

BBFW (Broadband Fixed Wireless) is one of the most commonly used terms in the fixed wireless industry. In general, it implies data transfers in excess of 1.5 Mbps, or having the capacity to do so.

Broadband In general a data system is deemed *broadband* if it has a constant data rate at or in excess of 1.5 Mbps. Its corresponding opposite is *narrowband.*

Broadcast In general, *broadcast* means the opposite of *narrowcast* and implies that a signal is sent to many points at the same time and/or is transmitted in an omnidirectional pattern. In the radio world, "broadcast" is a term of art; that is, it has a special meaning for radio. A "broadcast" signal is one that's intended for reception by the general public.

Carrier Frequency is the frequency of a transmitted signal that would be transmitted if it were not modulated. Some BBFW systems also have intermediate frequencies that reside between the indoor equipment and the outdoor equipment. Carrier "frequency" can be either a single frequency or a range of frequencies carried at one time between the transmitter and the receiver.

CDMA (Code Division Multiple Access) is a transmission scheme that allows multiple users to share the same RF range of frequencies. In effect, the system divides a small range of frequencies out of a larger set and divides the data transmission among them. The transmitting device divides the data among a preselected set of nonsequential frequencies. The receiver then collates the various data "pieces" from the disparate frequencies into a coherent data stream. As part of the RF system setup, the receiver components are "advised" of the scrambled order of the incoming frequencies.

An important aspect of this scheme is that the receiver system filters out any signal other than the ones specified for a given transmission.

Channel is a communications path.

Coaxial Cable is the type of cable used to connect BBFW equipment to antennas and indoor/outdoor gear. Coaxial cable usually consists of a center wire surrounded by a metal shield with an insulator separating the two. The "axis" of the cable is located down the center of the cable. "Coaxial" means that there is more than one conductor oriented around a common axis for the length of the cable. Coaxial cable, or "coax," is one of the primary means for transporting cable TV and radio signals.

Converter is also referred to as an "up/down converter" or "transverter." Some RF systems have two fundamental frequencies, one which is sent over the air (carrier frequency) and the other of which is sent back and forth between the indoor equipment and the outdoor equipment (intermediate frequency). The intermediate frequencies are converted to and from the carrier frequency by this converter.

dB (Decibel) is a unit for expressing a ratio of power or voltage in terms of gain or loss. Units are expressed logarithmically and typically expressed in watts. A dB is not an absolute value; rather, it is the measure of power loss or gain between two devices. For example, –3dB indicates a 50 percent loss in power, and a +3dB reading represents a doubling of power. The rule of thumb to remember is that 10 dB indicates an increase (or loss) by a factor of 10; 20 dB indicates an increase (or loss) of a factor of 100, 30 dB indicates an increase (or loss) by a factor of 1000. Gain or loss is expressed with a "+" or "–" sign before the number. As antennas and other RF devices/systems commonly have power gains or losses of up to even four orders of magnitude, "dB" is a more easily used expression.

dBi Values for decibels of antenna gain referenced to the gain of an isotropic antenna (hence the *i*) are stated in dBi. An isotropic antenna is a theoretical antenna that radiates with perfect symmetry in all three dimensions. Real-world antennas have radiation patterns that are far from symmetric, but this effect is generally used to the advantage by the system designer.

dBm represents decibels of power referenced to a milliwatt; 0 dBm is 1 mW.

dBW represents decibels of power referenced to 1 watt.

Demodulator is the part of a receiver used for assembling signals from the radio into a format usable by the network or device attached to the radio. The corresponding device on the transmission side of a system is a *modulator.*

EIRP (Effective Isotropic Radiated Power) is a term that expresses the performance of a transmitting system in a given direction. EIRP is the power that a system using an isotropic antenna would use to send the same amount of power in a given direction that a system with a directional antenna uses. EIRP is usually expressed in watts or dBW. EIRP is the sum of the power at the antenna input plus antenna gain in dBi.

Electromagnetic Spectrum is the full range of electromagnetic (same as magnetic) frequencies, a subset of which is used in commercial RF systems. Commercial RF systems are typically classified as microwave (3–30 GHz) and millimeter wave, which is approximately 30 GHz and above. Military and other RF systems may also use VLF (Very Low Frequency, 3-30 kHz) and LF (Low Frequency, 30-300 kHz) frequency ranges.

FDM (Frequency Division Multiplexing) is the modulation scheme that divides the total available spectrum into subsets, which are commonly used in parallel across one or more links.

Fixed Wireless is the type of wireless in which both the transmitter and receiver are nonmobile. BBFW wireless is always broadband wireless, i.e., capable of data rates in excess of 1.5 Mbps, although the links can be throttled to data rates below that, typically not less than 256 Kpbs downstream and 128 Kbps upstream.

Footprint is the geographical area covered by a transmitter.

Frequency Reuse is one of the fundamental concepts on which commercial wireless systems are based; it involves partitioning an RF radiating area (cell) into segments of a cell, which for BBFW purposes means the cell can be broken into up to 13 or more equal segments. Notably, most RF cells are segmented into either three or four segments. One segment of the cell uses a frequency that is far enough away from the frequency in the bordering segment that it does not create interference problems.

Frequency reuse in mobile cellular systems means that each cell has a frequency that is far enough away from the frequency in the bordering cell that it does not create interference problems. Identical frequencies are used at least two cells apart from each other. This practice enables cellular providers to have many more customers for a given site license.

Fresnel Zones are theoretically ellipsoid-shaped volumes that reside in the space between transmitting and receiving antennas. The industry rule of thumb for line-of-sight links is to leave 60 percent of the centermost part of the first Fresnel zone free from obstruction. There are many Fresnel zones within an RF link, and they are often referred to as the "first Fresnel zone," "second Fresnel zone," and so on as the area referred to becomes farther from the center of the theoretically straight line between the transmitter and receiver.

Gain For an amplifier, "gain" is the ratio of the output amplitude of a signal to the input amplitude of a signal. This ratio is typically expressed in decibels. For an antenna, "gain" is the ratio of its directivity in a given direction compared to some reference antenna. The higher the gain, the more directional the antenna pattern.

IS-IS (Intermediate System-to-Intermediate System) OSI link-state hierarchical routing protocol based on DECnet Phase V routing, whereby ISs (routers) exchange routing information based on a single metric to determine network topology.

License is the purchased right to transmit RF waves over a given BTA on certain frequencies for a certain period of time. The license tightly governs the design parameters of an RF system and its use. Licenses are usually issued in a way that ensures a greatly reduced probability of interference from other users of the same spectrum. Depending on the licensed service and the country in which the license is issued, the license may be issued as the result of an auction, or as the result of a "beauty contest" in which the regulator evaluates the merits of proposals to use the spectrum.

The theory behind auctions is that they use free market forces to ensure that the spectrum is put to its best use, as well as to set requirements for spectral efficiency, which is another way of stating efficient use of the RF spectrum.

LMDS (Local Multipoint Distribution Service) is a relatively low-power license for transmitting voice, video, and data. In the United States, typically, two licenses are granted in three frequencies each to separate entities within a BTA. These licenses are known as Block A or Block B licenses. In the U.S., Block A licenses are from 27.5 to 28.35 GHz, 29.10 to 29.25 GHz and 31.075 to 31.225 GHz for a total of 1.159 GHz of bandwidth. Block B licenses operate from 31.00 to 31.075 GHz and 31.225 to 31.300 GHz for a total of 150 MHz of bandwidth. LMDS systems have a typical maximum transmission range of approximately 3 miles as opposed to the transmission range of an MMDS system, which is typically 25 miles. This difference in range is primarily a function of absorption due to precipitation and other physics phenomena as well as FCC allocated output power limits.

LOS (Line of Sight) refers to the fact that there must be a clear, unobstructed path between a transmitter and a receiver. This is essential for millimeter wave products such as LMDS and most microwave products lacking modulation and other schemes specifically designed to overcome the effects of a partially occluded (blocked) beam path. Having an LOS path as opposed to a partially obstructed data path enhances general performance in every RF deployment. The opposite to LOS is NLOS, or Non Line of Sight (also referred to as Near Line of Sight).

MMDS (Multichannel Multipoint Distribution Service) is a licensed service in the U.S. The FCC has allocated two bands of frequencies to this service, which are 2.15 to 2.161 GHz and 2.5 to 2.686 GHz. Licenses have been assigned by BTA.

Mobile Wireless is the type of wireless utilized in mobile phones, PDAs, pagers, and other small portable, battery-powered devices that can either transmit and or receive information by radio.

NLOS (Non–Line of Sight) Also known as an obstructed RF path or pathway, this condition is often also referred to as "near line of sight". See "Fresnel Zone" and "LOS."

Oversubscription is the method of having more users on a network than the network could accommodate if all the users used the network simultaneously. What makes this work is the premise that rarely if ever do all users actually use the network at the same time. Oversubscription is mission critical to the financial models used by Internet Service Providers (ISPs) and other entities, and in many cases, oversubscription is what keeps an entity solvent. Oversubscription rates can be anywhere from a factor of 6 to a factor of 50 or more depending on the class of service the subscriber has agreed to and other factors, including how much bandwidth the subscribers use.

Parabolic Antenna is a dish-like antenna that sends and receives radio waves in a highly focused manner. Such antennas provide very large antenna gains and are highly efficient. This antenna is typical to most point-to-point RF systems but is not the only design available or appropriate for a given RF link. The primary task of an antenna is to provide gain (signal boost) and to radiate in particular directions in accordance with the network's intended use, i.e., point-to-multipoint or point-to-point, or to cover a prescribed geographic area.

Passband represents the range of frequencies that a radio allows to pass from its input to its output. If a receiver or transmitter uses filters with narrow passbands, then only the desired frequency and close-by frequencies are of concern to the system designer. If a receiver or transmitter uses filters with wide passbands, then many more frequencies in the vicinity of desired frequency are of concern to the system designer. In a frequency division multiplexing (FDM) system, the transmit and receive passbands will be different. In a time division multiplexing (TDM) system, the transmit and receive passbands are the same.

Path Loss is the power loss that occurs when RF waves are transmitted through the air is called *path loss.* This loss occurs because RF waves expand as they travel through the air and the receiver antenna only captures a small portion of the total radiated energy. In addition, a significant amount of energy may be absorbed by molecules in the atmosphere or by precipitation when the carrier frequencies are above 10 GHz. The amount of absorption due to precipitation depends on the amount of precipitation and is usually a factor only for systems that operate at frequencies above 10 GHz.

The amount of atmospheric absorption depends greatly on the particular frequency used. At 12 GHz, water vapor absorbs a great deal of energy, and at 60 GHz oxygen molecules absorb even more energy. Systems that operate at those frequencies have very limited ranges.

PTM (Point-to-Multipoint) This term, the more correct acronym for which is pt-mpt, generally refers to the communication between a group of sites with interfaces to a single hub site. A pt-mpt connection is commonly set up in three or four segments to a cell to enable frequency reuse, but such a system can be designed for as many as a dozen or more segments within a single cell.

PTP (Point-to-Point) Point-to-point systems (the more correct acronym is P2P) provide communication between just two endpoints. In the U.S., point-to-point systems are typically found in the ISM, U-NII, and LMDS bands.

QoS (Quality of Service) is a feature of certain networking protocols that treats different types of network traffic differently to ensure required levels of reliability and latency according to type of traffic. Certain kinds of traffic such as voice and video are more sensitive to transmission delays and are therefore given priority over data that is less sensitive to delay.

As an example, on the Cisco Systems pt-mpt BBFW systems there are traditionally 4 levels of QoS, but some systems have as many as 13 levels of QoS depending on how many bits are used to prioritize the traffic. Most systems use either 3 or 4 levels of QoS and are commonly referred to as *Unsolicited Grant Service (USG), CBR* (*Committed Bit Rate*—sometimes referred to as *CIR* or *Committed Information Rate*), and *BER (Best Effort Rate)*. USG has priority over CIR/CBR, which has priority over BER. QoS levels are set in Layer 2 (the data link layer) of the OSI reference stack.

RF (Radio Frequency) generally refers to wireless communications with frequencies below 300 GHz. The term RF is commonly used to cover all types of wireless.

TDMA (Time Division Multiple Access) is a technique for splitting transmissions on a common frequency into time slots, enabling a greater number of users to use a given frequency. It is now more commonly used than CDMA in BBFW systems and is used in state of the art systems with frequency division multiplexing (FDMA) in one data direction and TDMA in the opposite direction

Truck Roll is the concept of "rolling" trucks to the installation site in order to install, repair, or upgrade equipment.

U-NII (Unlicensed National Information Infrastructure) primarily refers to a U.S. frequency band. The wireless products for this are in the 5.725 GHz to 5.825 GHz frequency range for outdoor use. There are two other U-NII bands that are used indoors; 5.25 GHz to 5.35 GHz and 5.5 GHz to 5.6 GHz. The indoor U-NII frequencies are transmitted at lower power levels than the 5.725-5.825 GHz frequencies.

These frequencies do not require the use or purchase of a site license, but the gear does require certification by the FCC and strict compliance to its regulations. The U-NII is a term coined by federal regulators to describe access to an information network by citizens and businesses. Equivalent to the term "information superhighway," it does not describe system architecture, protocol, or topology.

Wireless Access Protocol is a language used for writing Web pages that uses far less overhead than HTML or XML, a fact that makes it preferable for low-bandwidth wireless access to the Internet from devices with small viewing screens such as PDAs and cellular phones. WAP's corresponding OS is that created by 3Com for their Palm Pilot. Nokia has recently adopted the Palm OS for their Web-capable cellular phone.

WAP is based on XML, which is Extensible Markup Language. XML dictates *how* data is shown, while HTML determines *where* data is located within a browser page.

Wireline is the use of copper phone lines, cable lines, or fiber. Wireline advantages include high reliability, high tolerance to interference, and, generally, ease of troubleshooting. In the case of fiber, wireline also has exceptionally high bandwidth. Wireline is the technological opposite of wireless.

NETWORK AND BROADBAND FIXED WIRELESS FUNDAMENTALS

Welcome to the world of broadband fixed wireless! This book will assist you in your journey as a networking professional, as a prospective customer, or as a person considering a career in information technology. You may be relatively new to the industry, and if so, this book is intended to guide you through the intricacies of broadband fixed wireless (BBFW) systems; you may also appreciate that it leads off with a primer on information networks in general.

Acronyms, while often the bane of those new to a technology field, are the express lane for verbal and written communication to those who are more experienced in the industry. Where technical terms are used, you'll find a clear and concise explanation as well as some context to assist you in your learning curve.

If you are a grizzled network system engineer or other technology professional, this work will provide context and architectural information on one of the newest and least understood access mediums. If you are an account manager or individual with strategic responsibility in your organization such as a CIO, you'll also find context within this book on issues such as market drivers, BBFW competitive strategies, and a working knowledge of common BBFW architectures and deployment challenges.

If you are currently an Internet professional, you know from your experience that finding concise and real-world information on technology is very much a game of "Find Waldo." You know what you are looking for is out there, but it's not easy to find. This book is a summation of the author's extensive library of BBFW and Internetworking information, the preponderance of which was assembled as a solutions engineer at Cisco Systems' World Wide Service Provider Technical Operations.

It should be noted that this work is not limited to the solutions provided by Cisco but is an industry-wide "best of the best" that the author has collected in the course of more than ten years of wireless industry experience. You'll find this book reflective of the state-of-the-art systems being offered at the time of publication.

It would be very difficult at the least to be highly technically fluent in all five of the primary network elements: core, aggregation, access, customer premises equipment, and network management systems. Each of these disparate elements is enormously complex, and many people enjoy successful and rewarding careers by focusing on only a small portion of the total network solution.

Increasingly, the most valuable persons on a network team or company are those who are generalists, as they commonly have the best view of a network from an "end to end" perspective. Having possession of the network in its entirety gives one a view to the two largest issues facing companies that provide networking equipment: accommodating unpredictable data traffic growth and the integration of disparate network elements.

While there is an enormous influx of information technology professionals, very few have either the time or the bravado to attempt to understand a network in its entirety. For that reason, the industry has found itself highly populated with individuals who are deep in their technical expertise but not very broad. John Chambers, CEO of Cisco Systems, has said, "There are two great equalizers today, education and the Internet." This book will help broaden your technical horizons to include one of the most intense, fast-paced, and economically rewarding work experiences available today: building the BBFW portion of the Internet.

OSI Reference Stack

State-of-the-art communications networks are based on the fundamentals expressed in the Open Systems Interconnection (OSI) Reference Model (sometimes called OSI reference stack). While for years the OSI model has been considered archaic, it has proved to be quite useful with the resurgence in *Intermediate System-to-Intermediate Systems* (IS-IS).

The OSI stack enables a layered network architecture. The process of sending e-mail or other forms of data between two points is a highly sophisticated process.

The function of a network is to transmit voice, video, and/or data between servers. While electronics communication is fundamentally the science of converting information into ones and zeros, which are then converted to waves sent across various mediums like cables, fiber, or BBFW links, a considerable number of tasks are involved.

Above all else, the sending and receiving ends of a communications link need to be symmetric in terms of the various protocols used and required. Each end of the link must be synchronized with regard to the following tasks:

- Format the information into commonly understood signals (modulation)
- Package the voice, video, and/or data
- Determine the route the data will take
- Regulate the rate of data transfer
- Transmit the data
- Receive the data

- Assemble the data from packets, cells, or frames
- Ensure all of the data has been received
- Reassemble missing data or have the data resent
- Place the data "pieces" in the correct order
- Notify the sender that the transmission was successful
- Make the data available to the correct application

The OSI reference model is made up of seven layers, but it is important to note that they are not always used in all networks. In other words, in virtually all networks, each of the two ends of a data link will need to negotiate each of the seven layers as the mechanisms which enable the communication. TCP/IP is an architecture has fewer layers than the traditional OSI stack; however, all the seven functions are accommodated.

The seven layers are:

- 7. Application
- 6. Presentation
- 5. Session
- 4. Transport
- 3. Network
- 2. Data Link (sometimes referred to as the Link Layer)
- 1. Physical

The layers just indicated are not mistakenly numbered in reverse but are often referred to in the industry by their number as opposed to their name. The easiest way to remember all seven layers is to use the mnemonic "Please Do Not Throw Sausage Pizza Away." There are other mnemonics associated with the stack, but this is more commonly used than others.

Both of the communications devices at each end of the link have identical sets of these layers, and each of the layers communicates only to the corresponding layer on the opposite end of the link. This is called the "peer process," and each of the peers operates independently of the others. The peers also primarily communicate with each other.

The OSI stack or model is shown in Table 1-1.

Each of the layers provides *services* to the layer above it. For example, the network layers delivers packets of data across the network at the request of the transport layer. The exception is layer 1, as it defines the physical and electrical characteristics, i.e., what the connector plug looks like and what constitutes a one or a zero. Layer 2 assembles these ones and zeros into frames and hands them on up the reference stack.

OSI Layer	Function	Units of Data
7. Application	Not an application like Word, Excel, or Eudora, this layer interfaces Word (etc.) to network resources via Telnet, SMTP	Messages or files
6. Presentation	Performs syntax translations between computers of different types	Messages or files
5. Session	Initiates and maintains the dialog between two communicating computers	Messages or files
4. Transport	Transports data between two devices at a specified quality of service	Messages or files
3. Network	Routes data through the network and manages congestion, i.e., maintains "flow control"	Packets
2. Data Link	Transmits data from one node to another and controls the connection between routers. Controls access to the medium (copper, fiber, wireless, etc.) allowing multiple systems to use a common medium	Frames
1. Physical	Consists of the "physical" interface and hardware. Turns information into electrical impulses or other signals. In BBFW, this OSI layer is where the radios exist	Bits

Table 1-1. OSI Reference Stack Layers

Originally the OSI stack was intended to become a standard, which it has, but commercial systems commonly do not conform exactly to this model but use various modified forms of it (i.e., TCP/IP). However, a working knowledge of the OSI model is important, as it's a common tool for understanding networking principles. An understanding of the OSI model provides the foundation for understanding TCP/IP.

Transmission Control Protocol/Internet Protocol (TCP/IP)

TCP/IP is a deep and highly sophisticated subject, well worthy of a book-length description, and indeed many books exist on this subject, a good example of which is *Cisco TCP/IP Routing Professional Reference* by Chris Lewis (McGraw Hill, 1998). As this book is focused on BBFW, TCP/IP will not be dealt with to the depth it deserves but will be reviewed by way of summary.

TCP protocols are also divided into layers, although it has four as compared to the seven in the OSI model.

The four TCP/IP layers are shown in Table 1-2.

The TCP/IP stack does not provide all of the protocols of the OSI model, nor do they neatly correspond to the OSI model. The Internet and transport layers of the TCP/IP model provide approximately the same functions as the network layer and the session and transport layers, respectively, in the OSI model, but the TCP/IP's link layer is different than the OSI model. An approximation between the two models is shown as follows:

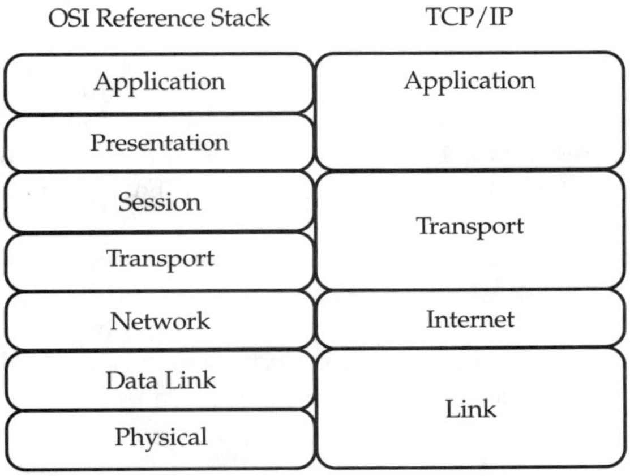

OSI Reference Stack	TCP/IP
Application	Application
Presentation	
Session	Transport
Transport	
Network	Internet
Data Link	Link
Physical	

Layer	Function
4. Application	Some TCP/IP application layers are full-fledged applications that generate network service requests, such as Telnet, FTP, Domain Name Server (DNS), and routing protocols.
3. Transport	Transports data between two devices at a specified quality of service in the form of two protocols, TCP and the User Datagram Protocol (UDP)
2. Internet	Contains the IP protocol, responsible for packaging, addressing, and routing data
1. Link	Contains protocols that provide an interface between TCP/IP stacks and the physical layer of the network

Table 1-2. TCP/IP Layers

The link layer in TCP/IP is special because it's designed for use on a wide range of hardware and network types. In the OSI model, the data link and physical layers respectively control access to the network and the actual signaling. On today's LANs, these are almost always Ethernet or token ring protocols.

Ethernet and token ring protocols are not part of the TCP/IP model and were developed by a different standards body. These protocols are implemented by the network interface hardware and device drivers installed into their respective operating systems. The TCP/IP protocols do not define data and physical protocols; rather, TCP/IP is designed to work with various protocols like Ethernet and token ring. This provides important functionality to various parts of the OSI Data Link layer but doesn't replace it.

Data Encapsulation

The control information is also commonly referred to as *overhead.*

Data encapsulation is the process by which each protocol in the OSI or TCP/IP (etc.) stack applies its own control data. Data encapsulation adds protocols as it moves "down" the stack, which is to say from Layer 7 towards Layer 1. Each protocol adds a new header to the payload.

The definition of "payload" depends on where the information is at any one time. When the data from the application (such as Word or Excel, etc.) has the Application application (Layer 7 of the OSI stack), the "payload" is the combination of the two. As the payload moves down through the stack, the payload is the control data plus the "Word" etc. information. The illustration shown here demonstrates this fact.

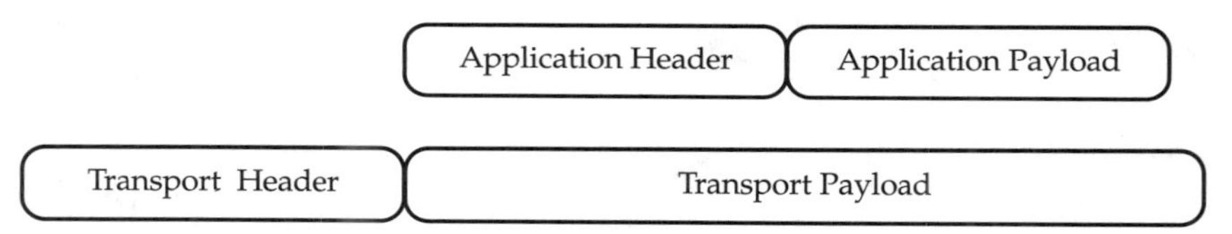

We can see from this illustration that as each protocol in the stack adds its header to the proceeding layer, it identifies the payload as its own.

A key concept then comes into play: following the addition of all the protocol layers, the entire structure is then called a *packet.*

The term "packet" can mean more than one thing in the Internetworking industry, though incorrectly. In IP networks, the term "packet" is used generically when referring to the units transmitted across packet switching networks, when in fact the fundamental unit sent across IP networks is actually a *frame* (in most cases). Having said that, I must note that numerous disagreements remain about what "packets" and "frames" indicate. One well-publicized version is that "frame" and "packet" mean the same thing and that historically, LAN people spoke of packets and WAN people spoke of frames, however when one uses IP they both become the same thing, just using different Layer 2 encapsulations for the LAN or WAN media.

An additional concept is that everything at the IP layer is a packet. Anything going to layer 2 and up through the stack is a frame, a cell, etc. Thus, you have an Ethernet frame in an IP network, but at no point do you not have a packet in IP.

Each header contains information specifically designed for the corresponding stack layer of the receiving system or device. The headers are stripped off as the packet is passed "up" the stack of the receiving system or device ("up" means from Layer 1 to Layer 7 or as appropriate). Specifically, the transport layer strips off the transport header, and so forth. The process of adding headers is called *multiplexing*, while the removal of headers is called *demultiplexing*.

The final header/protocol occurs at the data link layer. This protocol is typically either Ethernet or token ring in a LAN (but could be SONET, ATM, or the like in a WAN). It is the only protocol that adds a trailer (a form of a header located at the tail end of the payload as opposed to the front end of the transmission). The trailer is used to verify that the information sent was received with a high degree of accuracy.

Every network transmission requires a certain amount of control data in addition to the payload. There are networks where the control planes and data planes are one and the same, but these are very simple networks with very small payloads. These networks are typically used in monitoring and control networks such as those used by alarm systems and industrial monitoring and control (on/off, high/low, near/far, hot/cold, etc.).

There are a number of ways to add control data to a payload; the general premise is to have a comprehensive data control capability while maintaining the least amount of overhead. Limiting overhead is important because the larger the overhead-to-payload ratio, the smaller the throughput. A common method for resolving this is to separate the control data in their own packets and move them independently from the data between routers.

At the top of the protocol stack (Layer 7, application), the data encapsulated consists only of the original request for transmission by the application (e.g., Word, Excel, Eudora). This payload/protocol combination is then consecutively passed on down the protocol stack to Layer 1, which is where the data is sent by the device controlling the medium, e.g., the radio for BBFW, or a cable modem for a cable transmission, etc.

Addressing

One of the most important functions of the information in the protocol headers is to direct the transmitted data to an exact address. The importance of this task grows clear when one realizes that the Internet spans the entire globe with literally millions of exit and entrance points, which are often referred to as "nodes."

Similar in manner to the way a common letter is addressed with references to countries, states/provinces, towns, streets, houses, and then occupants, the TCP/IP addressing scheme is specified by the information contained in the headers, which are provided by successive protocol layers.

However, in order for information to arrive at its correct destination, the transmission process involved goes well beyond simple address location, as the routers and other network devices commonly process an array of protocols. To accommodate this, each layer of the stack must provide a sign to point the way through these protocols. Each header contains a field specifically designed to accomplish this purpose.

To add further complexity to the network, some payloads are simply too large to be sent contiguously. When this occurs, the payload is split, or *fragmented*, into smaller segments. That the IP protocol is capable of doing this is important, as some or many networks limit the size of the payload due to various necessary constraints and data transmission priorities.

Also, high-priority data loads can be split up and routed along numerous paths before arriving at the destination. This capability is enormously important, as networks need to manage unpredictable growth rates. Simply put, this architecture can provide a large number of new routes to accommodate unpredictable growth rates.

The segments are each given a number, and this number is stored in the data link protocol header and transmitted with the data. The fragments are transmitted individually and commonly take different routes to the destination.

Following assembly of the segments, the transmission is checked for accuracy. This is called *error detection* and *error correction.* Error detection is accomplished in a number of different ways, but essentially it involves having the sending system or device store a number in the protocol header that represents the size and order in which the segments are sent. Error correction is accomplished in a number of ways, ranging from having the data resent from the point of origin, to interpolating missing or corrupt packets (see Chapter 9).

One other thing to note is that in a frame relay (FR) network, error correction does not happen here. For efficiency, FR only detects errors and relies upon the upper layers of the OSI or TCP/IP stack to perform error correction.

However, not all Layer 4 protocols do error detection or correction, and it is important here to differentiate between error detection and error correction. Layer 2 protocols like Ethernet do error detection by throwing away damaged packets. Layer 4 protocols can do error correction by re-requesting damaged or missing frames. Some schemes used in BBFW will interpolate for missing packets and accurately reconstruct data on this premise (see Chapter 9).

Not only can segments be out of order or lost in transmission, but information can also be damaged, or "corrupt," within each segment. Accordingly, additional protection or insurance is provided through more thorough checks such as the *cyclical redundancy check (CRC)* used in Ethernet. CRC is delivered in the Ethernet header and is highly accurate in terms of ensuring segment or packet integrity.

Error correction is also accomplished through various means but again is fundamentally similar from technique to technique. The data, which is always in ones and zeros, can be replaced either by having the entire segment or packet resent from the originating source or by having the ones or zeros replaced by interpolation. Interpolation is generally used only by very sophisticated modulation schemes such as OFDM and is the process by which the missing or corrupted data is "repaired" through a statistical analysis.

IP Address Registration

Because any computer on the globe can send data to any other computer on the Internet, it is vitally important that every computer have its own IP address. IP addresses are assigned by one of two committees that keep track of the addresses on a worldwide basis. Network managers can add what are called *subnets* to the registered IP address, but the network at large does not "see" the subnet until it crosses into a specific network, i.e., the network owned and operated by a specific business or organization such as a school.

IP addresses have three classes, which are dependent on the size of the business registering the address to be used. Most businesses register as a class C organization, as they are small to medium in size and they can register up to 254 computers or routers. Large business will more likely use a class A or B. At this writing, all that remains available are class C addresses, but there are also class D addresses for multicast and even a class E, though it is undefined at this writing.

A class A license can have up to 16 million nodes, which means it's useful for Internet service providers (ISPs). Examples of ISPs include companies like Earthlink, AOL, and AT&T Cable. At the inception of this class, there were less than 128 of these addresses.

Comparing IP to ATM

The relative merits of IP and ATM are heavily debated, with ILECs generally in favor of ATM networks, and CLECs, ISPs and many equipment providers, most notably Cisco Systems, favoring the IP protocol.

ATM versus IP is often referred to as comparing "clouds vs strings," the analogy being that ATM networks require a dedicated circuit whereas IP networks use unique circuits for each transmission and indeed multiple and often changing circuits for a single transmission.

To better understand these two protocols, a brief review of their legacy is in order. ATM, rapidly adopted by the ILECs prior to the breakup of AT&T, is very well suited to networks that have predictable growth curves and predominantly one form of traffic, in this case voice. The IP protocol came into prominence in the 1990s as networks began to experience very rapid scaling and data took its prominent place on the telco networks.

ATM networks with their fixed-size 53-byte cells can not only transport both voice and data but also differentiate between various priorities of data traffic. This service differentiation is mission-critical to ISPs, as they need to offer at least three different classes of service. These classes of service have different cost rates to the end user and different oversubscription rates.

As an example, a BBFW link sent to the top of a *Multi-Dwelling Unit* (multi-business building) or *Multi-Tenant Unit* (condominium or apartment building) also known as *MxU* using ATM must provision separate end-to-end permanent virtual circuits (PVCs), which are electrical circuits that must be set up by a technician at each node of the transmission path or switched virtual circuit (SVC). An SVC is virtual and logical (having a network address) as opposed to an electrical circuit. The difference between an SVC and a PVC is that the PVC is permanently active whereas the SVC is only active when needed. ATM cells require both an address and the establishment of a circuit that is created prior to the transmission.

One disadvantage of an IP system is that, as packets are sent over disparate routes over a network, they can be subject to packet delay, loss, misdirection, or delivery/assembly that is out of sequence.

Error correction is performed by TCP modules at both the originating and destination sites. These modules communicate with each other and provide packet sequence numbers, acknowledgments, and time-based responses that in essence resend the packets to replace those which were corrupt. Error detection and correction can add an additional 10 percent or more of overhead to the network (additional traffic), which translates to a nearly identical correspondence in performance reduction.

One advantage of IP over ATM according to some is that IP network devices can make decisions based on QoS and readily prioritize traffic according to information primarily at Layer 3 of the network, although there is also important traffic management information in the headers at Layers 2, 4 and even 7.

When one briefly evaluates the oversubscription and scaling issues, both of which are critical in an era of unpredictable network growth, IP gains even more credibility. Take, for example, a BBFW link that has 100 subscribers. With an ATM-based link, the ISP must set up 100 circuits, each of which must remain in place regardless of whether actual traffic is being sent over the circuit or not. The circuit remains "tacked up" until the session is terminated.

In an IP BBFW link, a single circuit will carry IP packets from numerous sources and with different levels of priority. If bandwidth is available over the circuit, it can and will be used to capacity before another circuit is established, thereby significantly increasing the overall bandwidth of the link.

Efficient use of bandwidth has significant cost savings for the network along with other significant advantages such as greater oversubscription and higher availability to the end user. For a network utilizing differentiated service, an ATM-based system can cost up to seven times as much as an IP-based version. While the preponderance of contemporary radios use an ATM interface, it is the author's opinion that IP radios will eventually prevail when used as access devices on a network.

Another advantage of IP is that IP networks do not require the use of Class 5 switches but instead can use what are called soft switches, and Media Gateway Control Protocol (MGCP), to provide voice routing and call control if the user base requires voice capability over the network. While at this writing, toll bypass for voice isn't dominant, the voice traffic bypassing ILEC local loops is expected to increase steadily over the next few years.

BBFW links commonly reside on the edge of an existing ATM backbone, as many network owners are in the process of migrating their networks to IP. In this case, an appropriate architecture is to use an IP-based BBFW link with Multi-Protocol Label Switching (MPLS) to enable the different priorities of traffic. The traffic at the core will remain ATM and therefore require tacked-up circuits, but as the network migrates to an end-to-end IP protocol, network costs will reduce, revenues will increase as the service provider will be able to have higher oversubscription rates for a given amount of capital invested, and scaling will be much more readily accommodated.

Most of the technology and financial markets believe that IP will rule the day. Indeed, research carried out by Synergy Research Group indicates that sales of ATM based equipment slowed while IP based gear sales grew at double and even triple digit rates in the year 2000.

In turbulent financial times, cost savings are important in order to retain higher company valuations that bring in growth capital at a better price. Cost savings are best put into place during a bull market and best enjoyed during the bear portion of the market cycle.

IP networks cost less to deploy and run, and they generate more revenue on a per-dollar basis, making an IP solution a greater value. The downside is that truck rolls for this migration are expensive and must be planned and executed with precision.

This cost effectiveness is especially significant as cash-constrained CLECs continue to compete against much better funded ILECs. The equalizer in this competition is to have lower-cost gear that has better inherent value, scaling, and revenue generation profiles.

Asynchronous Transfer Mode (ATM)

ATM is a cell-oriented protocol that uses fixed-length packets to carry different types of traffic, primarily voice, but it is also well suited for video and applications where the highest levels of quality are required. ATM cells consist of packets that hold exactly 48 bytes of information and a 5-byte header, thus comprising 53 bytes per cell. ATM cells are constant in size.

ATM has its roots in the 1970s and 1980s, when ISDN was the predominant broadband protocol. It requires either a virtual or physical circuit to be established between two points prior to communication and is deemed *connection oriented*, which in this case means that different routes are selected to provide different classes of service and that the transmission source must "see" the other end of the link prior to sending data. Each circuit carries multiple traffic streams as well as having cells sent over multiple links simultaneously, a feature that enables ATM to scale for unpredictable traffic growth rates and even more unpredictable network usage.

A group of virtual circuits can be multiplexed (grouped) into what is called a *virtual path connection (VPC)*. While in theory ATM traffic occurs over the same circuit each and every time, a transmission even takes place while IP traffic can be routed over a number of different circuits according to a number of factors including network load, shortest path, and traffic priority. Once an IP network has converged, traffic will always follow the same path until a network disruption like a link down occurs. At that point a new path is calculated. No routing protocol today truly dynamically routes packets depending on load. The very new traffic engineering features of MPLS like Auto route can approach this, but it's not truly implemented in a real environment at this writing.

Almost all commercially deployed ATM networks use PVC. The efforts of various protocols to support SVC never gained major market share, as they were too complicated and had high rates of failure.

ATM circuits are usually connected with switches as opposed to routers, which connect IP traffic; however, Cisco routers commonly include the ATM protocol. Two ATM switches commonly have a large array of PVCs between them, and these PVCs are commonly owned by an array of different users.

ATM Classes of Service

There are five classes of service in an ATM network, and they greatly facilitate the prioritization of traffic. These classes of service are outlined in Table 1-3.

Service Class	Quality of Service Parameter
Constant Bit Rate (CBR)	This class is used for applications like voice, video conferencing, and television that are highly sensitive to cell delay. Therefore cells are introduced into the network and forwarded along ATM circuits at a constant time rate (not all cells are sent on constant time rates). Please note the following references in this table to Variable Bit Rates, Available Bit Rates, and Unspecified Bit Rates.
Variable Bit Rate–Non-Real Time (VBR-NRT)	This level of service is typically reserved for e-mail and other applications that are not time sensitive with respect to the timing of the cells arriving at the destination; i.e., the bit rate is variable and is dependent on the availability of user information. Groups of this traffic are sent along the circuits to maximize network efficiency.
Variable Bit Rate–Real Time (VBR-RT)	This class is similar to VBR-NRT but is designed for applications that have the highest rate of time sensitivity, like interactive video.
Available Bit Rate (ABR)	Depending on network congestion, non-time critical applications like file transfer and e-mail are sent along the network and "flow controlled," which means the cells are sent along no faster than the receiving end can accept them in a buffer. There are few metrics used to ensure this level of service.
Unspecified Bit Rate (UBR)	This class of service allows for traffic that is at the bottom of the priority list. UBR has evolved into TCP/IP.

Table 1-3. ATM Classes of Service

The classes of service are monitored by a combination of some or all of six different metrics that are part of the ATM protocol. These various forms of measuring an ATM network performance is indicated in Table 1-4.

Metric	Definition
Cell Loss Ratio (CLR)	Cell loss ratio is the percentage of cells that fail to arrive at their destination. Cells are typically lost due to transmission errors, network congestion, and buffer overflow at the receiving end of a transmission link.
Cell Transfer Delay (CTD)	Cells are commonly delayed in their introduction to a circuit by the transmitting side of a link from variables like processor loading and even power fluctuations. Other such events include cueing delays at switches along the transmission path and intended delay due to the prescribed level of service attached to a cell.
Cell Delay Variation (CDV)	A measurement of the variance of CTD. High levels of variation can imply large amounts of buffering for highly time-sensitive traffic like voice and video.
Peak Cell Rate (PCR)	PCR is the rate at which cells are transmitted. PCR can occur at any point in the circuit and is not necessarily limited to the circuit located closest to the transmission source. PCR is the inverse of the minimum cell arrival time between any two points in the network, considering that there may be many intermediate points in a network.
Sustained Cell Rate (SCR)	SCR is the average rate of cell transfer over the entire circuit between the origination and final destination points.
Burst Tolerance (BT)	BT is the maximum rate at which cells can be transferred between two points and is the primary parameter for controlling the traffic entering the network, or between any two points in a network.

Table 1-4. ATM Performance Metrics

Not all classes of service use the same set of metrics to monitor and manage performance between the point of origination and the final destination of a cell. Table 1-5 associates the various metrics with the five classes of ATM service.

ATM Traffic Management

ATM traffic management software has two primary objectives:

- To deliver quality of service guarantees for various applications and or customers
- To optimize network utilization

There are typically three elements to an ATM management system:

- Node Controls
- Near Real Time Network Controls
- Non-Real Time Network Controls

Node-level controls operate in real time and are typically implemented in the switch hardware. This type of control includes managing data cues and also involves the metrics of cell loss and delay. There are also controls that involve traffic policing, which is the concept of ensuring that users do not have access to more bandwidth than they are allocated. Some traffic policing policies allow a preset of data rates in excess of that which is agreed before charging the customer a higher rate for transmission of data during the excess period. This assumes, of course, that there is sufficient bandwidth to allow customers burst rates that are in excess of their service level agreement.

Node controls also include dynamically allocated buffers to accommodate the bursty nature of network traffic.

Network-level controls operate in near real time. These typically include admission control for new connections, network routing and rerouting systems, and flow control.

	CBR	VBR-NRT	VBR-RT	ABR	UBR
Cell loss ratio	Yes	Yes	Yes	Yes	No
Cell transfer delay	Yes	No	Yes	No	No
Cell delay variation	Yes	Yes	Yes	No	No
Peak cell rate	Yes	Yes	Yes	No	Yes
Sustained cell rate	No	Yes	Yes	No	No
Burst tolerance	No	Yes	Yes	No	No
Flow control	No	No	No	Yes	No

Table 1-5. Various ATM Class of Service Metrics

The management of admitting new traffic requires sophisticated monitoring of current traffic levels and also manages re-routing based on the failure of links in the PVC as well as the dynamic allocation of data loads sent to very large data trunks.

Flow control involves the adjusting of cell rates at the source in response to network congestion along with congestion at the final destination site. CBR traffic is generally immune to flow control, but there is a philosophy among highly experienced network engineers that flow control is included as part of CBR traffic in order to ensure the lowest rate of dropped packets. Flow control is of course a key element for traffic in the lower service guarantee echelons, as this is one method for ensuring available bandwidth for the higher classes of service.

Non-real time network controls are primarily used for configuration management and network planning tools.

Table 1-6, courtesy of the International Engineering Consortium, provides a snapshot comparison of IP and ATM.

Snapshot Comparison of IP vs. ATM Technologies
Scale of 1–5, 1 = Worst, 5 = Best

	IP	ATM
Installed LAN base	5	1
Installed WAN base	5	1
QoS maturity	2	5
Technical elegance	3	4
Vendor support	5	3
Service provider support	2	4
Market perception	5	3
Standards maturity	2	4
Standards activity	5	2
R&D activity	5	2
Investment activity	5	2
CPE pricing	4	3
Scope of applications	5	3
Simplicity	2	2
OVERALL AVERAGE SCORE	3.9	2.7

Table 1-6. A Snapshot Comparison of IP Versus ATM

Both IP and ATM have their advantages and disadvantages, and no area of technology is exempt from this premise. At the same time, however, while there is a considerable amount of ATM gear still being sold and no predictions are being made for its early demise, the market at every segment and level has stated that IP will eventually prevail over ATM.

This prevalence will not likely occur anytime before approximately 2005, and it may take as long as 2010 before ATM is placed at the wayside of technological evolution. By the same token, no one should imagine that IP will reign eternally as the protocol of choice, as the market migrates to the next level of its Internetworking phase.

Layer 3 BBFW Solution Versus Layer 2

As arguably the company providing the most state-of-the-art BBFW, Cisco offers a Layer 3 solution with its IP wireless and WLAN wireless solutions. At this writing, this level of solution provides key competitive advantages, described in the text that follows.

Full Integrated Wireless Link

Layer 3 RF links have a distinct industry advantage, as the RF elements are fully integrated to the backbone network. Most BBFW systems have a relative element value of approximately 25 percent (cost of the RF link compared to the total cost of the solution), while the backbone portion cost of the system (the non-RF portion of the network) typically comprises approximately 55 percent of the total network cost. The balance of the cost incurred is typically deployment.

An appropriate design approach therefore dictates that BBFW access links and methods be fully compliant with backbone as opposed to a "patched" network of disparate boxes (e.g., radios and their attendant management software from one vendor and the routers along with their attendant management software from another). The provision of a fully integrated RF link ensures compliance with backbone performance and parameters and allows the network owner to provide an end-to-end "solution" as opposed to a patched collection of electronic boxes.

Layer 3 RF links appear to the network and management systems as simply another pair of routers with specific management features relative to the radio links. Again as an example, Cisco RF links are managed by SNMP, CiscoWorks, and Cisco IOS in the same manner as any of the company's other network products. This is arguably the best known and most widely distributed network management software.

Differentiated Services

With a Layer 3 solution, the BBFW solution enables the provision of differentiated services to the end user. The encoding within the header of each packet set differentiates the payload from voice, video, or data. Specifically, this means a single transmitter can provide voice, video, and data to customers with a single transmitter across multiple users in a radiating environment. In point-to-multipoint applications, this means a single hub can provide differentiated services within a single sector (given that IP wireless cells have three or four relatively equal sectors).

Further, it means the end user can set up what is called a *grant and request policy* that is fully flexible. Grant and request policies are established by the end user to determine guaranteed bandwidth and prioritization of certain kinds of traffic in terms of traffic type, customer, or a combination of the two QoS criteria.

Differentiated services are set along the lines per the following graph:

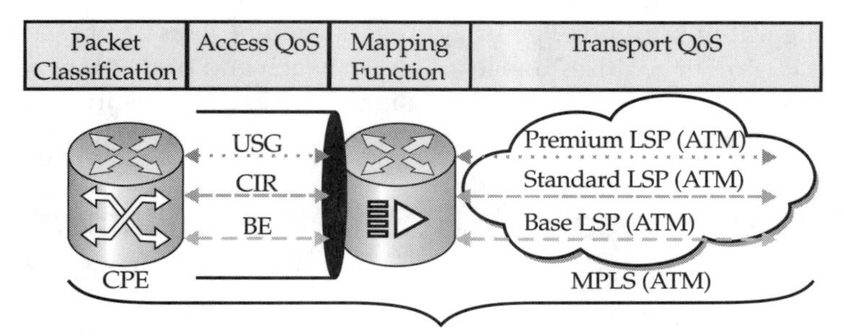

The preceding graph indicates the three levels of service that are supported by grant policies. The *UGS* policy is an *Unsolicited Grant Service* and is generally reserved for voice. This service guarantees maximum bandwidth at all times and does not require a "grant" to a "request" by an end user.

The *CIR* is the next level of service and represents *Committed Information Rate* which is a guarantee of bandwidth, but not at the same level as UGS; this is typically reserved for critical data pipes or telecommuters.

BE represents *Best Effort* and is most likely to be used by residential users. This access is supplied on the basis of its name, "best effort," which implies there may be minor delays in service if the service levels above are fully subscribed, but not to the dissatisfaction of the end user.

One of the premier features of a Layer 3 integration is that a user can upgrade or change their usage profile in real time, without intervention of a person at the host. This reduces cost to the ISP through lower required staffing levels, provides real time service to the end user, and enables the ISP system manager to allocate resources appropriately based on real time network demands.

RF Links "Look Like a Pair of Routers"

Importantly, because of the Layer 3 integration, all IP wireless links operate and appear to the network as simply a pair of routers. In other words, the wireless links provide any service or content that a fully featured pair of routers can perform in an IP network. This enables the ISP to have wireless access or extensions to an IP network with the wireless portions of the network remaining access and multiservice agnostic.

Access Link Include Routers

State of the art BBFW providers always quote RF data links with routers or switches included. Companies that only provide RF gear as opposed to end-to-end solutions commonly only quote the RF gear, which only represents approximately 25 percent of the deliverable. When comparing quotes from RF companies as opposed to network companies, it's important to have a full comprehensive quote.

Layer 3 BBFW Wireless Solutions Have Fewer Components

In addition to the preceding point, each end of a Layer 3 RF data link has routers with RF components integrated into the edge or aggregate routers. This approach is innovative and key, as the customer requires one less "box" at each end of the data link than in the case of most competitors. This also leads to significant reductions in deployment complexity and cost, as at the same time it increases Mean Time Between Failures (MTBF).

Operational System Support (OSS) Selected Fundamentals

OSS is one of the least understood and yet one of the most essential network elements because it's the software that manages the hardware. The network in its entirety is composed of core, aggregation, access, and CPE elements, and the OSS element cuts across them all; in other words, the OSS monitors, controls, and manipulates the hardware, establishing trouble tickets when elements operate out of their prescribed parameters.

The basic elements of the OSS element are shown in Figure 1-1.

The OSS is also the point on the network that actually handles the revenue generation through billing and credit card management.

An operational support system is enormously complex and is generally deployed in one of two variants:

- Proprietary to the customer
- Third party

Proprietary OSS

Proprietary OSS is generally substantially smaller in scope than OSS elements provided by third parties and tends to be provided by owners of BBFW networks that either are relatively small, i.e., have fewer than 10,000 customers, or are very large, like those used by telco-type entities such as Sprint, MCI, or AT&T.

Proprietary OSS systems owned by small companies often cost on the order of $500,000 to $1M USD to assemble. OSS systems used by very large firms that intend to have customer bases in excess of 10,000 and into the hundreds of thousands begin at about $5M USD and can readily cost in excess of $25M USD. Third-party OSS systems are provided by only a few companies, most notably Price Waterhouse Coopers and KPMG. These programs require enormous inherent logistics and program management capabilities, as the large programs include approximately a dozen major software elements, all of which require "gluing" together in a seamless manner. A typical third-party OSS development team is often composed of fifty or more individuals, most of whom are senior programmers, and development management teams.

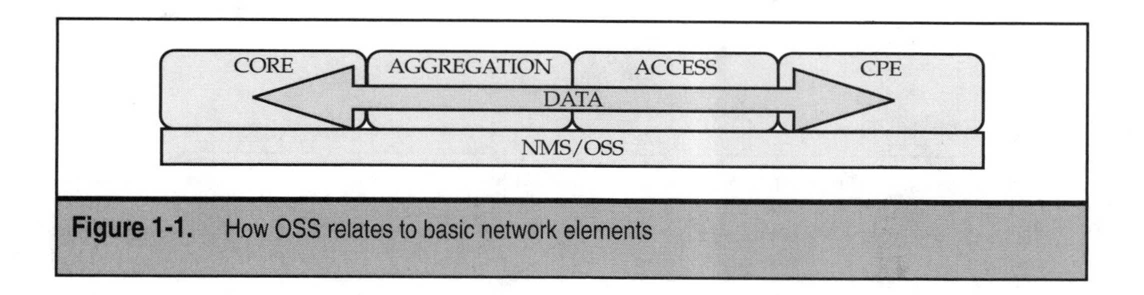

Figure 1-1. How OSS relates to basic network elements

The fundamental elements of a proprietary system are illustrated in Figure 1-2.

Customers who deploy a proprietary OSS system can migrate to a third-party system if their scaling needs eventually require it. One of the advantages to going to a third-party-supplied system is that with an effective transition plan, the customer can migrate from a system suitable for managing a relatively small BBFW customer base of less than 10,000 to being able to support a customer base in excess of a hundred thousand or more.

Third-party OSS systems may also be more suitable for network owners that have multiple access mediums such as BBFW, cable, optical, and DSL, as shown in Figure 1-3.

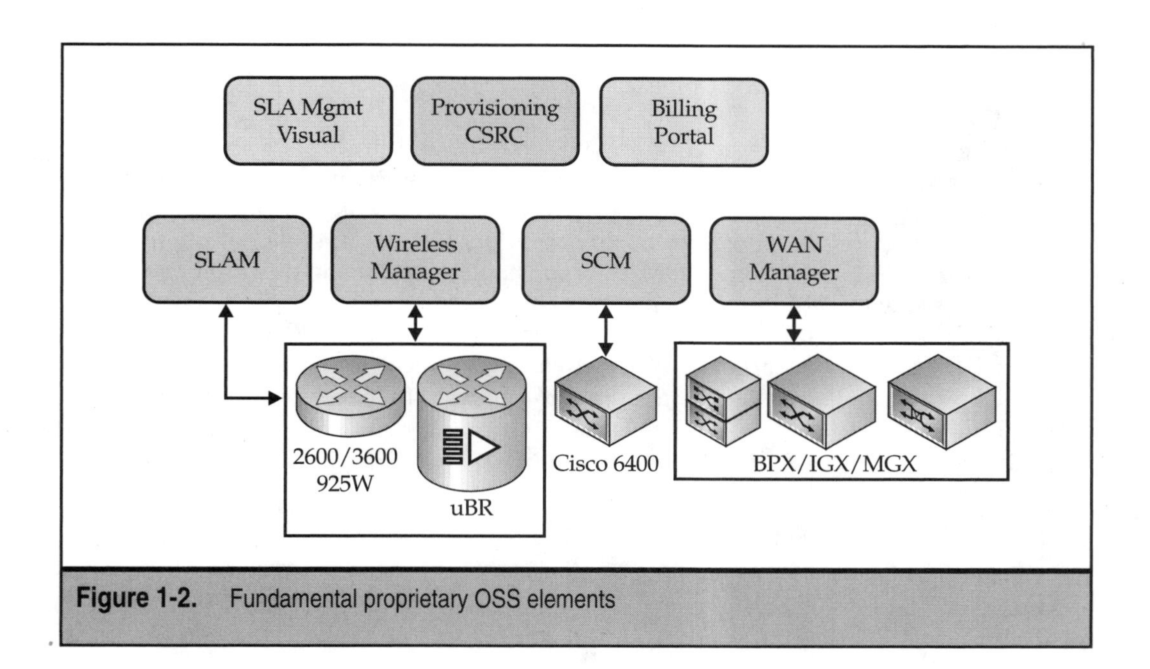

Figure 1-2. Fundamental proprietary OSS elements

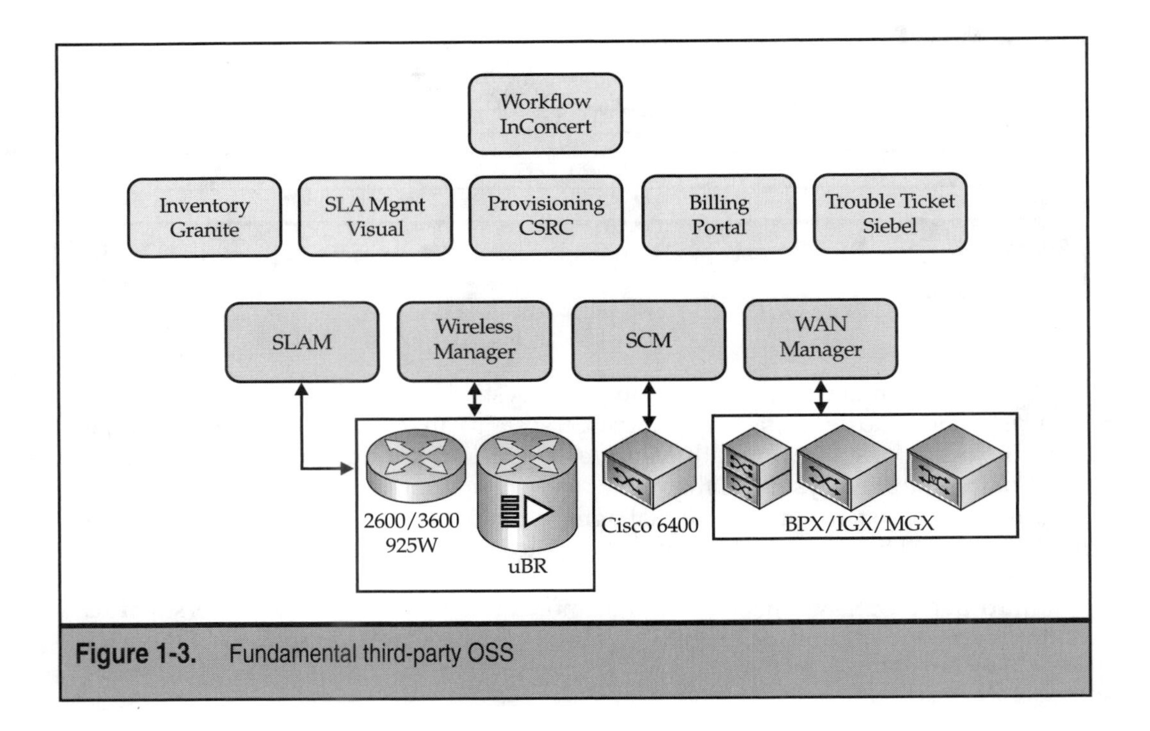

Figure 1-3. Fundamental third-party OSS

Third-Party OSS

One of the challenges in deploying an OSS system with multiple access methods is that each of the access methods tends to require unique software. Cisco Systems therefore offers what is called Cisco Powered Networks, which means that all of the network elements from core to aggregation to access to CPE run on a single platform (Internetworking Open System—IOS), thus substantially reducing the number of different types of software required to manage a network.

In newer networks, BBFW is often selected as the first access method, as it can often be deployed the fastest. In larger new deployments, optical is usually the first access method, but it's more often used as the ramp to and from the Internet, as it has very high bandwidth.

Table 1-7 provides a snapshot of each of the two systems.

Elements of a Total Network Solution

Today's networks must be viewed from end to end, or in other words, network providers must plan and deploy each and every element necessary for the BBFW customer to generate a profit through the provision of services and content and to accommodate.

	Proprietary OSS	Third-Party OSS
Target customer	Small number of customers Wants to tie OSS cost to existing base of subscribers Wants basic functionality	Mass initial rollout Has legacy applications Wants business processes tied to OSS
Functionality	Limited, no sales order customer care or inventory management Requires manual update of all databases	Extensive, includes sales order, customer care, and inventory management Automated databases updated with workflow management
Interfaces	Proprietary	Standard, TIBCO-based interface Allows easier integration to large-scale legacy systems
Customer Responsibility	Understand the biz flow	Partner provides turnkey deliverable

Table 1-7. Proprietary Versus Third-Party OSS

unpredictable network growth. Importantly, these elements must be tested for integration prior to deployment to the customer site. A fully comprehensive wireless solution must also include the issues of financing, deployment, maintainance, legacy, migration, and value propositions.

Commonly, the term "solution" includes the following primary technical elements:

- Customer Premises Equipment (CPE)
- Access
- Aggregation
- Network Management System (NMS)
- Billing/Operational Support System (OSS)

Customer Premises Equipment (CPE) Premises networks are the access and distribution networks that exist at the subscriber premises, which can be either a large business or a campus such as that found at a university, or with dorms, military barracks, etc. A Multi-Tenant Unit (MTU) is a building in which multiple businesses rent space.

A Multi-Dwelling Unit (MDU) is a condominium or apartment building, Small Or Medium Business (SMB), or Small Office/Home Office (SOHO).

Access The *access* portion of the network consists of the transport elements that bridge between the CPE gear and the aggregation points of the core. A core network can have multiple access technologies such as BBFW, dial, DSL, or cable.

Aggregation This includes the network elements that aggregate the numerous access ports such as with DSL or cable. The "front" end of an aggregation device faces the access lines (ports), while the back end of the device interfaces the core of the network.

Core The *core networks* are the public or private backbone networks that—in a general sense—will be utilized by the access network operators to connect their multitude of regionally dispersed POPs and to interconnect to public service provider network elements. For purposes of this discussion, the point of demarcation between the access network and the core network is a core switch that serves as an upstream destination point for a multitude of access network branches or elements. In general, the closer to the core one gets, the larger the router or switch. The current prevailing theory is that smarter, smaller boxes are placed at the edge of the network, with the smallest ones generally residing at the CPE site while the largest and "dumbest" ones reside at the center of the core.

Network Management System (NMS) The glue that ties the entire network together is the NMS. In its full implementation, the NMS is an exceptionally complex set of moderately to highly integrated software platforms.

Operational Support System (OSS) Virtually all BBFW customers will utilize an ecosystem (OSS) of deployment partners for systems covering *Basic Trading Area (BTA)*, a *Metropolitan Trading Area (MTA)*, or nationwide. This is due less to the inherent design of the RF components than to the matter of scaling deployment for dozens or hundreds of sites on a scheduled (and often simultaneous) basis.

Major companies like Cisco Systems that provide BBFW gear have an internal Professional Services group that is primarily responsible for managing and coordinating deployment partners. Cisco's primary ecosystem partners are Bechtel Engineering, Price Waterhouse Coopers, and KPMG. Bechtel, PWC, and KPMG are not exclusively dedicated to Cisco.

Bechtel, as an example, is recognized worldwide over nearly five decades for very large wireless system deployments. They have extensive expertise in site evaluations, construction, logistics, provisioning, finance, and project management.

Price Waterhouse Coopers and KPMG are world-class entities with extensive track record in logistics and complex project management. PWC essentially subcontracts the appropriate skill sets while retaining the role as prime contractor and providing or arranging customer financing.

There are three major providers of BBFW: Cisco Systems, Nortel Networks, and Lucent Technologies, and there are at least twenty or thirty smaller providers of RF components. Each of these companies uses its respective ecosystem partners for deployment and maintenance to Tier 1 customers like Sprint, MCI, and AT&T. Tier 2 and Tier 3 customers typically use smaller deployment entities and occasionally have captive deployment resources. Most of the Tier 2 and 3 BBFW customers are agile in managing the group of companies required to deploy BBFW and core networks.

The components and services required to fully install a BBFW network include:

- RF site survey
- Cables
- Mast or tower
- Lightning protection
- Mechanical assembly
- Construction
- Final assembly and test
- Antenna
- Connectors
- Routers/switches
- Spare components
- Negotiate rooftop or tower access.

The scope of issues involved in selecting subcontractors is nontrivial and includes the following:

- Generally must be FCC certified (if installing licensed frequencies)
- Financial solvency
- Local environmental experience
- Depth of resources
- Exceptional logistics and project management
- Established relationships with sub-subcontractors
- Site analysis
- Fresnel and multipath analysis
- Determine whether the RF links are LOS or non-LOS
- Incumbent propagation

- Proposed propagation
- Local permits
- Negotiate access to WAPs (wireless access points)
- Project management
- Prime Contractor
- Finance
- System integration
- ▲ Forward-going risk profile

There are generally two kinds of BBFW equipment providers: network companies and RF companies. Each have their pros and cons, but it's important to understand that the bulk of the deliverable even to a wireless customer consists of non-RF components. Table 1-8 illustrates the relative elements by ratio.

With the bulk of the deliverable being the core, selling RF components has what is called a "pull through" effect to their core. Another way of stating this is that selling RF gear has the added effect of selling more routers, switches, and software to support the RF components.

Smaller RF companies often have a better understanding of RF networks and technologies, as major companies tend to reassign non-RF personnel to augment the RF personnel within the large company. Smaller RF companies generally do not have this luxury, but the advantage there is that the smaller companies often have a higher level of expertise in RF technologies and deployment.

Just as all sophisticated products today require software either directly or indirectly, all BBFW deployments require networks, and indeed the BBFW components are considered network elements. Generally speaking, therefore, the Big 3 are better suited overall to deliver BBFW networks. This is especially true for major contracts.

Element	Typical Percent of Network
RF gear	25
Deployment/integration	25
Core	50

Table 1-8. BBFW Relative Element Values

Having stated that, however, I should note that companies that do RF exclusively play a vital role in the BBFW industry, as they service many smaller contracts, generally have leading-edge gear, and are more responsive to customer and market needs.

Table 1-9 summarizes the differences between the Big 3 and pure RF companies.

In summary, an excellent definition of a network solution was authored by Matt Irvin of Cisco Systems, who stated, "A Solution is a completely defined and operable service(s) that runs over a given architecture and satisfies a specific business need for one or more customers."

Characteristic	Cisco/Nortel/Lucent	Small RF Companies
RF literacy	Often limited	Superb
Highly nimble	No	Yes
Market-driven products	Generally late to market	Generally lead the market
Cutting-edge gear	Generally no	Generally yes
Network literacy	Superb	Generally lacking
Can deliver end-to-end network	Generally yes	Generally no
Will manage end-to-end integration	Generally yes	Generally no
Will finance a contract	Yes	Generally no
Can service major contracts	Yes	Generally no
Deep personnel resources	Yes	No
Require eco-partners	Yes	Yes

Table 1-9. BBFW Big 3 Versus Pure RF Companies

CHAPTER 2

Types of Wireless

Perhaps one of the most common technical terms used in highly industrialized countries is "wireless." This term has been around since the turn of the twentieth century, when Guillermo Marconi, Thomas Edison, and Nicola Tesla fathered modern wireless technologies in general. We owe a great debt to these men because they developed concepts and equipment based on the fundamentals discovered by Hertz and Maxwell in the mid-1800s.

Thomas Edison was the driving force behind the first commercially deployed wireless systems, which became the foundation of one of the largest and most successful companies in the world: General Electric. His work was based on that of Marconi and one of Edison's own former employees, Tesla, who had recently arrived from Europe as a brilliant immigrant with meager resources. Interestingly, after many years of controversy, Tesla was finally given formal recognition in 1998 by the U.S. Patent Office as the inventor of the radio, a distinction formerly held by Marconi.

WIRELESS TECHNOLOGY TODAY

The world of wireless technology has come a very long way from the days of Tesla, Marconi, and others. From the crude but brilliant devices built in their labs, different types of wireless technologies have proliferated on every continent and are deemed mission-critical business tools for efficiency as well as important items for personal safety and convenience.

In today's world of technology, wireless devices range from tiny crystal radios one can purchase at a hobby shop for a few dollars to commercial radio systems costing a million U.S. dollars or more per link. While the actual number of different types of "wireless" may be debatable, the key point is that there are many, a subset of which follows:

- Mobile (cellular telephones, Personal Data Assistants, pagers)
- Ham (long range devices used by private citizens)
- Citizens Band (radios used in transportation and for private citizens)
- Telemetry (remote control and measurement)
- Satellite
- WLAN (wireless local area networks)
- Broadband fixed (BBFW)

Clearly, an enormous number of these various types of radios are in use in the world today. It is therefore appropriate to be more specific about what type of "wireless" we are referring to in our discussions with customers, industry peers, and colleagues. A true story follows that highlights the need to be specific about what type of radio is being considered.

Not long before this work was written, a very large technology company was organizing its annual sales meeting. Key elements of this annual meeting were dozens of breakout sessions that provided tutorials on the various technologies sold by the company. The preliminary meeting agenda contained no less than five references to "wireless sessions," a fact that caused no small amount of confusion among the organizers and attendees of the sales conference. A few phone calls were quickly placed before the conference, and each of the breakout session owners was persuaded to give his or her sessions more specific names. The attendees were then able to attend the specific wireless sessions they were interested in.

BBFW systems can be appropriately referred to as "broadband wireless," "fixed wireless," or simply "BBFW."

Table 2-1 provides a snapshot of various "wireless" technologies.

With a brief review of competing types of wireless technologies, it grows clear that, while there are many different types, few types of wireless are competitive with each other. In general, the wireless medium is far more competitive to wireline technologies.

Indeed, at the writing of this book, BBFW has an increasingly rich opportunity to resolve the issue of bringing high speed network access into commercial buildings, multi-tenant units (MTUs), and especially the residential market. While the fundamentals

Type	Typical Use	U.S. Frequency	Bandwidth	Competitive To
Mobile	Cell phones, PDA, pagers	800 MHz/1.9GHz	----	CB, wireline telephone
Commercial AM Radio	Public broadcast	---	---	Satellite, television, internet broadcast
Commercial FM Radio	Public broadcast	---	---	Satellite, television, internet broadcast
Ham	Hobby/Rescue	---	---	None, wireline telephone, mail, fax
Citizens Band	Transportation, recreation, hobby	---	---	Mobile, wireline telephone
Telemetry	Industrial	300/900MHz	35 MHz	WLAN, wireline monitor/ measurement
Satellite	TV/	---	---	BBFW, mobile
WLAN	LAN	2.4GHz	85MHz	BBFW
BBFW-UNII	data/voice	5.7 GHz	100 MHz	WLAN/Satellite/ wireline
BBFW-MMDS	data/voice	2.5 GHz	198 MHz (U.S.)	WLAN/Satellite/ wireline

Table 2-1. Snapshot Comparison of Various Wireless Types

of BBFW remain relatively constant across the spectrum of users from large commercial campuses to the users in residences, there are key differences in the cost and uses of the radios.

RESIDENTIAL VS. COMMERCIAL SYSTEMS

The use/cost model from commercial BBFW radios is inverse to that used in the residential market. Figure 2-1 indicates these differences.

Businesses demand BBFW access and networks to be very reliable and to provide sufficient bandwidth such that there is little to no loss of efficiency. Applications like email, web browsing, VPN's, order entry, human resources, and data storage are some of the typical uses of this type of BBFW link. While during 1999 and 2000, the primary focus for a BBFW system was to bring in new customers, the latest deployments also feature increased business efficiency through broadband access to other offices and a reduction in the cost of doing business by taking advantage of bypassing the local phone loop for voice applications.

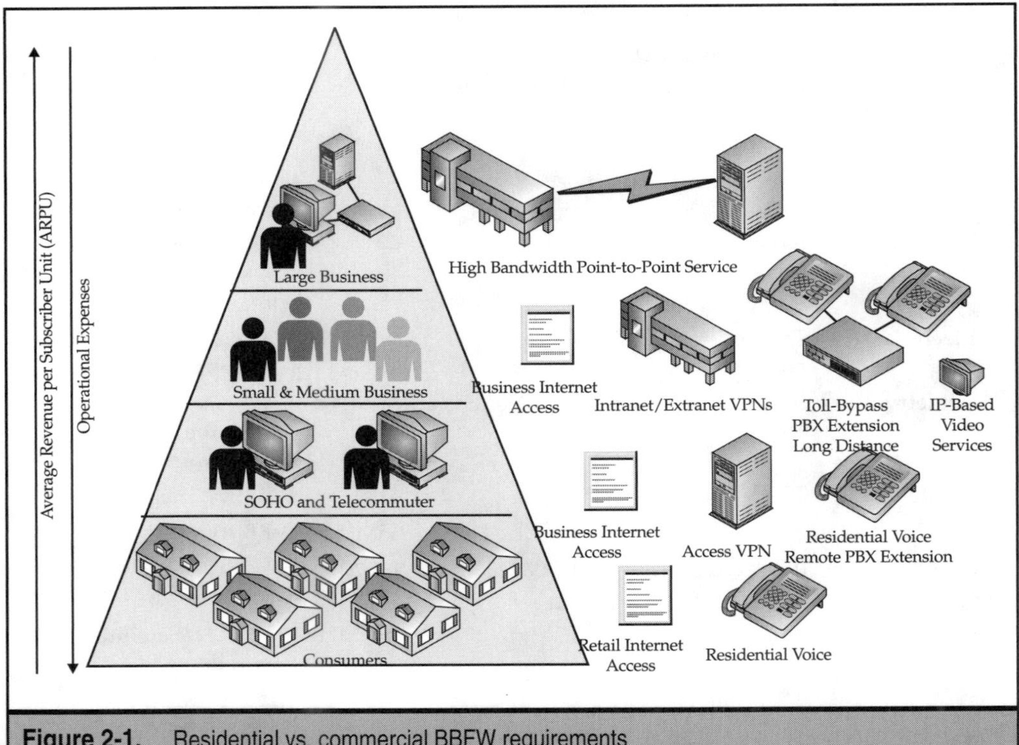

Figure 2-1. Residential vs. commercial BBFW requirements

Businesses are willing to pay a premium for this resource, while residential users tend to be more casual in their use of a network. The most common uses of a residential link are email, web browsing and online chat rooms. Thanks in large part to the level of service provided by internet services providers over the last five to ten years, most residential users are used to delays in service and minimal bandwidth. At the same time, residential users have shown a reluctance to pay much more than ten or twenty dollars a month USD for access through dial-up modems, while broadband users with cable modems or DSL are showing a reluctance to pay much more than about forty dollars per month.

One of the other key differences between these two markets is that business users generally pay tens of thousands of dollars for equipment, installation, and maintenance. Residential users, on the other hand, show a strong reluctance to purchase expensive gear, and most dial-up modems are commonly included in both desktop and laptop computers. Some laptops such as those provided by Dell and Apple have an optional wireless access device although this device is nearly always only for connecting that laptop to the LAN. Today, however, the cost of equipment required for broadband services is almost always bundled into the service charge and provided to residential users by the service provider. At this writing, residential broadband gear is primarily limited to cable and DSL modems, although there is a very substantial effort being put forth to provide BBFW residential services to compete against existing solutions, most notably by Sprint and MCI Worldcom, among others.

Never has BBFW had a better opportunity to enter the market in a major way since the DSL market shakeout in mid-2000. Layoffs at major players in the U.S. such as Rhythms and Covad were part of a generally weak technology sector performance. Verizon announced plans to stop DSL installations in 53 of its major sites, and other DSL providers reported severe network problems causing outages and customer churn. This state of affairs was preceded by many DSL providers aggressively advertising even to areas where they did not have facilities.

At the same time, residential broadband use soared by 230 percent in 2000. That phenomenon in conjunction with the generally poor performance of DSL providers has made for a fertile opportunity for BBFW providers, though this opportunity is not without risk or hurdles.

The challenging side of this opportunity for BBFW providers is primarily centered on the issue of scaling. To be specific, while it is relatively easy to perform a field trial with a limited number of subscribers, it is quite another matter to install devices at residences at a rate that must exceed 2000 homes or more—*per week.*

To have an effective installation rate, no element is more critical to the formula than that of having a sufficient number of trained personnel who can install BBFW devices in to homes. This installation often takes up to two hours per two-person team if any cabling needs to be run from a wall or roof-mounted dish to the interior of the home, which is virtually always the case.

The installation of BBFW is very similar to the installation procedure for a television satellite dish. The primary components are the dish, the cabling, and a modem inside the residence. Signal strength must be measured, the installation provisioned (installation of software that manages the connection), and bandwidth throttled. At present, only professional installers can provide the service to homes when the system uses a licensed frequency. There is a substantial effort to have an installation scenario where the resident can self-install the device, but this will be limited to the U-NII frequency of 5.8 GHz or some other nonlicensed frequency.

There are five primary BBFW alternatives to DSL and cable modem installation in the United States, and comparable services for the European and Asian markets. These services are indicated in Table 2-2.

A BBFW service to a home has significant advantages over broadband carried by copper. One of the advantages is the rapid rate at which a geographic area can be surveyed and therefore qualified. This is commonly done with a truck or van that has a transmitter and a receiver linked to a head-end location. Measurements taken from a number of locations within a neighborhood, town, or county constitute a site survey of the geographic area. This is done to ensure the link will be strong enough to have a good probability of remaining reliable and capable of carrying a reasonable amount of bandwidth.

One of the other advantages of BBFW over copper into the residence is that trenching is not required. This provides substantial savings in terms of cost and lead time to the service provider. Trenching is also a resource with a substantial backlog from the fiber optics industry, and therefore can have substantial lead-time issues. This is partly because the trenching companies are finding it more profitable to lay copper or fiber in areas already

Technology	Data Rates	Pricing
LMDS	600–800 Kbps Symmetric	$29.95 Modem and NIC card included
MMDS	1–2 Mbps down, 256 Kbps up	$50.00 Installation, $299.00, equipment included
U-NII	1 Mbps down, 256 Kbps up	$50.00 Installation, $299.00, equipment included
ISM	128 Kbps down, 64 Kbps up	$75.00 $299 for equipment, $30 activation fee, self-install
Satellite	500 Kbps down, 156 Kbps up	$60.00–$99.00 Installation $199, equipment $399

Table 2-2. Snapshot Comparison of Various Types of BBFW Residential Service

under construction, such as new residential tracts and new office parks. Trenching in these areas requires fewer, if any permits, and does not require special equipment and additional costs for ripping up and replacing paved streets. It's also safer for crews to work in an area relatively devoid of commercial traffic.

Having stated this, there is little reason to believe that the BBFW industry will not experience similar scaling problems as those found in the DSL and cable modem industries, especially if there is a strong increase in demand for BBFW to the home.

One of the more cost-effective ways to leverage installation resources is to establish the Multi-Dwelling Units (MDUs) as a higher installation priority over single-family dwellings. This is because a BBFW link can be readily installed on the roof of an MDU and data can be distributed more cheaply than to a single-family dwelling. This is because one radio assembly on a single rooftop can service a large number of living units inside the building.

Although it is not uncommon for service providers to run into hurdles set by building owners, such as revenue sharing demands, denial of access to roofs, and limited or restricted access to internal facilities such as conduits, walls, and power, the MDU presents a significant opportunity to provide high-speed access to many users in one stop, and is typically underserved. And although other obstacles exist to wireless service provider deployments, such as restrictions from homeowners' associations and local ordinances relative to installing radio dishes, they are minimized in the MDU environment. For these reasons, the MDU market space is attractive to service providers, as they can gain access to a dense population with a single round of approvals. A typical MDU architecture is shown in Figure 2-2.

A second BBFW configuration for MDUs is to bring the broadband access to the edge of the property and from there use existing copper that resides in the ground on the MDU campus. The rationale for this type of architecture is that some MDU campuses have a large number of buildings (twenty or more) with varying building heights that effectively block some of the beam path to certain buildings. Figure 2-3 indicates this architecture.

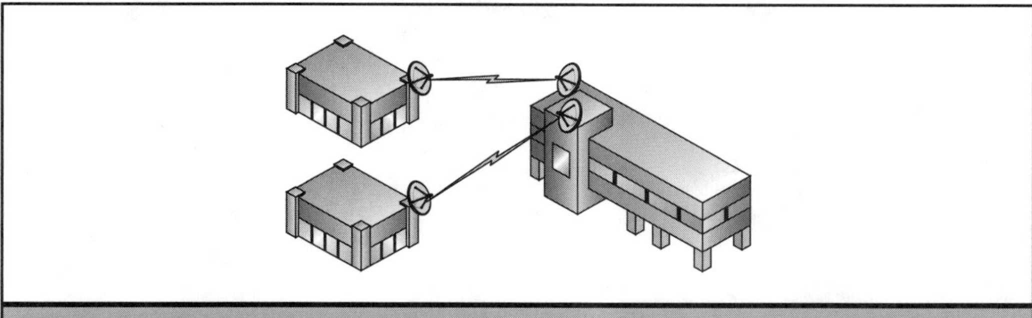

Figure 2-2. MDU point to multipoint BBFW architecture

There are other hurdles that must be overcome by service providers who endeavor to provide BBFW links, and they center on the issue of arranging a suitable "ecosystem" of partners. Not even the largest companies like Cisco or Lucent can provide a truly end-to-end BBFW product and service line. If this is true, then it is even more difficult for the midsize and smaller companies to adequately service this market.

Ecosystems must include the following entities:

- ASIC and RF chipset providers
- Development labs
- Integrated solution testing labs
- Mass production, test and verification
- Federal testing and approval (for licensed and unlicensed systems)
- Sales and distribution
- Finance and leasing
- Site survey
- Installation including construction and integration
- Network Operations Centers
- Technical assistance centers
- Network components compatible with the network core

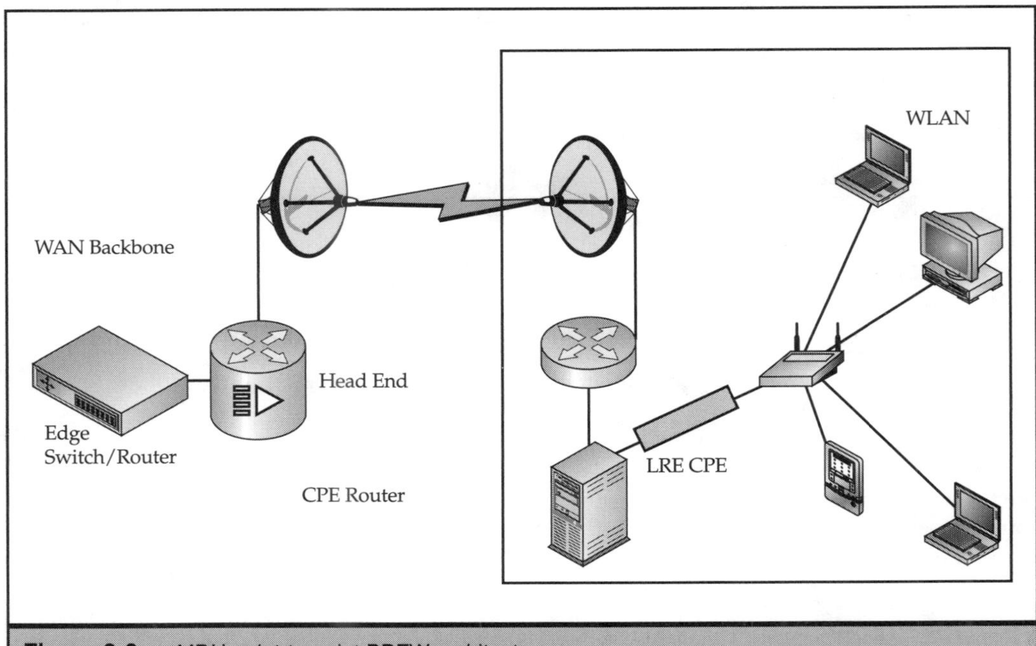

Figure 2-3. MDU point-to-point BBFW architecture

TYPICAL BBFW SYSTEM BLOCK DIAGRAM

There are a number of fundamental BBFW system architectures. The one best representing the state of the art is outlined in Figure 2-4.

The primary elements of a state-of-the-art BBFW link are the following components, which make up the "access" portion of the network. These elements are:

- Router/Switch
- Wireless Line Card
- Intermediate Cabling
- Outdoor Unit
- Antenna

A wireless link is essentially a link between two networks. It therefore requires routing and switching functions at each end of the link to manage and shape the traffic on the link. Most wireless links in use today are used as bridges extending local area networks (LANs). Performance across these bridges is exactly as would be expected in a LAN, except that it is wireless. A state-of-the art BBFW link operates at one level higher in the OSI stack, which is a routing layer as opposed to the layer where bridging occurs. They typically connect different LANs or provide WAN access to the LAN.

This requires the incorporation of routing into the wireless network. In spite of the considerable publicity for BBFW, the bulk of the equipment installed on BBFW networks is non-radio gear, including routers that reside at the customer premises, as well as aggregation and core devices. In fact, the non-radio gear generally accounts for approximately 50 percent of the entire expenditure by the customer. Deployment labor for construction, site survey, and integration will cost about 30 percent of the total money spent, and the actual radios will only cost about 20 percent of the total money spent by the end customer. This is

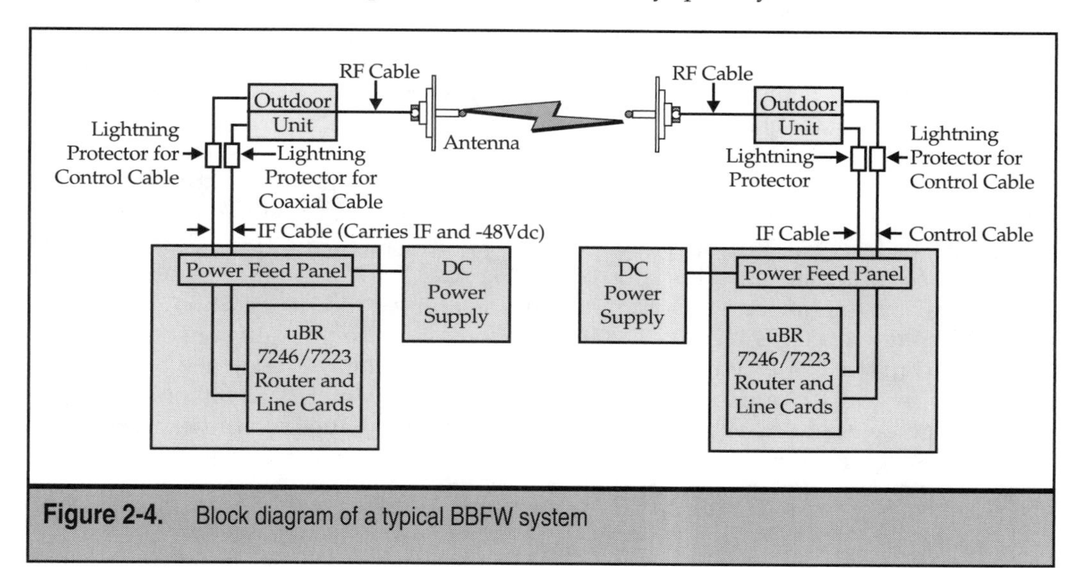

Figure 2-4. Block diagram of a typical BBFW system

not to say that the wireless portion of the network is insignificant, only that in reality, the wireless link is merely a serial extension of a routed network.

This low relative cost of the radio as a portion of the entire network is due in part to the fact that access devices such as BBFW are not as expensive as the aggregation and core devices at the center of a network.

The general rule is that the closer to the core of the network, the more expensive the device and the larger the device. While access devices such as a BBFW link connect the core to other networks, it is the aggregation and core elements of the network that do most of the processing of data transmissions. This is especially true of networks that have multiple access mediums such as optical, DSL, cable, and BBFW. Each of these network elements will be covered in detail as part of this work, as a fundamental understanding of these aspects is necessary to gain an appropriate understanding of the role of BBFW.

Increasingly there is less and less design isolation between the access, aggregation, and core elements of a network. Businesses are rightfully demanding "solution level" support from companies that supply and maintain their networks, which means in part that businesses are looking at their networks from end to end as opposed to focusing on the unique network elements such as access or aggregation.

However, it can probably be argued successfully that relatively few individuals in the networking business truly understand the concept of considering a network from a solution perspective, even though the term "solution" is commonly used. As a network must transmit information from end to end, the better those associated with networking understand the true solution perspective, the better the planning, deployment, integration, usage, and migration phases will be. See Table 2-3, keeping in mind that there are exceptions to the information found in the table and that this is intended as a general snapshot only.

Two of the most common use frequencies for BBFW solutions in the U.S. are the Multichannel Multipoint Distribution System (MMDS) at 2.5 GHz and the Unlicensed National Information Infrastructure (U-NII) at 5.8 GHz. In the United States, MMDS is a licensed band; it originated in the United States as an analog, one-way video distribution system for cable television and was originally known as "wireless cable TV." The FCC amended its rules in September 1998 to allow the spectrum to be used for two-way digital voice, video, and data services.

The reader should note that the terms MMDS and U-NII, as well as other band names such as LMDS, may have different frequency allocations in different countries. For the sake of discussion, when this book refers to MMDS, it is referencing the frequency band 2.500–2.690 GHz. For U-NII, the reference band is 5.725–5.825 GHz. The U.S. MMDS band for data traffic is actually a reallocation of the original MMDS band (2.596–2.690) and the ITFS (2.50–2.596 GHz) band, which make up the entire band. In addition, 12 MHz of bandwidth known as MDS (2.150–2.162) has become a part of this auctioned band.

Approximately 83 percent of the MMDS spectrum in the U.S. is owned by two entities, Sprint and MCI Worldcom. The two firms paid approximately $2 billion for their licenses, which were primarily purchased from the original auction winners. Nucentrix owns a considerable portion of the spectrum, and judging by various reports, there are approximately 50 to 100 other MMDS license holders in the U.S. Sprint's publicly stated

	MMDS	U-NII	WLAN	ISM	LMDS
Frequency (GHz)	2.5	5.7	2.4	2.4	28 (U.S.)
PTP bandwidth Mbps (up/down)	44/44	44/44	1–11/1–11	11/11	155/155
PTM bandwidth Mbps (up/down)	19/17	19/17	1–11/1–11	3/3	45/10
Carrier equivalent PTP	DS3	DS3	N/A	DS2+	OC3
PTP range LOS	25 mi	20 mi	400+ ft	25 mi	5 mi
PTP range NLOS	~4 mi	~4 mi	N/A	N/A	N/A
Over the air protocol	IP/ATM	IP/ATM	IP/ATM	IP/ATM	ATM
Approximate cost per link	$150K	$150K	$5K	$5K	$200K—$1M
FCC license req'd?	Yes	No	No	No	Yes
Preferred deployment	ISP, SMB, enterprise	ISP, SMB, enterprise	Indoor, SOHO, SMB	SOHO, SMB	SP

Table 2-3. Two Common IP BBFW Products—MMDS & U-NII

market is residential, small office/home office (SOHO), and telecommuters, while MCI is primarily going after the MDU/SOHO/telecommuter market.

UNII is an unlicensed spectrum in the U.S. and is a highly desirable spectrum for those who do not own MMDS or some other portion of the spectrum. Contemporary industry wisdom is such that viable business models can be constructed and sustained without the overhead of amortizing a very expensive license. This is especially true given that state-of-the-art U-NII systems are highly robust and able to resolve for what is known as "co-channel interference," which is the type of interference generated by transmissions in the same frequencies. While MMDS relies primarily on the legal system to ensure there is no co-channel interference, U-NII technology relies on advanced techniques to resolve this issue.

For further context regarding the issue of MMDS versus U-NII, it needs to be stated that the U.S. MMDS band is a reallocated band, which means it was occupied by a former owner. In this case the incumbents are narrow-band spectrum owners providing one-way analog television. This deployment issue must be dealt with by the current MMDS license holder because the FCC mandates that the two-way MMDS broadband solution must not interfere with the existing narrow-band signals already in the spectrum. Therefore, the technology that enables robust communications in the U-NII band is also valued in the MMDS band.

IP BBFW SERVICES

Most service providers have discovered that the primary service the end user requires is simply high-speed access. While this is an appropriate tactic to generate revenues for the service provider, customer turnover, or "churn," is not desirable, particularly when the cost of installing a BBFW link is relatively high compared to a simple dial-up modem cost. To reduce the risk and affect of churn, service providers will be required to do one of three things within the first twelve months of installing a BBFW link to a customer. These options are:

- Increase bandwidth
- Decrease monthly charge
- Provide added services

Increase Bandwidth

Increasing bandwidth is a network issue that is relatively easy to resolve, and in state-of-the-art systems the customers can request increases in bandwidth. Whether or not the customer can acquire additional bandwidth is called "policy management" and is set at the head end or NOC. Networks in healthy business entities always require additional bandwidth and this has obvious limitations if too many network subscribers desire additional bandwidth. The ultimate extension of this growth is that the service provider is required to install and maintain a larger network with bigger "pipes." Translation: increased cost to the service provider.

Decrease Monthly Charge

Decreased charges are clearly not desirable from the service provider's perspective, as this immediately cuts into profit margins, erodes the ability to upgrade the network and add additional customers, and can help drive down the stock price if the company is publicly traded. At the same time, service providers are unable in the present market to raise prices substantially because of competition. In the case of ILECs, the FTC prohibits charging different prices in different geographic locations, a fact that prohibits charging a different (higher) price to new customers in new areas.

Provide Added Services

After Internet access and e-mail, the most requested add-on services are second-line voice for long-distance bypass, virtual private network (VPN) capability, and service level agreements (guaranteed data rated) Additional value-added services can include:

- Toll bypass
- Wireless LAN
- Hosted applications

- Web Hosting
- Gaming
- VPNs
- IP Security for credit card and VPN transactions
- Service level agreements (insurance of a minimum level of Internet access)
- Firewall services
- Security services (intrusion detection, anti-hacker measures, etc.)
- Content hosting and delivery
- Remote Storage Transit
- Video
- IP TV
- Telelearning
- Telecommuting
- E-business
- Telemedicine/Tele-imaging
- Teleshopping
- Remote training
- Infotainment
- Digital (Web) TV

The subscriber can be given a "menu" of options from which to select any number or combination of services that the service provider wishes to include. It is not inconceivable that the service provider may be able to require a premium payment for those services that require substantial increases in bandwidth or that add significant value, such as in telemedicine.

As an example of a broadband service, during the writing of this book, the author's wife was taken to the emergency room of a local hospital due to abdominal pain. The doctors and technicians ran a number of tests, including X-rays and CAT scans. These images were sent by broadband to the homes of the on-call staff, who evaluated the tests and determined that her appendix needed to be removed. Instead of additional medical staff being summoned to the hospital, which would have incurred further treatment delay, the results were obtained in minutes in the middle of the night to the satisfaction of all concerned.

POINT-TO-MULTIPOINT ARCHITECTURE

While point-to-point architectures are used for backhaul and for providing dedicated, high-bandwidth access to a specific customer or building, a point-to-multipoint architecture is used to cover a large geographic area from a single point. Typically, this area, referred to as a *cell*, is divided into sectors to enable greater bandwidth coverage for the subscriber and to allow for improved distances. The most common sector sizes today are 60, 90, and 120 degrees wide. These can support distances of 2–10 miles (more in some situations) depending on several factors. Issues such as frequency band used, topology, and amount of obstruction can dictate the distance covered effectively. This distance becomes part of the business tradeoff a service provider must make when looking at the return-on-investment model for a BBFW network. A representation appears in Figure 2-5.

Point-to-multipoint sectors have different ranges from the head ènd than point-to-point links, in part because as the power transmitted from the head end remains covers a much larger geographic area. Think of a common flashlight; you get longer range when the beam is small and focused, and the range decreases as you widen the beam. Therefore, as this power is spread over a wider area, the effective distance drops.

There are two relationships that apply here: 1) Frequency is inversely proportional to distance. Simply stated, as the frequency increases, the effective distance decreases. 2) Distance is directly proportional to power. As the power drops (for a given path) the distance must also decrease to maintain a given signal-to-noise ratio.

There can be as many as 4000 or more subscribers to a single router at the head end, although that number would have to be carefully managed to provide an acceptable level of service. To be specific, if a service provider had four sectors in a cell and a maximum data rate of 20 Mbps per sector downstream, 1000 subscribers per sector would receive an

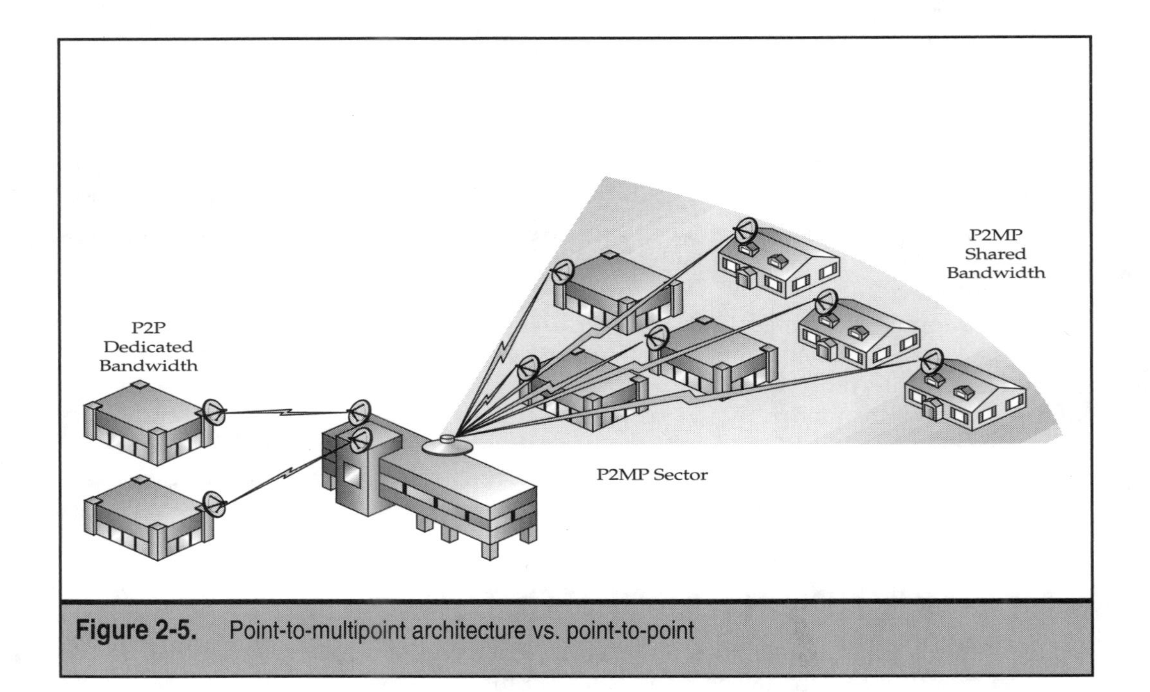

Figure 2-5. Point-to-multipoint architecture vs. point-to-point

average data rate of 20 Kbps, which is barely competitive to dial-up access. However, this can be improved through statistical modeling.

Statistically, the probability of all subscribers in a sector going online at the same time is very low. In a residential service, this fact enables the concept of what is called *oversubscription.* Oversubscription rates of 4:1, 6:1, and even as high as 10:1 have been used for this variable. If we use an oversubscription rate of 6:1 in our example, the average available data rate goes from 10 Kbps to 120 Kbps for 1000 subscribers per sector. These numbers must be used with some caution but have proved effective in business modeling and have been used by service providers for many years.

Point-to-multipoint cells are typically located in clusters, which means that the selection of frequencies used is important. A single cell is composed of anywhere from one to a dozen or more sectors (usually three or four), which are situated around the head end. Each sector uses its own set of frequencies, some for downstream traffic (from the head end) and others for upstream traffic (toward the head end). In order to prevent interference from two or more sectors using the same frequency, the sectors are kept at least one cell apart. There is an array of different frequency reuse strategies and methods that are based on number of subscribers, projected growth rate of number of subscribers, and total amount of spectrum available.

The mapping of a metropolitan cellular plan can be a very complex process, especially after taking frequency reuse and sectorization into consideration. Without a comprehensive plan in place before the first cell is deployed, system scalability can be dramatically limited. Cell planning involves a mathematical model that uses the limited frequency range over and over to provide contiguous coverage over a specific geographic area. Issues such as adjacent channel separation, co-channel interference, and terrain can further complicate the model. All of this said, a top-level reuse plan can be developed in a relatively short period of time with a few simple tools and some ground rules. Advanced software tools such as CellPlan are commercially available for this purpose. The ground rules include:

- Adjacent channels must be separated by at least one sector.
- Similar channels must be separated by at least one sector (co-channel interference).
- Adjacent sectors must be a minimum of 24 MHz apart in frequency (adjacent channel interference).
- In opposing cells, no similar frequencies can be spaced 0 degrees or 180 degrees from each other. (The risk here is that a subscriber or head end in one cell may create co-channel interference to another cell.)

A common frequency reuse pattern is shown in Figure 2-6; "D" represents a single radio channel.

Figure 2-6 is theoretical; in the real world, the actual shapes of the cells are not symmetric and there is overlapping coverage from cell to cell and from sector to sector, though with different frequencies.

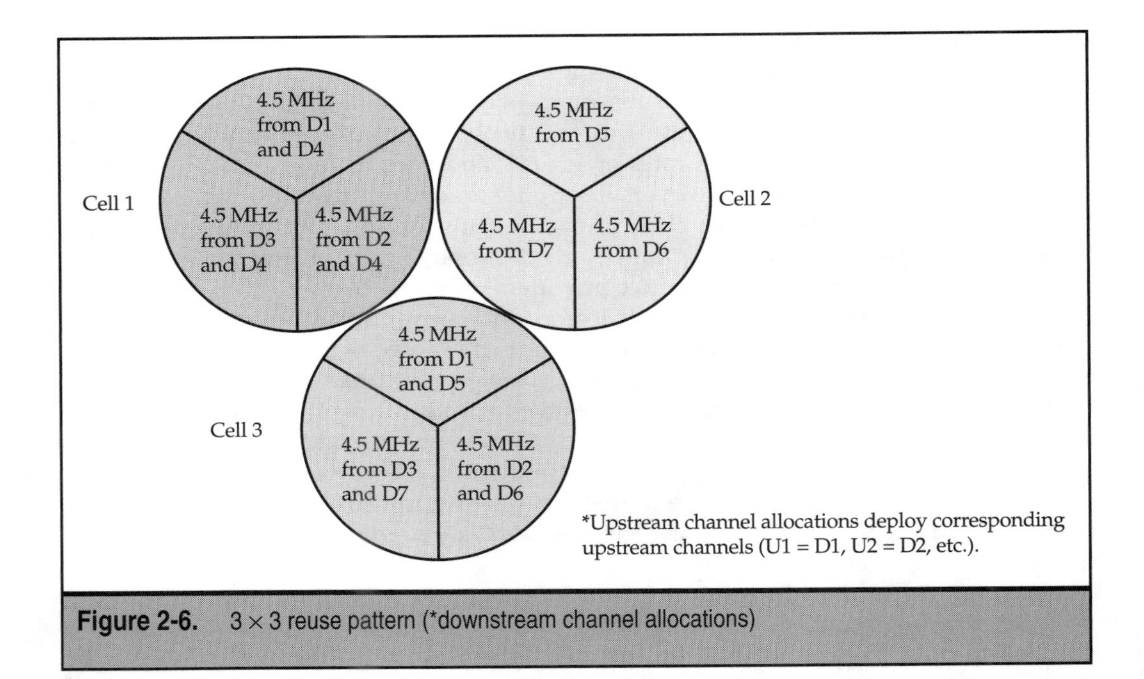

Figure 2-6. 3 × 3 reuse pattern (*downstream channel allocations)

Once a cell model is developed such as the one shown in Figure 2-6, geographic coverage can be accomplished as needed, as the basic pattern can be repeated over a theoretically infinite area. As stated at the beginning of this section, without a solid, well-thought-out frequency reuse and sector plan, a BBFW deployment will not scale. With a comprehensive model, a BBFW network can scale almost indefinitely as demand increases and metro areas grow.

SUPERCELL VS. MICROCELL NETWORKS

A *supercell* is a BBFW architecture that radiates a large geographic area and has low overall capacity. (Supercells typically have three or more sectors and a radius of 16 to 20 miles.) However, they may be attractive for initial deployments as they are inexpensive and easy to deploy, although they do require a tall tower, typically on the order of 500 feet or more. No frequency reuse is performed within the supercell, as the isolation factor required from sector to sector over the radius of a supercell is greater than can be provided with an antenna.

As the number of subscribers increases, additional sectors can be added along with narrow-beam antennas to aid in coverage at the far reaches of the cell. It is assumed that all of the subscribers are in a line of sight to the head end.

Supercells are best used in areas where minimal subscriber growth is projected and the total addressable market is well understood. The obvious reason for this is the data rate limitation in a given sector of the supercell. Once this number is achieved, no more subscribers can be taken in that sector. Therefore, revenue from that sector will go flat unless services or content is later provided over the BBFW link.

Microcells, on the other hand, are cells with radii of approximately 2–5 miles that can have a dozen or more sectors. Frequency reuse is mandatory on this architecture, and the relatively short range lends itself well to sectorizing via antennas that restrict the beam to anywhere from approximately 30 to 120 degrees. Microcells require more equipment, but the sectors and cells have higher rates of return because there is generally a high number of subscribers.

Additionally, as the number of subscribers increases, additional sectors and cells can be deployed to continue to take new customers. Compared with the supercell model, this keeps the revenue increasing for a substantially longer time, especially as content and services are later provided over the links. This type of architecture is well favored by service providers using unlicensed bands such as U-NII and 2.4 GHz in the ISM band. Most cells in the real world are closer in design to the microcell with a radius of 3–7 miles, which is an appropriate compromise between the high cost of a true microcell and the low performance of a supercell.

CHAPTER 3

Selected RF Fundamentals

In order to understand RF fundamentals, the first item to consider is electromagnetic energy, which is pervasive throughout the planet and intrinsically part of every electrical system. Every electrical system or device that carries electricity has an electromagnetic field.

The fundamental element of electromagnetism is the activity of electrons. Electrons are the particles that orbit around a nucleus of protons and neutrons. The essential art of RF is to get enough electrons to move in relative unison at the transmitting antenna such that they have a detectable effect on electrons at the receiving antenna.

James Clerk Maxwell (Scottish theoretical physicist, 1831–1879) predicted that electric and magnetic fields move through space as waves. His theory was based on the following four concepts:

1. Electric lines originate on positive charges and terminate on negative charges.
2. Magnetic lines always form closed loops- that is, they do not begin or end at any specific point.
3. A varying magnetic field induces a magnetic force and therefore an electrical field.
4. Magnetic fields are generated by moving charges (currents).

Maxwell's scientific work is often considered among the most important of the nineteenth century. He supported these four concepts with a corresponding mathematical framework to prove two phenomena: 1) electrical and magnetic fields are symmetric to each other, and 2) by rapidly changing the electrical field, one can produce a magnetic field well beyond the electrical field.

Maxwell discovered that when an electrical field was rapidly changed, magnetic waves were sent through space much like waves generated by a stone when tossed into water. Importantly, he also calculated that these waves travel at the speed of light.

What is also of considerable interest, Maxwell discovered that visible light, electrical waves, and magnetic waves have tightly linked relationships. His discovery that magnetic and electrical fields each generate the other is one of the greatest discoveries in science and had a profound influence on later developments in science and industry.

Both magnetic and electrical fields are created when energy is transferred from one point to another. It is this shift in energy that enables the phenomenon of magnetism and, by extension, radio frequency propagation. Therefore, to transmit information between two points, the energy must not be in a constant state; that is, it must change in either amplitude or frequency. Amplitude is the magnitude or strength of a waveform, while frequency is how often a complete wave passes a given point in space.

This change of energy is called modulation, of which there are many types, but all stem from either changes in amplitude or frequency. The issue of modulation will be treated more fully later in this book.

The prevailing concept is that RF and electron behavior are inseparable in that any time you have an electrical current in place, you also have an electromagnetic field in place—and vice versa.

The full electromagnetic spectrum is considerable in scope, and the portion used for commercial wireless is a small subset of the available spectrum currently understood. Interestingly, the portion of the electromagnetic spectrum occupied by visible light (light perceived by human eyes) is less than one percent of the total spectrum.

Electromagnetic fields affect conductors. Antennas are, in simplified terms, electrical conductors of a prescribed size and shape. To radiate a signal, the power from a transmitter must alternately push electrons into an antenna and then pull them out again. One *cycle* is one completed push and pull of electrons.

For each particular RF transmission and reception, there will be a certain number of cycles completed each second. This phenomenon is called frequency, and it occurs at very high rates. Table 3-1 indicates the relative cycles, or frequency, expressed in common terms. While there are no commercial terahertz radios at the time of this writing, experimental work is being performed in this area.

For general reference—and there is a fair amount of misuse of the terms—the spectrum is divided into large segments of frequency bands. The lowest bands, usually below 1 GHz, are referred to as the RF spectrum. The band between approximately 1 GHz and 10 GHz is known as the microwave spectrum, and the band between approximately 10 and 100 GHz is referred as the millimeter wave spectrum.

Certain electrical components commonly found on printed circuit boards are relatively frequency dependent. Put another way, these components and circuits operate at higher levels of performance when carrying certain frequencies. By designing circuits and selecting certain components that are more responsive to specific frequencies and less responsive to other frequencies, circuit designers can improve that system's performance by a factor of 10 to 50 times.

Unit of Frequency	Symbol	Cycles per Second
Hertz	Hz	Once
Kilohertz	kHz	1000
Megahertz	MHz	1 million
Gigahertz	GHz	1 billion
Terahertz	THz	1 trillion

Table 3-1. Units of Frequency

FREQUENCY

The concept of frequency is most easily understood through the analogy of the wave patterns created after dropping a pebble into a body of water, first considered in the context of radio frequency by Maxwell. The waves emanate outward with a specific shape from the point where the pebble entered the water. The waves weaken in energy as they move farther from the point the pebble entered the water, and they also change shape as they move away from the center. These waves are called *sine waves* because of their shape.

Sine waves depart a transmitting antenna in much the same manner as the waves that emanate from the point at which the rock enters the water. The number of times per second a complete sine wave departs from a transmitting antenna (or is received by a receiving antenna) is its frequency. The unit of time used is "hertz" (see Table 3-1).

The broadband commercial wireless portion of the electromagnetic spectrum begins at about the 1 GHz frequency and continues up through approximately 32 GHz, though experimentation and product development for commercial wireless communications is being explored at 60 GHz and higher. Currently, several countries are exploring the possibility of allocating (and auctioning) these millimeter wave bands now.

Sine waves are "managed" in terms of amplitude, frequency, and phase. This management is known as *modulation,* which has a tremendous effect on data output and other key RF attributes. This "management" is accomplished through the execution of very specific engineering designs, which entail the selection of specific electrical components. When considered individually, these components are sometimes referred to as "discrete components" or "discretes," a term that typically refers to their unique electronic characteristics. It is important to note that application-specific integrated circuits (ASICs) are replacing the use of discrete components to simplify design and reduce manufacturing cost and complexity. ASICs combine large numbers of discretes and circuits into a single component.

Discretes communicate with each other via electrical signals, which are typically transmitted over copper or gold paths that reside on a printed circuit board (PCB)or thin-film substrates manufactured from ceramics, depending on the frequency of operation. PCBs are complex in commercial radios and can have as many as eleven layers or more of conducting paths connecting the components. Commercial radios consist of a series of boards. Connectors reside within the walls of the "box" that surrounds the PCBs; these connectors are used to connect the radio to other "boxes" such as power supplies that convert AC power to the DC power required by radios as well as other devices such as antennas.

Software is of course necessary to determine how some of the discretes interact with each other and how the user interacts with the "box." Some of this software is fairly easy to modify, while other types of software reside inside certain types of discretes and cannot be altered without changing the component itself.

The characteristics of an RF wave are quite different at different points of the electromagnetic spectrum. Typically, the higher the frequency, the shorter the transmission distance for a given power and the greater the issue of multipath (covered later in this book). Conversely, for a given modulation technique, in general terms, the higher the frequency, the more data can be transmitted in a given time. Modulation issues are covered later in this chapter.

Characteristic	Microwave	Millimeter Wave
Frequency range	<10 GHz	>10 GHz
Cost	Less than millimeter wave	Higher than microwave
Complexity	Less than millimeter wave	Greater than microwave
Nominal Range	5–20 miles	< 5 miles
Affected By Weather?	Usually No	Usually Yes
Typical Use	Multipoint	Point-to-point
Multipath An Issue?	Yes	Generally No
License Required?	Usually	Usually

Table 3-2. Snapshot Comparison of Microwave Versus Millimeter Wave

RF systems fall into two broad categories: wavelengths less than 10 GHz (often referred to as microwaves) and wavelengths greater than 10 GHz (often referred to as millimeter waves). Each of them have certain advantages and disadvantages that pertain to cost, complexity, and the manner in which the waves propagate along a beam path. The differences are summarized in Table 3-2.

Since microwave wireless transmission equipment was first developed, manufacturers and operators have tried to mitigate the effects of reflected signals associated with RF. These reflections are called *multipath signals.* State-of-the-art BBFW products currently in the MMDS and U-NII bands not only tolerate multipath signals but, due to the repeatability of some multipath signals in the microwave band, can actually take advantage of them.

Several bands exist below 10 GHz for high-speed transmissions. These may be licensed bands such as MMDS (2.5 GHz), or unlicensed bands such as U-NII (5.7 GHz). Bands below 10 GHz have propagation distances up to 30 miles, while millimeter wave products are generally limited to ranges of approximately 5 miles or less. For a given transmit power, the maximum transmit distance is inversely proportional to the transmit frequency.

POWER TRANSMISSION

RF transmission is a power or energy transfer process, although a very inefficient one. Fortunately, this very low efficiency is offset by very sensitive receivers. A reception of one-millionth of the signal sent by the transmitter is actually considered a good signal in terms of strength. A typical RF signal sent between two sites is often 10,000 times weaker than that, yet quite usable.

The ratio of the power loss between two sites is called *path loss* or *free space path loss.* This refers to the energy that is lost during the time in which it is transmitted between two points. An important factor is that the path loss over a given path is typically constant, regardless of the amount of power used at the transmitting site, so variations due to modulation are quite faithfully reproduced at the receiver. This factor should not be confused with the concept that the rate of loss over a given path is constant; rather, the total loss over a path is relatively constant for different levels of power or modulation types. Path loss through free space will occur at a rate different than that which occurs when a path is partially blocked.

Path loss is important to calculate because an RF link (often simply referred to as a "link") between two points must account for distances and obstructions between transmitters and receivers. An appropriate link design will provide parameters for *maximum allowable path loss.* If the proposed solution has a free space path loss in excess of the maximum allowable path loss, the system will either lack appreciable bandwidth, have excess unreliability, or simply not work at all.

To overcome excessive path loss, you must increase the amount of power received at the receiving antenna. This can be accomplished in several ways. The most obvious is to increase the transmitted power, up to the limit set by the regulatory authorities. Other techniques can include providing more directional antennas, increasing the gain (sensitivity) of the receiving antenna, or increasing the elevations of both the transmitting and receiving antennas to clear the obstructions causing the path loss. You can also, in some circumstances, use repeaters.

Because path loss numbers are typically orders of magnitude, they are typically expressed on a *decibel* (expressed as dB) scale, which is logarithmic. Simply stated, a loss of three decibals means that half of your transmit power has been lost. The common notation for expressing this loss is –3 dB. Table 3-3 shows commonly used dB values.

The area closest to the transmitting antenna has two fields of energy that reside in the same space: an electrical field and a magnetic field. This field is called the *induction* field. Outside of this field, the RF wave loses any identity from the original electrical field.

The RF wave then exists independently of the original current or voltage that created it, and it will continue to radiate through any space where there are no conductors or absorbers. When the wave comes near a conductor, some of the energy will be absorbed by that conductor, and will set up miniature copies of the currents and voltages that originally sent off the radiation in the first place. Again, these "copies" are so small that we use the dB scale to more easily express this change in strength. What does not change over time or during transfer through space is the original rate (frequency) at which the waves were sent.

However, in terms of mobile systems such as cellular phones or pagers, moving a receiver toward or away from a transmitter induces a phenomenon known as the Doppler effect, which changes the rate at which the waves are received. If the receiver is moving

Factor of	dB
2	3
4	6
10	10
100	20
1000	30

Table 3-3. Decibel (dB) Values

toward the transmitter, the waves are received at an increased rate; conversely, if the receiver is moving away from the transmitter, the rate at which the waves are received is reduced. This is typically not an issue in BBFW, since by definition it is "fixed."

When signals are sent through a medium such as the atmosphere, the speed of the transmission is slowed slightly from the original transmission speed of 186,000 miles per second (again, the speed of light). An interesting physical phenomenon then occurs, which is that the frequency remains constant though the waves are slowed down. This is due to the fact that there are many more atoms in the atmosphere than in a vacuum.

Each atom along the beam path is affected by the electrons, which must forward the energy on an atom-by-atom basis. The more atoms, the more times the energy is transferred. As there is a miniscule loss of energy each time the energy is transferred from one atom to the next, the wave not only decreases in energy, but is slowed down in speed. The question that arises then is: How can the wavelength shorten and yet the frequency remain constant? The answer is perhaps best understood with a simple analogy.

Imagine you are standing next to a railway where cars are passing you at the rate of exactly one per second. If the train is moving at 60 miles per hour, each car must be 88 feet long. If the train passes you at 30 miles per hour, the cars must be 44 feet long to pass at the same rate. The slower cars would be half the length because they're traveling at half the speed.

This is, in effect, what happens to radio waves. When they pass through a medium such as atmosphere or water, the length of the waves shorten but the frequency remains the same. Some of the wavelength is converted to heat, which is the prevailing principle behind the microwave ovens we are so familiar with. Microwave ovens operate at a frequency of 2.4 GHz, which is the frequency that has a particular reaction with water; that is, it excites the water molecules, which is another way of saying it heats water. It is the water content in foods that heats, which then heats the rest of the meal content.

The key point here is that as frequency is the most reliable aspect of RF radiation; it's the most accurate measurement of a signal in relation to others like amplitude, wavelength, and phase.

FREQUENCY BANDS

The U.S. government, like every other major government in the industrialized world, has set up groups of frequencies (bands) for use. There is no scientific or engineering reason for how these divisions are established, but the division and allocation of these bands is essential to efficient and reliable use by the public.

Not all of the frequencies shown in Table 3-4 are relative to the scope of this book, but all are shown for reference and context. It's also important to note that in general, most of these terms are no longer used. The wavelengths indicated are shown by general figures only (close estimates).

You'll note that these bands represent a 10:1 ratio of wavelength to frequency. Again, this is not driven by scientific or engineering principles, rather it's an easy way to keep frequencies in blocks that are generally consistent with the way the carrier frequency propagates through the air.

SHF is also generally known as millimeter wave frequency while UHF is also known as microwave.

Band	Frequency Range	Wavelength At Lower End Of Spectrum
Very Low Frequency (VLF)	0 kHz–30 kHz	100km
Low Frequency (LF)	30 kHz–300 kHz	10km
Medium Frequency (MF)	300 kHz–3 MHz	1km
High Frequency (HF)	3 MHz–30 MHz	100m
Very High Frequency (VHF)	30 MHz–300 MHz	10m
Ultra-High Frequency (UHF)	300 MHz–3 GHz	1m
Super-High Frequency (SHF)	3 GHz–30 GHz	100mm
Extremely High Frequency (EHF)	30 GHz–300 GHz	10mm

Table 3-4. Frequencies as Defined by the U.S. Government (Antiquated)

MODULATION SCHEMES

Any signal that can be translated into electrical form, such as audio, video and data, can be modulated and sent over the air. Data, such as e-mail, is the easiest to modulate, while video with voice is the most difficult for which to retain fidelity to the original source.

Modulation is the technique of turning bits into something carried by the carrier frequency over the air. Carrier frequency has no intelligence; the modulated data carries the intelligence (i.e., data) between two points.

The purpose of commercial RF systems is to carry modulated waves of energy, that is, information. Earlier in this section, it was stated that it is the *changes* in either amplitude, frequency, or phase that enable data to reside on a wave of radiated energy. Modulation is the difference between a steady-state RF signal (a signal that is nonchanging, referred to as a continuous wave [CW] tone) and one that carries information.

The selection of modulation schemes is based more or less on the compromise between maximizing bandwidth through high spectral efficiency and bit loss through complexity of scheme. Also, in general, the better the spectral efficiency, the worse the power efficiency and vice versa. Simpler systems such as phase-shift keying (PSK) are very robust and easy to implement because they have low data rates. In PSK modulation, the shape of the wave is modified in neither amplitude nor frequency, but rather in phase. The phase can be thought of as a shift in time.

At lower frequencies (such as 2.5 GHz), the selection of a modulation scheme is very important because there is inherently less bandwidth in general to work with than at higher frequencies such as LMDS uses (28 GHz in the U.S.). The proper term with respect to this fundamental is *spectral efficiency,* that is, how to get the most out of the available bandwidth. The common term used for spectral efficiency is bits per hertz.

There are many different modulation schemes. Table 3-5 includes a partial list.

While Table 3-5 includes eleven different types of modulation schemes, all of them are based on only three methods, although they can be used in combination with each other. It must be noted that there are many derivatives of these three basic types of modulation in contemporary systems:

- Amplitude (AM)
- Frequency (FM)
- Phase (PM)

Amplitude modulation changes are when the amount of power in a single sine wave sent to an antenna is changed or modulated. Frequency modulation is when the frequency is changed (up or down) to carry information, and phase modulation is when the timing of a specific portion of a sine wave is managed. Regarding phase modulation, all sine waves, when transmitted, have an associated phase between 0 and 360 degrees relative to the center of the beam path. For a true sine wave, the phases of each signal are constant. Changing the phase of a given sine wave allows the receiver to interpret this as information. The most common type of phase modulation is binary phase-shift keying

Symbol	Modulation Scheme
AM	Amplitude Modulation
FM	Frequency Modulation
SSB	Single Side Band
PM	Phase Modulation
CW	Continuous Wave (telegraphy)
PCM	Pulse Code Modulation
VSB	Vestigial Side Band
BMAC	Type B Multiplexed Analog Components
QAM	Quadrature Amplitude Modulation
BPSK	Binary Phase-Shift Keying
QPSK	Quadrature Phase-Shift Keying

Table 3-5. Various Contemporary Modulation Schemes

(BPSK). If a sine wave's phase is rotated from 0 to 180 degrees, the receiver can interpret the rotation as a shift in information from a 1 to a 0.

The greatest gains in radio efficiency are had when multiple types of modulation schemes are used in concert with each other. Combining AM with PM provides very robust modulations known as quadrature phase-shift keying (QPSK) and quadrature amplitude modulation (QAM). Different levels of QAM exist to provide higher and higher modulation capability. The true advantage of these advanced types of modulation is that they provide the ability to transmit more information in a given bandwidth of spectrum.

For instance, QPSK can provide twice the transmitted data as BPSK, while 16 QAM can provide four times more information than QPSK. Each, in turn, provides more and more information transmission capability relative to less complex modulation approaches or schemes. Modulations as high as 64 QAM are in use today in commercially available wireless systems, while modulations such as 256, 512, and 1024 QAM are being used and experimented with in closed systems such as cable. ("Closed" means the modulated information travels through coaxial copper as opposed to atmosphere.) The number in front of the term "QAM" indicates the number of different modulation states available for a given modulation.

These advanced modulation states also include specific compromises. As the modulation complexity increases, errors are more likely in the transmission. While it is relatively easy for a receiver to distinguish BPSK states, that is, the 180-degree differences in the received signals, the difference between two states of 64 QAM can be as small as a few tenths of a degree and a few millivolts in amplitude. Better, cleaner, and stronger signal paths (the technical term is signal to noise [SNR] or signal to interference plus noise [SINR]) are required to successfully transmit these advanced modulations.

Many modern fixed microwave communication systems are based on quadrature amplitude modulation (QAM). The 64 QAM level is relatively common for state-of-the-art microwave band products, while QPSK modulation is the most common modulation technique used in millimeter wave solutions.

In binary phase-shift keying (BPSK), only one bit is transmitted per cycle (a complete cycle with bits of information is called a symbol). In more complex modulation schemes, more than one bit is transmitted per symbol.

The modulation scheme QPSK (quadrature phase-shift keying) is similar to the BPSK. However, instead of only two separate phase states, QPSK carries two bits per symbol. Like BPSK, QPSK is used because of its robustness. However, since it only modulates two bits per symbol, it still is not very efficient for high-speed communications. Hence, higher bit rates require the use of significant bandwidth.

Even though QPSK does not make changes in amplitude, it is sometimes referred to as 4 QAM. When four levels of amplitude are combined with the four levels of phase, we get 16 QAM. In 16 QAM, two bits are encoded on phase changes and two bits are encoded on amplitude changes, yielding a total of four bits per symbol.

This approach to modulation can be expanded out to 64 QAM and 256 QAM or higher. The higher the density in QAM, the higher a signal-to-noise ratio (SNR) must be maintained in order to meet the required bit error rate (BER). BER is the ratio of corrupted bits to good ones and is generally expressed exponentially. State-of-the-art systems have BER as low as 10^{-11}.

As the modulation complexity increases, the risk of incorrectly interpreting the correct state also increases. In BPSK, the signal can be correctly identified despite a very high noise level. In QPSK, the possibility of error doubles (four states versus two states and so on) as the complexity increases. The level of modulation complexity used in a system must be traded off against several factors. Available bandwidth, SINR, frequency spectrum, distance traveled, required signal reliability, and more are issues that must be addressed before selecting a modulation for transmission.

How the data is encoded also plays an important part in the equation. The data is usually scrambled, and a significant amount of forward error correction (FEC) data is also transmitted. Therefore, the system can recover those bits that are lost because of noise, multipath, and interference. A significant improvement in BER is achieved using FEC for a given SNR at the receiver.

In wireless QAM systems, decision feedback equalization (DFE) is used to mitigate the effects of the intersymbol interference (ISI) caused by multipath. Delay spread is a condition when a replicated signal arrives later than the original signal. When this occurs, the echoes of previous symbols corrupt the current symbol.

Direct sequence spread spectrum (DSSS) is a signaling method that avoids the complexity and the need for equalization. Generally, a QPSK modulated signal is used. This narrowband signal is then multiplied (or spread) across a much wider bandwidth.

Code Division Multiple Access (CDMA) has been adopted for many advanced IP RF systems in the mobile telephone environment and is used to allow several simultaneous transmissions to occur. CDMA is currently being expanded from Narrowband CDMA (1.23 MHz) to Wideband CDMA (5–15 MHz) to allow for greater carrying capability. The value of this is that each data stream can be voice, video, or data and broadcast within the same transmission. In practical terms, it gives the ability to provide different services within a given transmission area. A single customer could send video broadcasts, perform CLEC services such as data traffic, and provide Web browsing capability all via their cell phone.

With CDMA, each data stream is multiplied with a pseudorandom noise code (PN code), which is another way of stating that each data stream is supplied with its own identification number. All users in a CDMA system use the same frequency band. The intended signal is recovered by using the PN code. Data transmitted by other users or a even a native transmitter is filtered out during the reception phase. It's important to note that all receivers within transmission range will receive the signal because it's on the same frequency, but only the intended receiver recognizes the specific code and allows the signal to pass the filters.

Another advantage of CMDA is that the amount of bandwidth required is now shared over several users, giving more users per transmitter. However, in systems where there are multiple transmitters and receivers, proper power management is needed to ensure that one transmitter does not overpower other users in the same spectrum. These power management issues are mainly confined to CMDA architectures. Frequency division multiple access (FDMA) and time division multiple access (TDMA) systems are more tolerant of fluctuations in power.

In a frequency division multiplexing (FDM) system, the available bandwidth is divided into multiple data carriers. The data to be transmitted is then divided between these subcarriers. Since each carrier is treated independent of the others, a frequency guard band must be placed around it, which is another way of saying that no other data will be carried on an adjacent frequency. This guard band lowers the bandwidth efficiency. The most common example of a FDM system in operation today is the Advanced Mobile Phone Service (AMPS).

It's interesting to note here that AMPS and CDMA typically operate in the same frequency spectrum, but because AMPS is in the frequency domain and narrowband by definition and CDMA is in the code domain (using spreading techniques), they do not interfere with each other.

In some FDM systems, up to 50 percent of the available bandwidth is used for guard bands, which prevents that portion of the bandwidth from carrying data. In most FDM systems, individual users are segmented to a particular subcarrier; therefore, their burst rate (short peaks of maximum data rates) cannot exceed the capacity of that subcarrier. If some subcarriers are idle, their bandwidth cannot be shared with other subcarriers.

Several techniques have been used to make digital modulation schemes more robust, including: QAM with decision feedback equalization (DFE), direct sequence spread spectrum (DSSS), frequency division multiplexing (FDM), and orthogonal frequency division multiplexing (OFDM). DSSS, FDM, and OFDM are called *spreading techniques.*

Spreading techniques carry one or more modulation schemes across a set of frequencies. Certain spreading techniques such as Direct Sequence Spread Spectrum (DSSS), Frequency Hopping Spread Spectrum (FHSS) and Orthogonal Frequency Division Multiplexing (OFDM) are becoming very popular because of their high spectral efficiency and ability to work in high-interference environments. The popular cellular telephone technique called CDMA (Code Domain Multiple Access) is a type of DSSS in use today.

Multi-Path

RF systems below 10 GHz are only mildly affected if at all by climatic changes such as rain and snow. These frequencies are generally not absorbed by objects in the atmosphere, such as dust and air pollution, but they are more readily absorbed by vegetation. On the other hand, products in these frequencies tend to reflect off various objects such as buildings, concrete, and water and thus can result in a high amount of multipath.

Multipath is generally not an issue with millimeter wave products, since most of the multipath energy is absorbed by the physical environment. However, products in this frequency range tend to have much shorter ranges for a given power level, as waves of this length are typically absorbed into solid objects or scattered into vegetation.

Energy transmitted from an antenna radiates outward in a direction perpendicular to the beam path, and the farther the distance from the transmitter, the wider the beam.

Multipath, as shown in Figure 3-1, is when the energy from these waves strikes objects such as buildings, water, roads, and even certain types of vegetation (under the right conditions) such as ripened wheat.

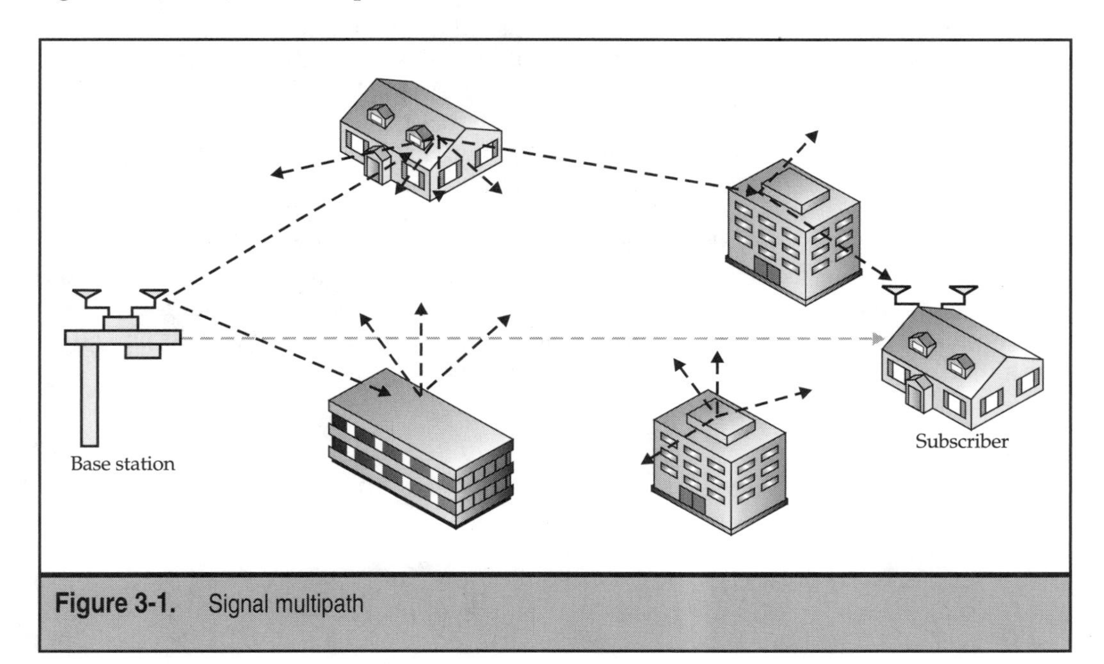

Figure 3-1. Signal multipath

Signals, when bounced off buildings, pavement, water, or the like, travel a longer path than the direct line of sight (called the incident signal). These signals containing the same information arrive at a given receiver at different periods in time. This offset in time results in a phase shift from the original, or incident, signal. Depending on the specific phase relationship between the multipath signal and the incident "out-of-phase" signals, there is generally a distortion of the signal at the receiver. This is because the received signal is a combination of an incident signal and one or more echoed signals; in other words, the incident signal and the multipath signal interact destructively. The time between the first received signal and the last echoed signal is called the *delay spread*. This delay spread confuses the receiver and results in what is termed *inter-signal interference* (ISI).

Further, the greater the distance of the echoed signals, the longer the delay and the lower the power of the echoed signal. One might think that the longer the delay, the better the reception would be. However, if the delay is too long, the reception of two or more echoed symbols can interact to the disadvantage of the receiver.

In analog systems such as television, you can see this multipath effect as a "ghost" image on your television. Occasionally, no matter how much you adjust the set, the image does not go away, though these days this primarily occurs with older systems. In these analog systems, this is an annoyance. In digital systems, it usually corrupts the data stream and causes loss of data or lower performance. Correction algorithms must be put in place to compensate for the multipath, resulting in a lower available data rate. Digital systems not designed to deal with ISI give up substantial performance and reliability.

In LOS (line-of-sight) environments with very short range, multipath can be minor and overcome easily. The amplitudes of the echoed signals are much smaller than the primary one and can be effectively filtered out using standard techniques. However, in obstructed environments where the incident signal is reduced in strength, echoed signals may have higher power levels or combine to have much stronger signals than the original signal. In this scenario, resolving the problem with conventional methods such as equalization is more difficult or impossible, thus limiting the deployment of these "line-of-sight" systems.

In all the previous discussions, the multipath occurs from stationary objects. However, multipath conditions can originate from objects that are mobile within the radiating environment. These mobile objects can initiate varying degrees of multipath from one sample period to the next. This is called *time variation.* Superior digital systems withstand rapid changes in the multipath conditions, referred to as *fast fading.* In order to deal with this condition, digital systems need fast automatic gain control (AGC) circuits and adaptive equalizers, as discussed in the following sections, to compensate.

OFDM

In Orthogonal Frequency Division Multiplexing (OFDM), multiple carrier frequencies (or tones) are used to divide the data across the available spectrum, similar to FDM. Unlike traditional FDM, however, OFDM systems spread a given piece of information over several subcarriers in a spreading technique. Additionally, in an OFDM system, each tone is considered to be orthogonal (independent or unrelated) to the adjacent tones. Each tone is a frequency integer (one whole number) apart from the adjacent frequency and does not therefore require a guard band around each tone. At any given point in time when one of the carriers is at maximum voltage (one), all other carriers are at minimum voltage (zero), thereby eliminating the need for guard bands between the individual carriers.

Since OFDM only requires guard bands around a set of tones, it is more efficient spectrally than FDM. Because OFDM is made up of many narrowband tones, narrowband interference will degrade only a small portion of the signal and has little or no effect on the remainder of the frequency components.

OFDM systems use bursts of data to minimize intersymbol interference (ISI) caused by delay spread. Again, delay spread is the difference in time from when the intended signal (the one that took the shortest path to the receiver) arrived to when another signal arrived a handful of microseconds later due to multipath (because it took a longer route to the receiver).

Data is transmitted in bursts, and each burst consists of a cyclic prefix followed by data symbols. For example, an OFDM signal occupying 6 MHz is made up of 512 individual carriers (or tones), each carrying a single QAM symbol per burst. The cyclic prefix is used to absorb late-arriving signals due to multipath (transients) from previous bursts. An additional 64 symbols are transmitted for the cyclic prefix. For each symbol period, a total of 576 symbols are transmitted, but there are only 512 unique QAM symbols per burst.

In general, by the time the cyclic prefix is over, the resulting waveform created by the combining multipath signals is not a function of any samples from the previous burst. Hence, there is no ISI. The cyclic prefix must be greater than the delay spread of the multipath signals. In a 6 MHz system, the individual sample rate is 0.16 μsecs. Therefore, the total time for the cyclic prefix is 10.24 μsecs, greater than the anticipated 4 μsec delay spread.

Some OFDM systems only use QPSK for the modulation scheme. If 16 or 64 QAM is used, then the amount of data transmitted significantly increases.

In addition to the standard OFDM principles, the use of spatial diversity can increase the system's tolerance to noise, interference, and multipath. Common throughout the microwave industry, spatial diversity has been used for years to reduce the impact from multipath interference. Spatial diversity is another way of stating "second antenna," which is used on the receive side of the transmission link only.

Diversity is based on the concept that a set of multipath signals arriving at a given receiver exhibits a definable and generally repeatable set of characteristics. If a second receive antenna is placed some distance away from the first (minimum distance between the antennas depends on the frequency used), it will receive a second, unrelated, set of signals. If deployed properly, one of the two paths will exhibit a stronger signal than the other path. A receiver can select and use the strongest signal from the two signal paths or combine the two paths for a composite signal. In the OFDM technique, this is referred to as "VOFDM," where the "V" stands for "vector," which means it is located either vertically or horizontally off to one side of the beam path.

A well-understood example of spatial diversity is when you stop your car at a red light and the radio broadcast in your car emits a great deal of static—or cuts out altogether. We have all moved our car a few feet (or even less than a foot in some instances) to regain the high-quality broadcast signal. By moving the car, you have relocated your antenna outside of a void or null in the local reception area.

Spatial diversity is a widely accepted technique for improving performance in multipath environments. Because multipath is a function of the collection of bounced signals, that collection is dependent on the location of the receiver antenna. If two or more antennas are placed in the system, each could have a different set of multipath signals. The effects of each channel would vary from one antenna to the next; therefore, carriers that may be unusable on one antenna may become usable on another. Spacing between the antennas is usually at least ten times the wavelength, and significant gains in the SINR are often obtained by using multiple antennas. Typically, a second antenna adds about 3 dB in LOS environments and up to 14 dB in non-LOS environments.

Neither QAM with exotic filters for multipath nor DSSS systems elegantly scale with increases in bandwidth. Designs for high-speed communication links using these techniques yield either very expensive and complex systems or ones that cannot guarantee the BER rates needed to consistently deliver data with the same relative quality of fiber optics in obstructed radio paths.

In comparing today's modulation techniques, a very strong case can be made for OFDM using diversity (VOFDM) as the only practical solution to resolve multipath issues existing in obstructed environments.

The ability to use wireless broadband products in obstructed environments opens up many possibilities that have been closed to non-OFDM technologies. One of the key elements is enhanced "harvest rate" over systems which cannot provide quality transmissions in obstructed environments, as the end customer can expect upward of 30 percent greater coverage area for a given transmitter.

Microwave systems can therefore be deployed in dense urban environments where multipath is pervasive. Instead of tall microwave towers, simple rooftop mounts can be used. In general, VOFDM allows the user to create simple, spectrally efficient, multipath-resistant, robust data communications links in areas where none have gone before.

WLAN SPREAD SPECTRUM

Most WLAN products utilize spread spectrum techniques. In the U.S. in the 2.4 GHz ISM band, the IEEE 802.11b standard requires signal spreading, and it has been adopted by the FCC. There are two kinds of spread spectrum available, direct sequence spread spectrum (DSSS) and frequency hopping spread spectrum (FHSS). DHSS typically has better data carrying capability, while FHSS is typically more resilient to interference. In state-of-the-art systems, DHSS has provided significant improvements in processing gain, which is another way of stating that bits are managed such that DHSS has the same relative performance of FHSS without the bandwidth loss attributed to FHSS due to extra processing.

A commonly used analogy to understand spread spectrum is that of a series of trains departing a station. The payload is distributed relatively equally among the trains, which all depart at the same time. Upon arrival at the destination, the payload is taken off each train and collated. Duplications of payload are common to spread spectrum, so that when data arrives excessively corrupted, or fails to arrive, the redundancies inherent to this architecture provide a more robust data link. Payloads are replicated from seven to eleven times; this replication is called "chipping."

With DSSS, all trains leave in an order beginning with Train 1 and ending with Train N, depending on how many channels the spread spectrum system allocates. In the DSSS architecture, the trains always leave in the same order, though the numbers of railroad tracks can be in the hundreds or even thousands.

With the FHSS architecture, the trains leave in a different order; that is, they do not depart sequentially from Train 1 to Train N. In the best of FHSS systems, trains that run into interference are not sent out again until the interference abates. In other words, in systems using FHSS, certain frequencies (channels) are avoided until the interference abates. It should be noted that not all FHSS systems have this capability. The FH system with its random hopping between different channels will, at times, hop into a band with sufficient interference to obliterate the signal.

In time, the system will hop out of that band into another clean area and can then retransmit the lost information. Channels with very high interference have payloads that arrive with excessive corruption. When the corruption is excessive, the channel is not used for a set period of time, at which point the channel is "tested" again. If the data arrives sufficiently intact, the channel is again used routinely. If the data still arrives with excessive corruption, it is not used for an even longer period of time until the channel is tested once again.

Interference tends to cover more than one channel at a time. Therefore, DSSS systems tend to lose more data from interference as the data is sent out over sequential channels. FHSS systems "hop" between channels in nonsequential order. Either approach is appropriate and dependent on customer requirements, the selection criteria primarily being whether severe multipath or RF interference exists.

TYPICAL BBFW PT-PT RF SYSTEM BLOCK DIAGRAM

A typical point-to-point RF system is shown in Figure 3-2.

Generally, state-of-the-art BBFW systems have several key elements on each end of a link, which are:

- ▼ A router or switch
- ■ A radio (generally called a transverter or up/down converter)
- ▲ An antenna

Router or Switch

Router selection is dependent on bandwidth requirements and whether the device is at the head end of a PT-MPT system or the customer premises. These devices range from the small routers for residential deployment that cost on the order of $500 to $1,000 on a point-to-multipoint basis all the way up to a Cisco 7200 series uBR router for IP protocol

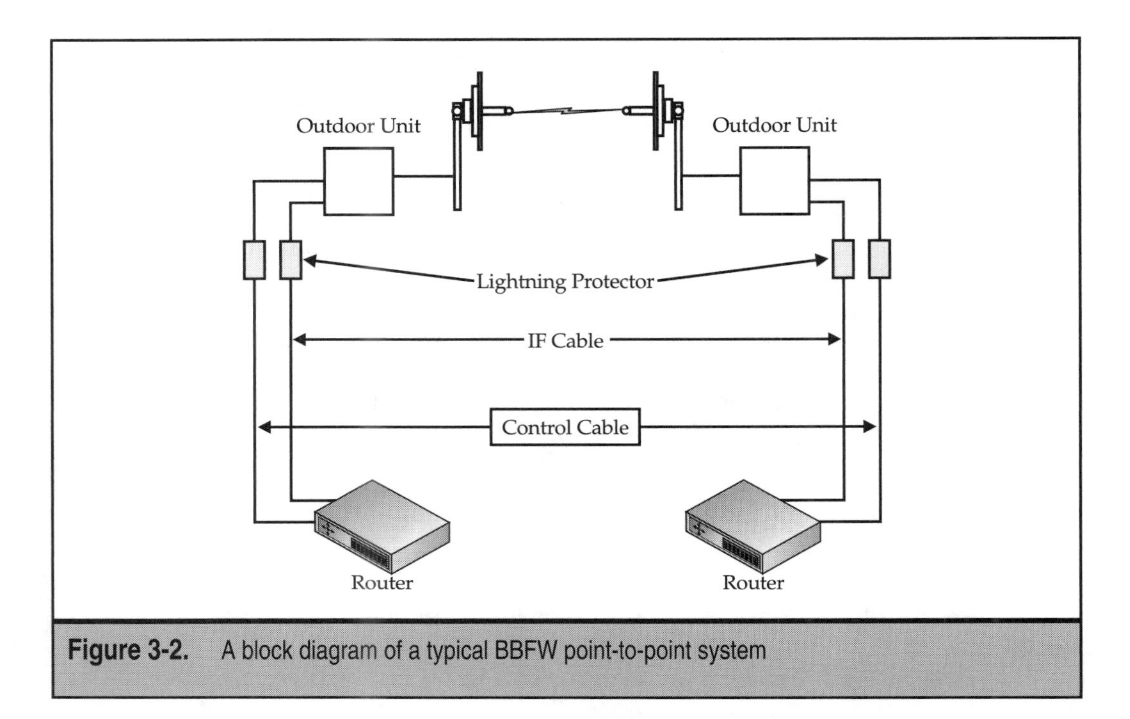

Figure 3-2. A block diagram of a typical BBFW point-to-point system

hub use, costing approximately $65,000, or an Cisco MGX 8850 for an ATM interface to the wireless transport gear, which fully loaded costs approximately $1 million.

Outdoor Unit (ODU)

The ODU is the device that converts the carrier frequency (the frequency directed between two or more points) to an intermediate frequency sent down to the router. The ODU also governs carrier frequency and includes a transverter in the IP system to select whether the top half or bottom half of a selected series of frequencies will be dedicated to transmission or reception. In some systems, the outdoor unit may take data directly from the modem, modulate it, and upconvert it to the microwave band. In other systems, there is no "outdoor" unit at all; rather, the modulation and upconversion are done directly at the modem, and the microwave signal is sent directly from the indoor unit to the antenna.

Miscellaneous Cables, Connectors, and Antenna Mast

Signal, power, and control are carried to and from the ODU via coax cables. The nominal length of these cables is approximately 200–400 feet from antenna to router (approximately the height of a 20-story building), but length can exceed 400 feet. By federal law, lightning arresters must be deployed inline at distances no greater than every 100 feet.

The antennas typically reside on a three-inch-diameter pole (mast) appropriately secured to the building or site. An antenna can also be attached by other mechanical means to elevated platforms.

Antenna

There are several types of the most common antennas, but where possible, most companies deploy a parabolic type (dish) antenna, as this antenna has the best performance and is most directional; however, patch and yagi antennas are very common.

Most BBFW systems carry more data than voice. State-of-the-art systems differentiate between the different priorities data traffic requires. These differentiated levels are managed by what is called Quality of Service (QoS), which topic will be treated in further detail later in this work.

PT-PT RF links are symmetric, meaning the radios at each end of the link are identical. PT-PT links typically have much greater range than PT-MPT links and have more bandwidth. When used in the last mile, directly to the subscriber, PT-PT links tend to generate greater amounts of revenue than PT-MPT links, as they are considered high-value high-speed trunks. More commonly, point-to-point links are deployed in the overhead capacity to provide backhaul service to an array of PT-MPT links.

PT-MPT links generally have what is called a head end (HE) radio and a smaller RF radio at the Customer Premises End (CPE) of each link. The CPE radio is situated at an MxU, an SMB, or a residential site. Residential sites have the smallest radios and the smallest bandwidth, at this writing, typically up to 1.5 Mbps.

PT-PT links don't share bandwidth with other radios, while PT-MPT links have a head end that distributes bandwidth among as many as several hundred (or more) CPE sites. As a very rough rule of thumb, PT-MPT links typically carry about half of what a PT-PT link can carry, although it's easy to configure links that don't adhere to this premise. This is generally for two reasons: 1) there is substantially more overhead associated with managing a large number of subscribers as opposed to a single, large-bandwidth subscriber, and 2) a similar amount of energy is used to cover a geographic area of dozens of miles or more as opposed to having a relatively tight beam that resides between two points.

PT-PT links have very narrow and long beam paths, while PT-MPT beam paths tend to be the opposite—relatively short and quite wide. A typical PT-PT link can be up to 20 miles (or more) and approximately 50 to 75 feet wide. PT-MPT cells are commonly made up of three or four "sectors," which are typically 90 to 120 degrees wide and three to seven miles in range.

BBFW networks are commonly an amalgam of PT-PT and PT-MPT links. A typical network is shown in Figure 3-3.

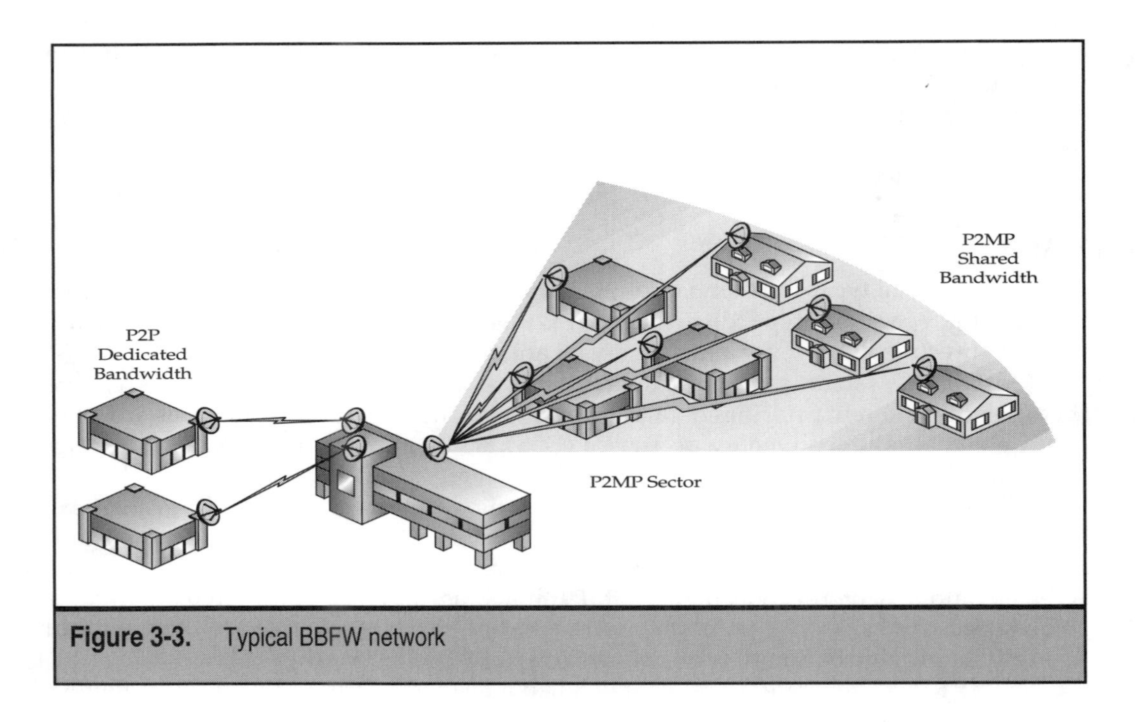

Figure 3-3. Typical BBFW network

EARTH CURVATURE CALCULATION FOR LINE-OF-SIGHT SYSTEMS

Line-of-sight systems that carry data over distances in excess of 8 miles will require additional care and calculations. Since curvature of the earth causes bulges at the approximate rate of 12 feet for every 18 miles, Table 3-6 can be used to maintain line-of-sight status.

While observing these calculations, it's important to remember that this accounts only for earth bulge. Vegetation, such as trees, and other objects, such as buildings, must have their elevations added into this formula. A reasonable rule of thumb is 125 feet of elevation at both ends of the data link for a distance of 25 miles, but this should be considered an approximation only.

That which must clear the earth bulge and other obstructions is the Fresnel zone (pronounced "frennel"), which is the ellipsoid-shaped area directly between the antennas. The center area in this zone is of the greatest importance and is called the "first Fresnel zone." While the entire Fresnel zone covers an area of appreciable diameter between the

Distance (Miles)	Earth Bulge (Miles)
8	8.0
10	12.5
12	18.0
14	24.5
16	32.0

Table 3-6. Earth Bulge Table

antennas, the first Fresnel zone is considered as a radius about the axis between the antennas. A calculation is required to determine the radius (in feet) that must remain free from obstruction for optimal data transfer rates.

The industry standard is to keep 60 percent of the first Fresnel zone clear from obstacles. Therefore, the result of this calculation can be reduced by up to 60 percent without appreciable interference. This calculation should be considered as a reference only, and does not account for the phenomenon of refraction from highly reflective surfaces.

PART II

Business and Market Strategies for BBFW Providers

CHAPTER 4

BBFW Versus Other Broadband Methodologies

BROADBAND COMPARISONS

The Strategis Group published a report in early 2001 on projections of broadband into the residential market; it states that BBFW will not play a major role in terms of total marketshare. This may be because BBFW providers have generally pursued the service provider, business to business (B2B), small and medium business (SMB), and multitenant/multidwelling unit (MxU) market segments to the exclusion of the residential market, or at least they have experienced a lack of success in penetrating that market. The findings of the Strategis report are probably justifiable given the dearth of appropriately designed and priced BBFW products for the residential market.

On the other hand, providers of cable and DSL equipment have focused on the residential market to the general disregard of the markets targeted by BBFW providers. The result is that both DSL and cable modems at this writing have captured the bulk of the residential market space for broadband communications. It is the author's belief that this will not change any time before 2004 at the earliest. Dominance by BBFW providers will rely on the combination of the DSL and cable modem markets fumbling away their market shares through execution problems in connection with a continually increasing demand for broadband.

According to the Strategis Group, in the year before this writing, broadband access to the residential market rose by 230 percent, with projections of approximately 36 million broadband subscribers by the year 2005, as Table 4-1 shows. If these projections bear out, it makes the residential market one that cannot be ignored. This will prove to be a significant challenge for BBFW providers, as the retooling of their products from ones geared toward the MxU and SMB market is nontrivial.

Importantly for BBFW equipment providers, these projections mean that there is a large customer base that will resell their equipment to the residential market. This fact should draw them into providing RF gear that is cost effective (at the retail price of approximately $300 U.S.), easy to install, and readily available from retail outlets such as Tandy Corporation (Radio Shack). There will also need to be a comprehensive technical

	1999	2000	2001	2002	2003	2004	2005
Cable	1.25	4.11	7.12	9.57	11.88	14.19	16.13
DSL	0.58	1.93	4.22	6.88	9.38	11.87	14.22
BBFW	0.01	0.06	0.19	00.95	1.96	3.04	4.71
Other	0.00	0.01	0.10	0.15	0.30	0.50	0.80

Table 4-1. Residential High-Speed Households by Technology (Millions), Report by Strategis Group, 2001

support hierarchy in place for the myriad of users who will require support to get online with BBFW gear.

Even so, significant hurdles will remain in place for a few years, as the same resources that install cable and DSL modems are generally the same ones that would provide BBFW installation into the residential market. While SPs, telcos, and cable companies claim they've made great strides in overcoming this hurdle, what the market is seeing on the street is a decreasing level of expertise and customer satisfaction as an increasing share of the deployment task falls to small local companies to provide installation and maintenance.

This problem demonstrates the necessity for gear to be self installed and provisioned remotely and simply. In other words, the gear has to be very user friendly, something that is achievable but not prevalent on the market just yet. While start-ups are providing DSL and cable modem interfaces to home networks, these small companies lack scaling and resources to provide any meaningful solution to the overall problem.

While an ever-increasing array of broadband choices is available to customers, each of the options has both pros and cons. Perhaps the greatest challenge for the service provider is offering an increasing array of choices to the user. The biggest problem is the issue of meeting the increasing demand, which has commonly outstripped the ability of service providers to install and maintain service. Rapid growth of end users also quickly elevates the issue of providing sufficient/growing bandwidth to customers while that bandwidth is divided between existing and new customers. These lessons were and are well understood by DSL and cable modem service providers.

There are, at this writing, four prevalent types of broadband access available in the world today. These types are:

- Fiber
- DSL (digital subscriber lines)
- Cable (cable modems)
- BBFW

Table 4-2 compares the common broadband access technologies.

Fiber

Less than three percent of the 750,000 commercial buildings in the U.S. are served by fiber. This number is even lower in most countries around the world, with the relatively rare exceptions of a few countries such as Singapore that are very extensively connected by fiber.

Fiber deployment is backlogged and is the source of an industry bottleneck. This is primarily due to shortages of the construction labor required to trench and install the fiber into the ground or to install it along power or telephone lines. Fiber is also relatively expensive to install—anywhere from ten to more than one hundred dollars U.S. per

Transport Method	BER	Maximum Service Range	Primary Advantages	Primary Disadvantages	Can Relocate?
BBFW	10^{-11}	25 miles	Quick install, scalable, low BER	Expensive, requires comprehensive installation procedure and link monitoring	Yes
DSL	10^{-6} at 25 Mbps (estimated BER)	3,000 ft from CO	Inexpensive	Bandwidth drops quickly after 3,000 feet, substantial install backlog	Modem yes, service to new site not guaranteed, requires fiber to the curb
Cable	N/A	N/A	Inexpensive	Incomplete service coverage	Modem yes, service to new site not guaranteed
Fiber	10^{-11}	20 miles without amplification	Excellent BER, technology independent, low loss of bandwidth for a given distance	Requires trenching or aerial for access, generally not a final-mile technology	No

Table 4-2. Broadband Comparisons

foot—and is slow to deploy in older urban areas. Fiber, however, is routinely installed at new construction sites both for businesses and in many of the newest residential home tracts. Because it is only an incremental cost relative to the ongoing construction, it can use existing construction permits, and appropriate construction equipment and labor to use the equipment are on hand.

For reference, there are four common ways in which fiber is installed. They are:

- Aerial (along existing power or light poles)
- Buried in existing conduit
- Jet fiber (fiber that is blown into conduits with compressed air)
- New trenching

There are commonly areas of terrain where fiber simply cannot be installed, or at least installed at cost-effective rates. These areas include deeply forested areas, steep hills, swamps or bogs, and dense metropolitan areas. The cost of trenching for copper or fiber can vary by as much as a factor of five on a per-foot basis depending on a number of variables, including area to be trenched, right of way costs, payment for trenching on an expedited basis, geographical difficulty, and competition for limited construction resources.

With the plethora of fiber providers, municipalities, businesses, and citizens commonly endure the aggravation of having high-traffic roadways dug up repeatedly as the new providers of fiber install their lines. This has led to the concept of *condominium fiber,* which is the concept of grouping a number of entities such as schools and businesses to purchase fiber collectively. The consortium then hires a management entity to maintain the links, with the overall benefit that there is less disruption to the community. An additional benefit is that the consortiums can purchase fiber in substantial excess of what they need at an incremental cost, as the bulk of the expenditure for fiber is labor, not the fiber lines themselves. The excess fiber is often referred to as *dark fiber* because it is not "lit." Dark fiber can be lit up at any point in the future and is relatively trivial to configure into networks.

DSL

Prior to deregulation in the U.S., DSL was exclusively sold by telcos, but it is now sold by a considerable number of small companies, many of whom are being consolidated through acquisition or are going out of business. DSL requires access to a copper plant; that is, integration into a telco facility, as it is carried by common telephone wires.

DSL is not typically broadband by definition (data rates greater than 1.5 Mbps) when transmitted over distances greater than 6,000 feet from the central office (CO), which is where the phone gear is located. Many communities are not served by DSL, as the cost of installing a DSL plant is very high, although the market rate of acceptance and installation of DSL is currently running in excess of 230 percent.

DSL has a history of problematic installations, as the telcos and SPs are required to subcontract the installation and maintenance of the modems. Smaller and smaller local companies are being contracted at an increasing rate just to accommodate the rate of sales incurred by the SP or telco, but at present, the great preponderance of residences and telecommuters use either DSL or cable modems.

At the time of this writing, the DSL market was enduring a very substantial shakeout, which included bankruptcies at NorthPoint and Flashcom in the U.S., along with major layoffs at other DSL providers such as Rhythms, NetConnections, and Covad. The DSL industry suffered a further blow when Verizon terminated DSL sales at 53 of its largest sites.

Another major provider, Telocity, reported severe network-wide problems due to challenges caused by hyper-rapid buildout schedules. Problematic stories abound surrounding the issue of DSL installation. The author of this book experienced a delay in excess of six months in getting DSL installed from the time of order, despite residing well within the range of service. Yet, interestingly enough, DSL providers continue aggressive advertising campaigns, even in areas where services are not even available.

DSL is used in two primary formats: 1) to residential homes, or 2) within large condominium or apartment buildings, both of which are categorized as MDUs (Multi-Dwelling Units). A common architecture is that either fiber, DSL, or BBFW is brought into the building and the existing copper twisted pair wiring carries the DSL services within the building.

DSL providers in general have not pursued the SMB or large business sector but rather have focused on SOHO or residential markets.

Cable

Cable modem providers have likewise pursued the residential market to the general exclusion of the large and SMB markets. Cable lines are laid to residential tracks and MDU sites but not to SMB or large business geographies. Traffic speeds are typically in excess of what DSL can provide, but both upstream and downstream traffic speeds are typically throttled to allow for more users on a given network. Again, as with DSL, deployment of cable modem services can be hampered by problems in developing next generation equipment, upgrading, and certifying two-way cable plants capable of digital communications.

Although the existing cable (the same type used for cable television) can be used for the final access from the street to the home, the trunk line is typically fiber, and replacing or upgrading the in-ground cable has the same cost implications that were discussed in the preceding "Fiber" section.

BBFW

BBFW reaches where cable, fiber, and DSL either will not or cannot. The largest providers of BBFW products, including Cisco Systems, Lucent Technologies, and Nortel Networks, have necessarily included BBFW into their broadband product lines in order to provide a fully comprehensive portfolio of broadband access devices.

Compared to other broadband technologies, BBFW can be deployed in a much shorter period of time, as it doesn't require trenching and can be deployed in virtually every metropolitan and urban area. Solutions can provide data rates of 155 Mbps or more, and the state-of-the-art systems include Layer 3 management of over-the-air links, enabling quality of service (QoS) features that are expected in a business-to-business broadband service.

BBFW is also portable, which in this case means that as a company grows and requires relocation to new facilities, an unlicensed system can also be relocated. In instances where the business relocates within the same basic trading area (BTA), the owners can also relocate the gear as long as they perform an additional RF engineering survey, which will be filed with the federal authorities.

THE PROS AND CONS OF BBFW

BBFW plays an important role in today's networks, and while many individuals and companies are passionate about this subject, an appropriate context includes the concept that

BBFW systems are an excellent tool or mechanism but that, like any good methodology, it has pros and cons.

BBFW Pros

BBFW has advantages that make it unique among the array of broadband access technologies. These items include the following:

- Shorter deployment time than other broadband mediums such as fiber and copper
- Bypass of incumbent telephone companies
- High data rates
- Links that can be installed where copper links (phone lines and/or coaxial line used by cable companies) cannot be installed

Short deployment time is a relative term, of course, but if there are no delays in right-of-way access (where customers will locate their antennas and equipment), a BBFW link can be up and online within about six weeks, which includes the lead time required to receive the equipment. Once the radio gear arrives at the deployment site, a link can be assembled in half a day or less (of course, more than one link has required half a day to simply solder connectors to LMR cable!).

Once the system is operational, or, during the period the system is being made operational, site surveys on the primary targets such as high-density office building complexes, condominiums and other campuses can be performed so that prequalification time of candidates for the wireless service is minimized. (The process is reasonably similar to the prequalification process for DSL or cable modems. If the same customer were to trench fiber or co-locate copper, the delay would and often does run into multiple months for a broadband link competitive to BBFW.)

Bypass of incumbent telephone companies is a good strategy to bypass the charges for local line lease as well as delays in the provision of that service from a telco. It is common for BBFW gear on a toll bypass implementation to amortize between nine and eighteen months, and this amortization rate is shorter where leased line for broadband access is more expensive. This is an especially attractive aspect of BBFW in countries where simply receiving phone equipment and service can take months. However, it should be noted that from a subscriber's perspective, they are trading one leased line service for another. Therefore, competitive advantages such as time to availability, reduced cost through competition, or better-provisioned services come into play.

BBFW systems can commonly provide point-to-point data rates on the order of 45 Mbps (DS-3) and beyond in any frequency band, licensed or unlicensed. Multimegabit services, T1 and greater, are achievable in the multipoint scenario.

BBFW Cons

Again, as with the pros associated with BBFW, many of the cons are unique to the medium of BBFW and include the following:

- Limited bandwidth due to FCC spectrum allocations
- Generally difficult to compete with wireline or fiber quality
- Multi-path in <10 GHz frequencies
- Short range in > 10 GHz frequencies
- Many systems require expensive governmental licensing
- Probability of link degradation increases over time from local environment
- More expensive than wireline

The general rule is that "fiber rules." There are, however, two exceptions to the end user, which are

- If they have to lease fiber
- If they have to trench for fiber

Fiber has such exceedingly comparatively high data rates that no BBFW system can even remotely approach the same level of data-carrying capacity. Indeed, no other broadband medium approaches fiber for capacity. With regard to BBFW this is primarily for two reasons, 1) because the federal authorities allocate relatively limited amounts of bandwidth on which to load traffic and 2) the modulation schemes don't exist to carry data at rates comparable to fiber. Table 4-3 illustrates the bandwidth allocated to the most common frequencies allocated in the U.S.

BBFW bandwidth capacity is diminished even further as distance is factored into the equation. Most BBFW systems are limited to transmission distances of less than 25 miles and cannot have physical obstructions within that distance. A single strand of fiber can carry more data than many thousands of BBFW links. BBFW accommodates this to the

Spectrum	Frequency	Speed
ISM	2.4 GHz	11 Mbps
MMDS	2.5 GHz	44 Mbps
UNII	5.7 GHz	44 Mbps
U.S. LMDS	28 GHz	192 Mbps

Table 4-3. Comparable Speeds For Common U.S. BBFW Frequencies

best of its ability by using more exotic modulation schemes, but the trade off is cost and bit loss, which requires equally exotic forward error correction (FEC), which can occupy precious bandwidth.

Where state of the art BBFW systems are comparable to fiber is in the area of bit error rate (BER). State of the art systems can provide broadband data rates in excess of OC-3 (155 Mbps) at relatively short and unobstructed ranges of approximately two miles at the same BER as fiber, which is 1×10^{-11}. BER is a measurement of how many bits fail to make it to through the demodulation phase on the reception side of a data link. A BER of 1×10^{-11} means that only 1 bit in 10,000,000,000 fails to make it through this phase of the data transmission. For the sake of context, a $1\mathrm{x}10^{-6}$ BER is considered sufficient for most voice calls.

Wireline data transfer has a number of advantages to BBFW, which include

- High tolerance to environmental interference
- Probability of link degradation increases at much slower rate than with BBFW
- Better understood by more network managers
- Easier to trouble shoot
- Can be less expensive to repair

While a customer can theoretically install a BBFW link almost anywhere the interference conditions are favorable, the link distance is limited and, typically, line of sight is required. Data carried on a wireline, however, is not generally subject to interference from sources outside the copper or fiber. Although, when comparing rate to rate in the last mile distribution network, BBFW is capable if sending multi-megabit services well in excess of the current DSL limitation of 18,000 ft from the point-of-presence.

A trade-off between microwave systems (1 GHz to 10 GHz) and millimeter wave systems (above 10 GHz) is duplication of signals, which is the result of the radio waves bouncing off local objects such as buildings, pavement, and water, and is called *multipath*.

The issue of multipath is addressed in more detail elsewhere in this work but the general concept is that the receiver acquires duplicates of packets. Although the duplicates are typically lower in power, they arrive at the receiver slightly later in time than the packets received via the line of sight. These multipath signals have the ability to combine with the original signal and cause sever degradation of the packets arriving at the receiver. This causes at least two problems for most RF networks: 1) how to deal with the packets that arrive late, and 2) how to construct the packets for suitable use in the system.

The first of the two problems is generally dealt with on most RF systems by filtering out the late-arriving packets, which also takes care of the second problem: degraded packets. State-of-the-art systems such as those provided by Cisco actually turn this challenge into a substantive competitive advantage through two methodologies: a modulation technology called Orthogonal Frequency Division Multiplexing (OFDM) and the concept of spatial diversity, which is the use of two antennas on the receiving end of the data link.

When both spatial diversity and OFDM are used at the same time, it is called Vector Orthogonal Frequency Division Multiplexing (VOFDM). The "Vector" merely refers to the second antenna as it is at a different vector to the angle of reception than a single receiving antenna.

On the other hand, millimeter wave systems have problems with the radio waves being either absorbed into the local environment or scattered due to precipitation. Millimeter wave systems typically have a PTP range on the order of two miles and must occur in a strict LOS environment. PTM millimeter wave architectures have ranges of less than one mile. The physics of why millimeter wave systems feature carrier waves that are readily absorbed or scattered is beyond the scope of this book, but millimeter band frequencies (>10GHz) with wavelengths shorter than about 100 mm in length provide predictable and exploitable multipath signals, such as those produced in the microwave band (< 10 GHz). The higher the frequency, the shorter the wavelength. This subject is discussed in more detail in the section "RF Fundamentals."

Data links degrade over time but at different rates, depending on the medium. While wireline links can have connectors that corrode or in the case of outdoor connectors take on water, RF links are typically degraded as the local environment fills with sources of interference.

The odds of a link degrading over time increase as the radiating environment includes an increasing amount of interference sources such as arc welders, competitive broadcasts in the same frequencies (known as co-channel interference), and even interference from transmitters from within its own network due to multipath (a subject treated later in this work).

BBFW systems tend to be considerably more expensive if the customer does not include the cost of trenching, which can be as high as $100 per foot in the United States. While some systems can be installed for as little as $30,000 (RF equipment only, necessary switches or routers not included), other systems cost five to eight times that amount. As in many instances, the lowest price does not always provide the best value or return on investment. In fact, the more expensive systems tend to provide more value than the less expensive systems when you include items such as truck roll.

An additional word by way of context is that BBFW systems are typically referred to as either *edge* or *access* devices, which means they connect customers to a core network of either fiber or large copper-based trunks. There are three primary kinds of BBFW systems: backhaul, Wireless Local Loop (WLL), and Wireless Local Area Networks (WLAN). Each of these types of BBFW will be considered in this work. Characteristics of common BBFW products are shown in Table 4-4.

NOTE: There are exceptions to the above; this is intended as a general snapshot only. Monthly charges may decrease over time if service providers do not add additional services or increase bandwidth.

One thing is certain about the provision of BBFW technology to various market segments, which is that the industry has an obligation to keep the transport as transparent as

	WLAN	ISM	MMDS	UNII	LMDS
Frequency (GHz)	2.4	2.4	2.5	5.7	28 (U.S.)
PTP Bandwidth Mbps (up/down)	1-11/1-11	11/11	44/44	44/44	155/155
PTM Bandwidth Mbps (up/down)	1-11/1-11	3/3	19/17	19/17	45/10
Carrier Equivalent PTP	N/A	DS2+	DS3	DS3	OC3
PTP Range LOS	400+ ft	25 mi	25 mi	20 mi	5 mi
PTP Range NLOS	N/A	N/A	~4 mi	~4 mi	N/A
Over The Air Protocol	IP/ATM	IP/ATM	IP/ATM	IP/ATM	ATM
Approx Cost Per Link	$5K	$5K	$150K	$150K	$200K-$1M
FCC License Req'd?	No	No	Yes	No	Yes
Preferred Deployment	Indoor, SOHO, SMB	SOHO, SMB	ISP, SMB, Enterprise	ISP, SMB, Enterprise	SP
Monthly Price	None	$75	$50	$50	$30–$90
Equipment Price	$1,500 retail	$200	None to end user	None to end user	None to end user

Table 4-4. Snapshot Comparison of Common BBFW Products

possible. End users in particular don't purchase DSL, cable modems or BBFW because of any inherent virtue or charm; rather what they seek is the broadband experience without the aggravations experienced with flawed processes in ordering, installation and maintenance. BBFW providers must provide their products and services seamlessly with the ability to scale rapidly.

CHAPTER 5

Primary Uses for BBFW

As stated previously, a BBFW link is used as either a Layer 2 bridge or with state-of-the art equipment as a Layer 3 link with a router interface to both ends of the link. At the time of this writing, virtually no service provider offers many services, if at all. This is primarily because most end users are demanding high speed access only for e-mail, presentations, file transfers, and communications applications like Instant Messaging from either Yahoo or AOL.

However, as users gain more time on broadband links, their needs will become more sophisticated as will their tools. Furthermore, BBFW service providers will need to offer services to help reduce customer churn. These services may be provided by the service providers themselves, at least initially, but as the Internet continues to mature, services will become increasingly specialized and therefore will require the content or service to be acquired and sent over the links by specialized entities, which will be segmented by industry such as medical, gaming, or storage area networks. The ability to offer and use services over a BBFW link takes the Internet experience to the next level, which is an exciting and practical development that will tie us all even closer to the links we use to communicate with each other.

IP BBFW SERVICES

The primary purpose of a BBFW link is to provide connectivity to end users. It stands to reason, therefore, that the most valuable aspect of the link is not the radios themselves but the traffic carried over the network. Many in the industry have focused on the prediction of a "rich" Internet experience and have accordingly assumed that end users would require numerous types of service and content. However, equipment builders and service providers are discovering that the primary service the end user requires is simply high-speed access. Little thought or planning is being given to the concept of even adding voice capabilities, much less other content and services.

While pursuing broadband as the primary deliverable to end users is an appropriate tactic for the service provider to establish a revenue generation base, over time as broadband access makes the transition from being a service of single-source suppliers to being a commodity available from an array of providers, service providers will need to plan in advance to induce customers to remain with their service.

VPNs

VPNs enable a remote user to access their corporate or other secured networks. VPNs are also used to connect remote offices or sites to the main campus. The fundamental architecture of a VPN is shown in Figure 5-1.

VPNs are one of the first services that will become well utilized on a BBFW link. They are one of the hottest selling broadband products at the time of this writing, in part because they increase efficiency and productivity while allaying a great concern about security. A VPN can be either an application as sold by companies like Cisco Systems or it can

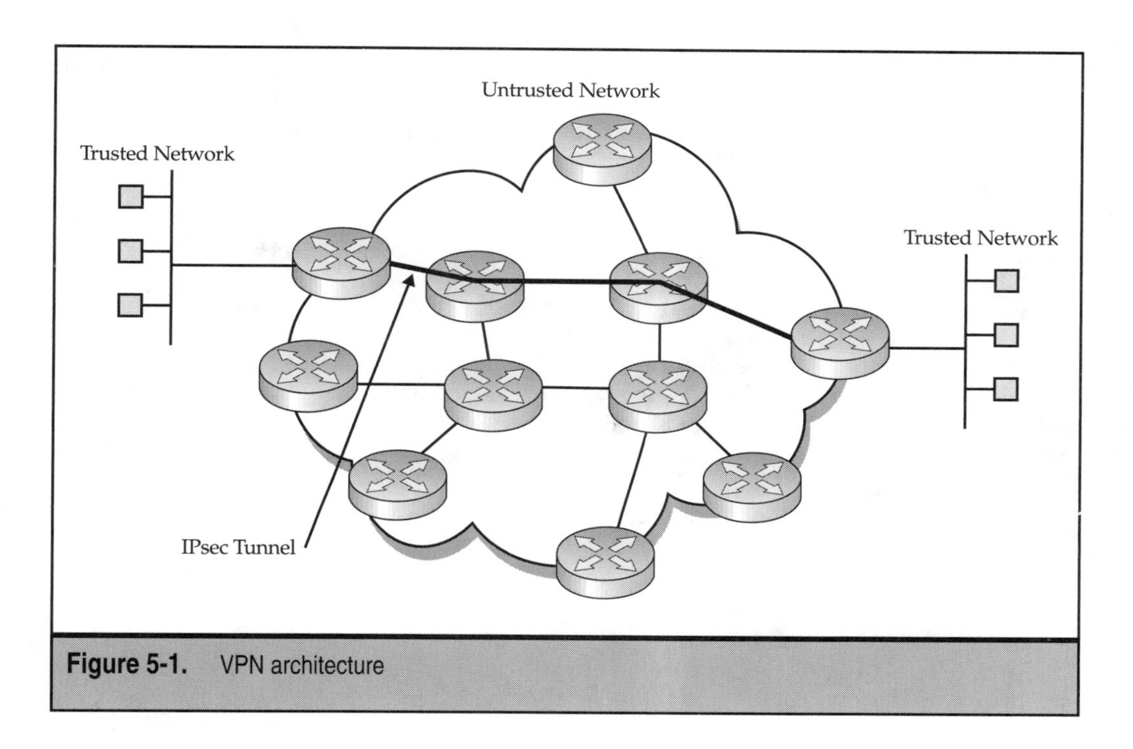

Figure 5-1. VPN architecture

be a service provided by third parties to service providers to ensure secure links between a campus and a remote office or an office and a telecommuting employee.

In today's world of "networks of networks," Internet traffic commonly traverses multiple networks. At some point, virtually all traffic crosses networks owned by the PSTNs (local or regional phone companies), IXCs (companies like Sprint or MCI WorldCom that connect disparate long-distance carrier networks), or long-distance carriers (like AT&T in the U.S.).

In many instances, this works sufficiently well and there are few if any security issues. However, a considerable and growing amount of network traffic requires at least minimal amounts of privacy, if not moderate to high levels of commercial-grade security. This is where virtual private networks (VPNs) come in.

While there are numerous types of VPNs, the most common are IPSec (a standardized architecture that defines a set of standards that can be used to secure the Internet Protocol, L2TP (Layer 2 Tunneling Protocol) and GRE (Generic Router Encapsulation). For the sake of brevity and to keep as close to the main scope of this work as possible, only IPSec VPNs will be summarized.

VPNs, regardless of type, provide secure paths (tunnels) over networks that carry public traffic. They connect mobile users (referring in this case to a person with a laptop as opposed to a cell phone) or telecommuters to their networks. VPNs also connect remote offices to the main networks. A VPN tunnel exists when a security agreement is established between two hosts to provide (and require) some security services on a flow of IP packets.

To accommodate this, the VPN tunnel between the end users must ensure:

- Privacy
- Data integrity
- Data origin authentication
- Nonrepudiation (receiver knows who the sender is and sender verifies that they sent data to the receiver)

The key to making a VPN work is standards-based security measures. The primary security measure is called Internet Key Exchange (IKE), which is used to negotiate cryptographic parameters, authenticate identities, and establish the associations between each end of the tunnel.

IKE is a hybrid protocol that enables key exchanges in conjunction with ISAKMP (Internet Security Association and Key Management Protocol), whose purpose is to provide authenticated keying material for, and secure negotiation of, security associations. ISAKMP defines a framework for managing SAs (security associations) and establishing cryptographic keys for the Internet.

An SA is a set of policies and keys between two parties used to protect information exchange between them. IKE uses ISAKMP SAs, which must include negotiations of the following attributes: encryption algorithm, hash algorithm, authentication method, and info using a set of keys known as the Diffie-Hellmen group. A minimum of three SAs are required for a bidirectional IPsec session: 1 transmit SA + 1 receive SA + 1 IKE SA.

A VPN fundamental architecture involves the following basic building blocks:

- A VPN gateway device that resides on the edge of a trusted network
- A software client that resides on a laptop computer, and/or
- A VPN gateway device that resides on the edge of the remote network

Two of the more common VPN architectures are called intranet VPNs and extranet VPNs. Intranet VPNs provide interconnections between telecommuters and remote offices and their corporate network, typically using dedicated connections over a shared (public traffic) network. Figure 5-2 illustrates a typical intranet architecture.

Extranet VPNs extend corporate intranet services to suppliers, customers, partners, or communities of interest over a shared infrastructure, typically using dedicated connections over a shared network. Extranet VPNs are architecturally comparable to intranet VPNs, with additional security and interoperability issues. Figure 5-3 illustrates a typical extranet architecture.

The advantages of having a VPN include having a connection between two sites without having to build or maintain the entire path between two sites, a process that can generate substantial costs, as the distances between the remote office or user and their corporate network can be considerable. VPNs are relatively easy to set up, and virtually all contemporary networks readily accommodate VPNs. VPNs also provide very high security for the entire data path without the need to hire, train, and retain network security experts, and

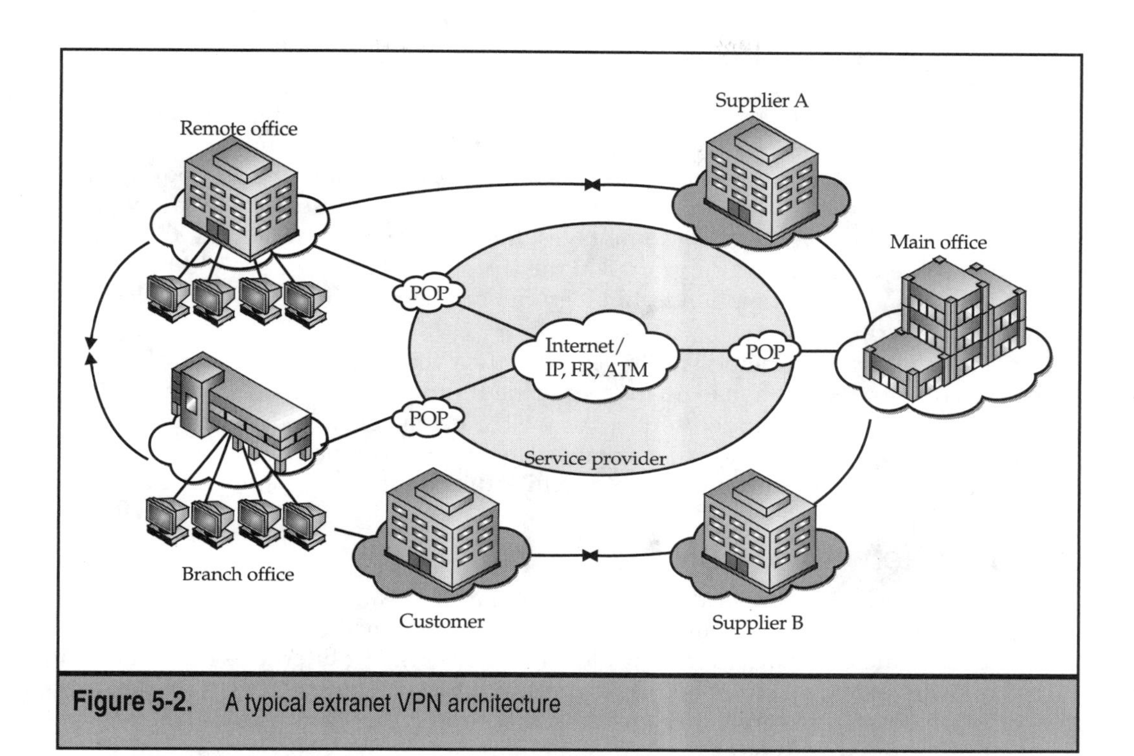

Figure 5-2. A typical extranet VPN architecture

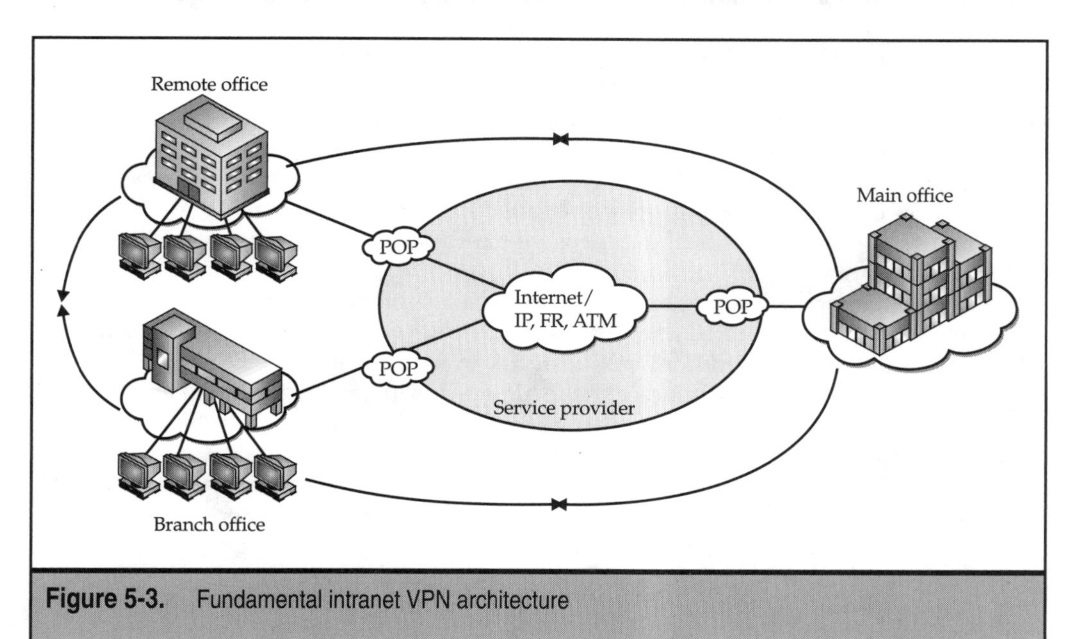

Figure 5-3. Fundamental intranet VPN architecture

they are very flexible, in that the remote users can be very mobile, needing only to connect their laptops to a standard phone jack from virtually anywhere in the world.

Some of the disadvantages of a VPN involve scalability issues, in that tunnels are not shared; that is, each user has his or her own separate tunnel. Encryption can affect the throughput of the tunnels, although this is being offset by increasing the power and capacity of processors in the network engines at the LANs and putting faster processors in laptop computers. Two other disadvantages, though relatively minor, are that client software must be installed and supported at the laptops and remote sites and that there are restrictions on what encryption technology can be exported from the U.S., at least at the time of this writing.

Service providers who include BBFW as one of their access technologies can also offer VPNs. The advantages for them to do so include the ability to generate additional IP revenue streams through value-added services as well as being able to reduce infrastructure costs because they won't have to build a network between all of their customer sites. Furthermore, VPNs have flexible and scalable architectures that can be deployed globally.

Content Data Networks (CDNs)

While the emphasis over the last five years or so has been on moving data across networks, the next few years will demand ever greater performance and efficiency from networks that have increased in size to truly global dimensions. The Internet is now well and truly a "network of networks." When the distances between networks cover thousands of miles or when clients and servers are great distances apart, there are delays in the receipt of the traffic, as the signals have to traverse a number of networks prior to arrival at the end destination. It should also be noted that while major network backbones have enormous capacity and are generally fiber based, the networks that connect major networks (cross Internet connections) typically have slower speeds. Final mile, or in other words, local loops typically have even less bandwidth and speed before the data is placed onto LANs. Figure 5-4 shows a nationwide CDN example.

The fundamental premise of a CDN is to deliver higher network performance at a lower cost. By eliminating cross-Internet connections, a network can improve performance by a factor of ten by trading expensive bandwidth for content engines, routers, and managers, as shown in Figure 5-5.

One of the primary principles of CDNs is to locate content, or copies of content, at locations that are closer to the end use and also to help servers scale through caching content in devices that enable commonly requested data to reside in a device next to a server. As by its nature, Internet traffic tends to have high peaks in demand, content engines and other CDN elements can assist in improving network performance at the client, reducing bandwidth requirements and easing server loads, as shown in Figure 5-6.

CDNs are ideal for applications such as distance learning, corporate communications, partner/supplier relationships, retail information applications, documentation delivery, accelerated network performance initiatives, load balancing, and applications where high security is an issue.

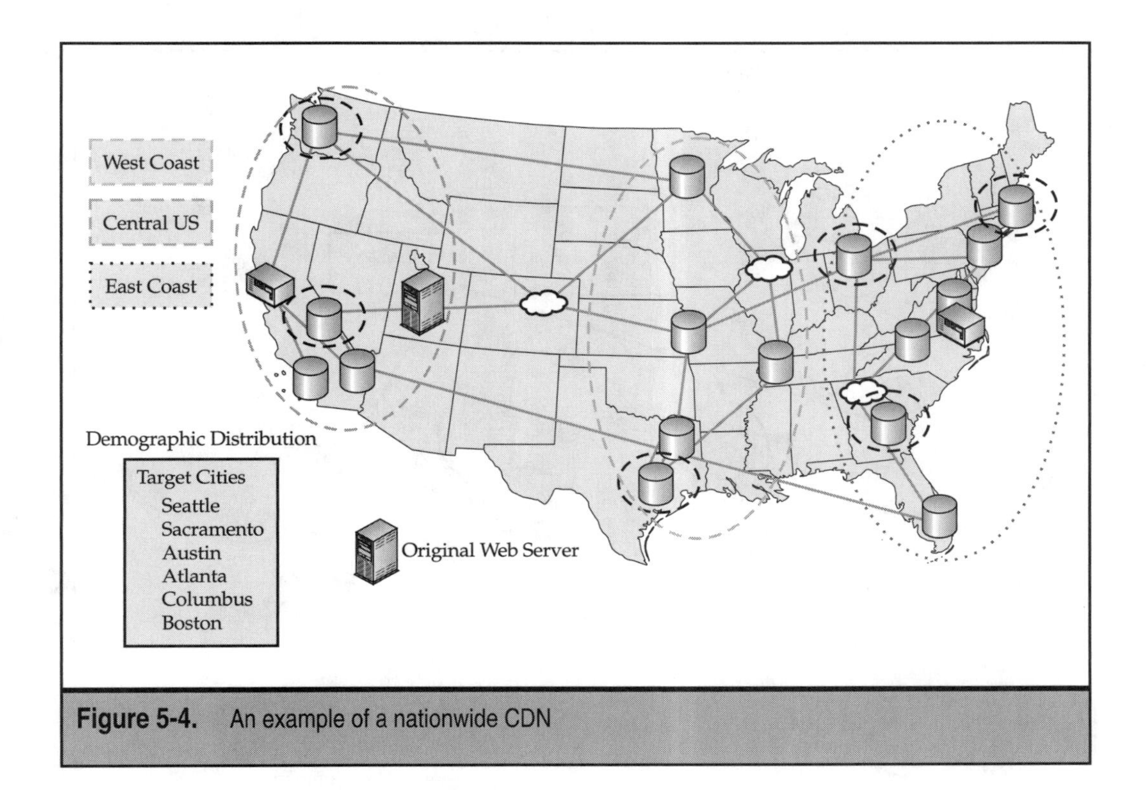

Figure 5-4. An example of a nationwide CDN

CDNs not only speed up response time and enable applications such as voice and video to perform more smoothly, they also balance the load across a network. In general, the content is "pushed" toward the edge of a network—that is, closer to the access part of the network on which the end user resides. Another key feature of CDNs is that they can insert "local content" such as advertisements, mapping, and other geographically sensitive information.

Networks that incorporate CDN are also more reliable in that if one portion of the network goes down, the CDN features can reroute traffic between clients and servers without delay. Also, if the server closest to the client is subscribed to the point where performance is diminished, the content can be retrieved from the next closest server.

CDN is one of the hottest Internet architectures on the market today and is a combination of both software and hardware. It enables service providers to construct CDN for the deployment of premium revenue-generating Internet services, and when deployed along with solution-based hardware from a CDN-specific product line, building a CDN network for applications such as acceleration of static Web content, streaming media to massive audiences, and accelerating applications or online advertising can be accomplished simply and affordably.

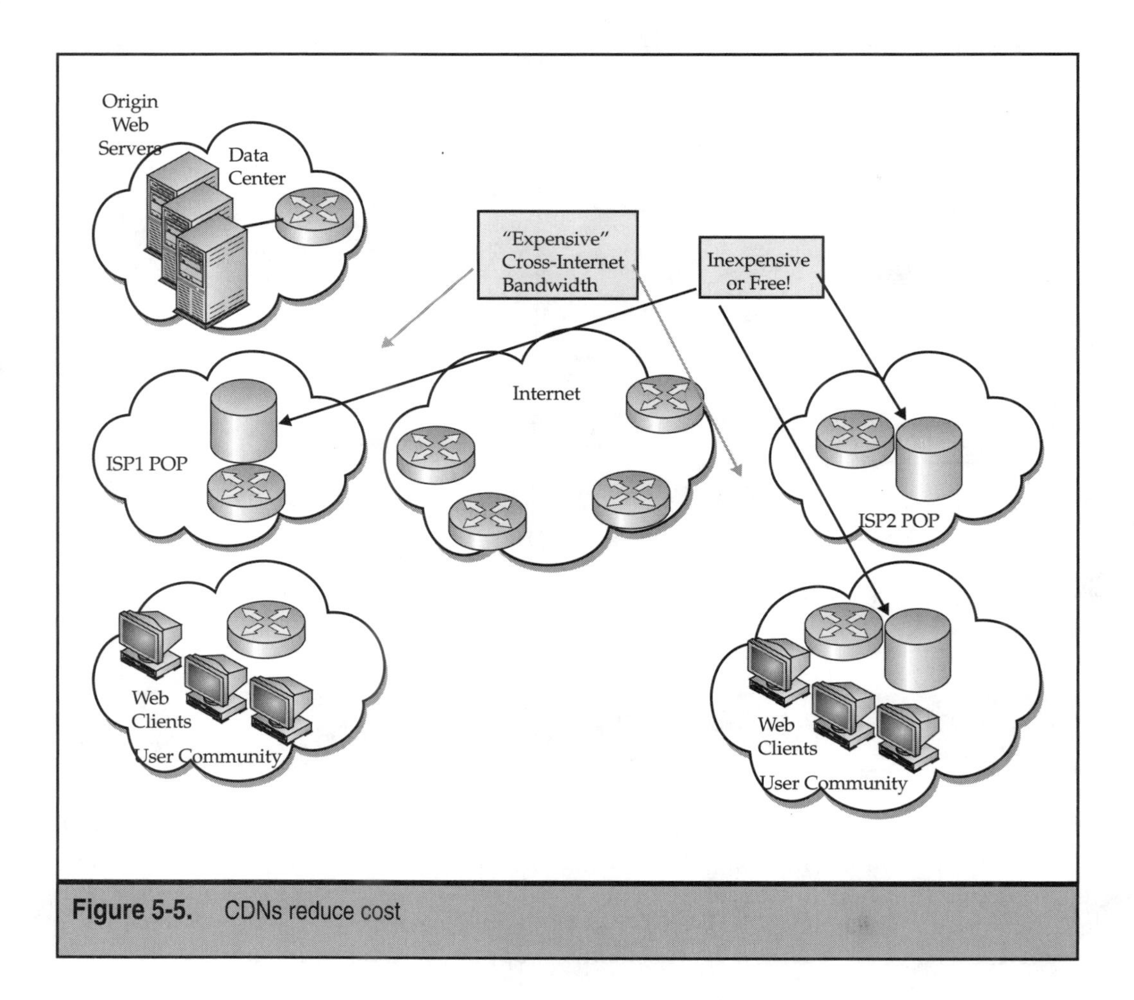

Figure 5-5. CDNs reduce cost

The five key elements to a CDN architecture are:

- Content distribution and management
- Content routing
- Content switching
- Content edge delivery
- Intelligent network services

Content distribution and management involves encoding the media for security, importing information, and distributing it to the appropriate network edge site. This element of CDN also configures the delivery engines or servers at the edge of the network, manages bandwidth, and measures CDN system performance and usage.

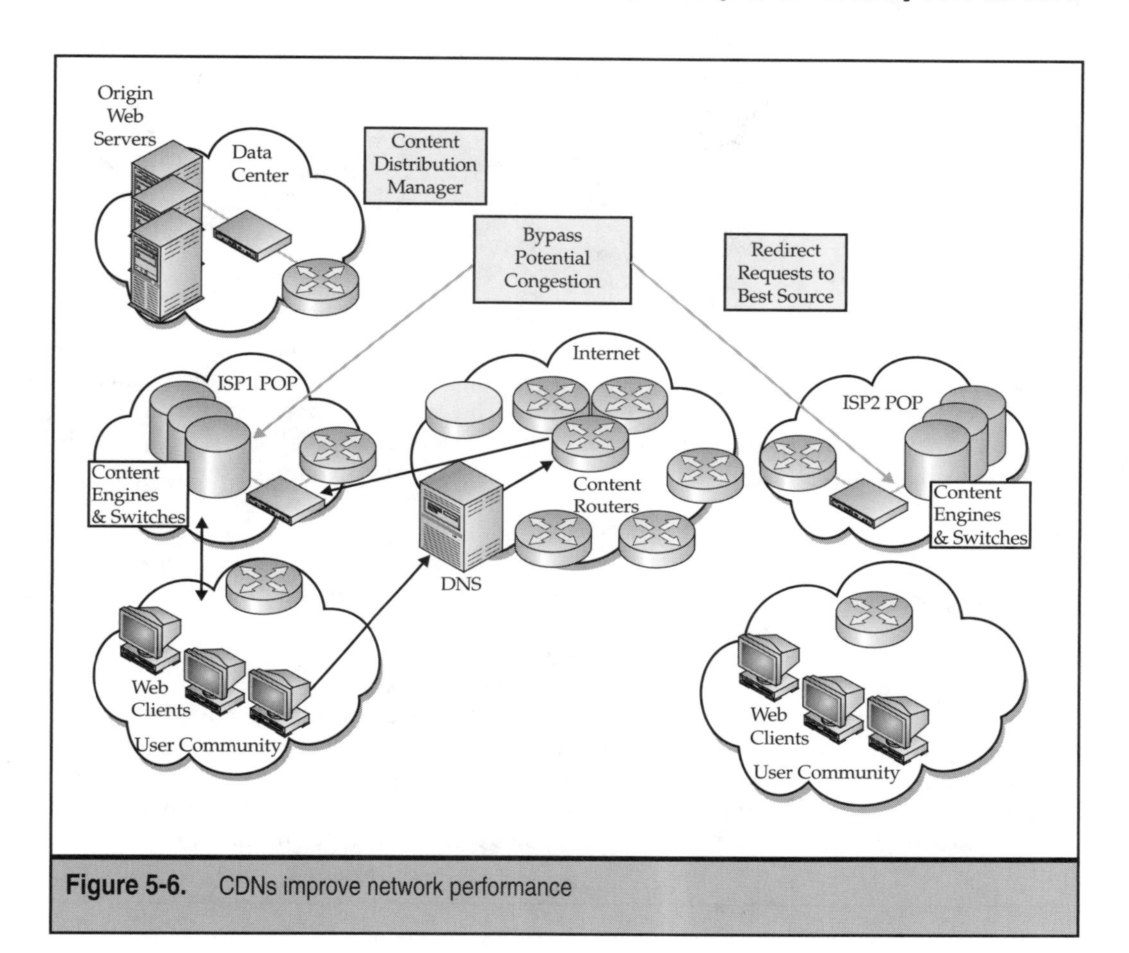

Figure 5-6. CDNs improve network performance

Content routing involves reliably routing user requests to the best server or cache engine on the CDN network. These decisions are based on which server or cache engine contains the requested information, geographic proximity of the server or cache engine, network conditions, point of presence load, and/or content engine load.

Content switching is different from content routing in that it intelligently switches traffic across origin servers for load balancing. It also measures and determines the availability of content servers and the availability of the content.

Content edge delivery is the element that seamlessly delivers the content to the final mile of the client's networks. It can provide content serving, streaming, and caching all in one device.

Intelligent network services manage the security, QoS, VPNs, and multicast functions.

In an enterprise network, CDN can be deployed when a network is monopolized by bandwidth-hungry yet noncritical applications, as opposed to more critical applications such as real-time inventory management. Therefore, instead of investing heavily in more bandwidth from end to end in their network, enterprise customers can take advantage of their native QoS ability and combine it with CDN for maximum performance. QoS will enable the network to differentiate traffic type, while the CDN architecture speeds critical applications by connecting clients and vendors to the most appropriate servers.

Differentiated Services

With the implementation of a Layer 3 solution using Quality of Service features, the BBFW solution enables the provision of differentiated services to the end user. The encoding at the header of each packet set differentiates the payload from voice, video, or data. Specifically, this means a single transmitter can provide voice, video, and data to customers with a single transmitter across multiple users in a radiated environment. In point to multipoint applications, this means a single hub can provide differentiated services within a single sector (IP wireless cells generally having three or four relatively equal sectors).

Furthermore, it means the service provider can set up bandwidth management using a grant/request policy that is fully flexible. These policies enable the service provider to establish the level of service, type of service, guaranteed bandwidth, and whether or not the end user can alter the type or amount of bandwidth.

Differentiated services are set along the lines per Figure 5-7.

Figure 5-7 indicates the three levels of service that are supported by grant policies. The USG policy is an "Unsolicited Grant Service" and is generally reserved for voice or time-sensitive services. This service guarantees maximum bandwidth at all times the original grant request is authorized for the duration of the session. No additional bandwidth grants are required. Committed Bit Rate (CBR, also referred to as Committed Access Rate

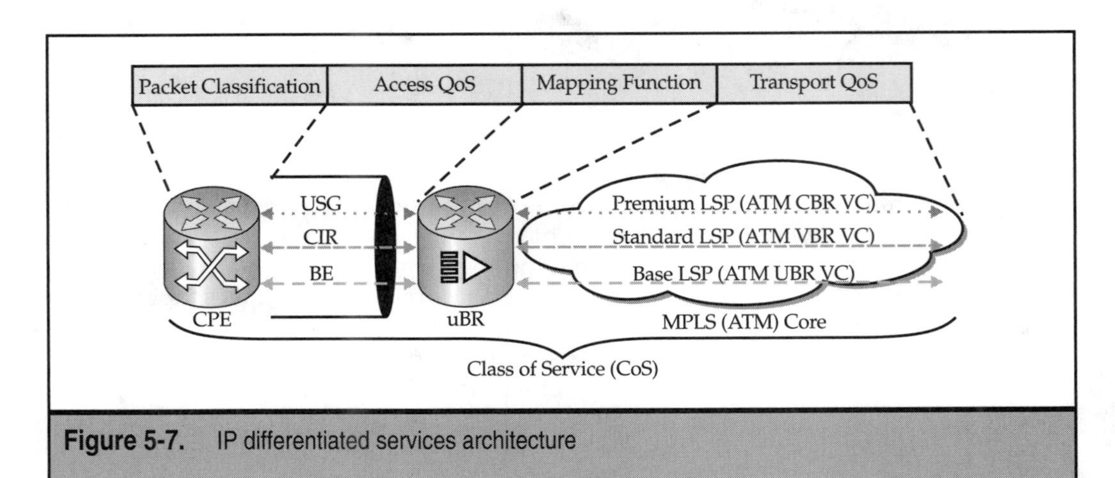

Figure 5-7. IP differentiated services architecture

or Committed Information Rate) is the policy set for users who want guaranteed bandwidth but who generally don't require voice applications. Best Effort Rate (BER) is the rate most end users require, as it has the lowest price of the three types of policies. It is generally well suited for e-mail and applications like file uploading and downloading where delays are quite tolerable. For more information on the grant policy process, refer to the Data Over Cable Interface Specification (DOCSIS) provided by Cable Labs.

In many IP systems that offer QoS, two or sometimes three levels of each QoS exist, but most service providers offer only two levels in order to keep management simpler. Furthermore, virtually all types of users fit nicely into one of the six QoS levels.

While there is a wide array of available services, odds are only a few options will be offered. These options will vary in importance from the residential user to the telecommuter to the business campus. The first group of services likely to be in demand will be Virtual Private Networks (VPNs) as this enables secure links between the residential area and the business site.

Service Delivery Architecture

The effort involved in providing services is nontrivial and multifaceted, requiring an array of different companies to interface with each other seamlessly to provide reliable and rich content and features to consumers, telecommuters, and business campuses. In the earlier days of the Web, service providers were a one-stop do-it-all entity. The ability to provide broadband speeds requires not only broadband links but traffic optimized for broadband service, as well as seamlessly integrated network elements from disparate companies.

Service delivery architectures will require increasing amounts of specialization as each element becomes more sophisticated. Furthermore, as each element increases in sophistication, it will also increase in cost and maintenance. This will drive service providers to subcontract these disparate elements or engage an array of vendors to provide the same. The final aspect of this optimized architecture will be the element of time; that is, service providers won't have years to design and ruggedize this architecture, further necessitating an array of eco-partners to focus tightly on specific elements and bring to bear appropriate levels of expertise and tools.

The primary elements of this architecture will include: Application Service Providers (ASPs), third-party application developers, application platforms, hosting companies, and service providers.

ASPs

ASPs will be responsible for building applications used by business or consumers such as instant messaging, paramutual or other gaming applications, telemedicine, and distance learning. These applications will require specific industry and customer expertise. Over time, ASPs will become dedicated to niches of applications such as those provided for education, gaming, and telemedicine.

Application Platform Providers

Application platform providers are at this writing one of the newest genre of start-ups, neither well known nor understood. Many of the founders and employees are from network companies such as Lucent, Nortel, and Cisco, which means they have strong networking backgrounds and perspectives. These types of companies assemble protocols and software toolkits for loading/downloading, reliability monitoring, and reporting, as well as the software to repair or provide redundancies. Some application platform developers will also include the hardware on which the application will reside.

Hosting Companies

Hosting companies enable the transport of the applications between the ASPs and the service providers via the PSTN. These companies generate revenue by purchasing PSTN bandwidth at wholesale rates and then selling the same bandwidth to the ASPs for a profit. In addition, they receive revenue from service providers for the applications themselves. Hosting companies own numerous types of servers, including conference servers, application servers, media servers, and database servers.

Service Providers

Service providers distribute the applications to the end users via last-mile access such as BBFW and connect the end users to the hosting companies. Service providers generate revenue by purchasing content and services from hosting companies and selling them at a profit to end user. Service providers also generate revenue from providing end users broadband access. Service provider routers initiate the QoS that enables the prioritization of traffic necessary where multiple types of traffic will be provided to end users. This prioritization is critical to end users because voice or video services must be transmitted on a priority basis and not be delayed by other traffic. Service providers provide the link between end users and the hosting companies. They will also provide the necessary billing and troubleshooting, maintenance, and other provisioning tasks in addition to owning the BBFW links.

Service providers tend to be regional in their coverage to end users as the result of the geographic limitations of BBFW. In theory, a service provider could have a worldwide subscriber base of customers, but even multistate or provincial coverage requires tens of millions of dollars, which effectively limits the number of service providers in existence.

WIRELESS LOCAL AREA NETWORKS (WLANS)

WLANs are one of the most important elements of a total BBFW solution, as they provide the final 400 or so feet between the BBFW link to the building and the actual users. WLAN provides two key aspects to BBFW, which are:

- Broadband mobility
- Pervasive computing

Broadband mobility refers to the concept of being able to move around within a building or from building to building while retaining connectivity to the network. This type of mobility should not be confused with the type of connectivity provided by mobile devices like cell phones. One of the key differences between these two types of product is that broadband mobility is primarily intended to work at pedestrian speeds, that is, at about 3 mph (5 kph), whereas a cell phone can work at relatively high speeds of 70 mph (100 kph) or higher. The other difference is that broadband mobility is indeed broadband, that is, having data rates in excess of 1.5 Mbps, whereas cellular or other data is, at this writing, much closer to 9600 baud. 3G mobile data rates are projected to be in the 128K range at pedestrian speeds, but it will be some time before the market sees ubiquity with that technology.

Pervasive computing, on the other hand, refers to the concept that users can experience broadband connectivity in a virtually limitless range of locations beginning at their work desks and from there extending to conference rooms, their automobiles, and home offices, as well as airports, train stations, libraries, customer sites, and so on. This type of ubiquitous connectivity is made possible through the use of standards. The Institute of Electrical and Electronics Engineers (IEEE) has developed a standards-based approach to Wireless LAN (WLAN) referred to as IEEE 802.11.

This ratified and widely accepted standard provides interoperability of the PC cards and the access points in the WLAN. The specific standard in use at this writing is IEEE 802.11b, enabling up to 11 Mbps data rates in the 2.4 GHz (ISM) spectrum. Other standards for increased data rates are in committee at this time and expected to be ratified soon. The concept of being able to upload and download broadband data in a very wide variety of locations has clear implications for productivity and professional effectiveness, as you can acquire mission-critical data virtually regardless of your present location.

802.11 WLAN equipment will likely be one of the common uses for BBFW and will certainly be the form of BBFW that most people will be acquainted with. Its ease of installation and use, relatively low cost, and the value it provides through broadband mobility and pervasive computing will ensure that this type of BBFW will remain popular for many years to come.

CHAPTER 6

Customer System Selection

The BBFW market has an array of customers, each representing certain segments of the BBFW market, which is truly no more than a subset of the Internet market. Although all of these customers have many technologically similar requirements, there are financial and architectural requirements unique to each of these market segments. Each of these markets focuses on specific areas of the Internet, although these boundaries are not always predictable and consistent. Figure 6-1 demonstrates the major Internet architectural elements and the market segments that primarily focus on these elements.

TYPES OF CUSTOMERS

There are generally five major types of BBFW customers:

- Incumbent Local Exchange Carriers (ILECs)
- Competitive Local Exchange Carriers (CLECs)
- Inter-Exchange Carriers (IXCs)
- Internet Service Providers (ISPs)
- Building Local Exchange Carriers (BLECs)

One of the similarities among these markets is that the opportunity is substantial, as indicated in Figure 6-2.

ILECs

An ILEC is often also referred to as the "incumbent" or the "RBOC" (Regional Bell Operating Company). ILECs are typically the legacy of the local or regional phone company that has commonly been providing telephone services for 100 years or more. While they have an interest in the final mile of voice service and provide considerable amounts of DSL, they are generally and primarily interested in backhaul; that is, they want to have large trunks servicing Class 5 switches that reside in neighborhoods. Class 5 switches are those that aggregate approximately 60,000 to 100,000 phone lines. A switch consists of a series of racks of gear that typically fill a room approximately 25 feet square. Although ILECs have provided phone service since the days of Alexander Bell, they understand the need to do data and are designing current networks to separate data from voice traffic. This is done in order to free up what are called voice ports on their various network elements. They recognize that to remain competitive and indeed to survive, they need to provide Internet service, especially given that revenues generated by voice traffic are steadily decreasing.

However, as the "baby" Bells decentralize, break up, reform, and form consortiums with other former Bell companies, it is getting increasingly difficult to tell an ILEC from its competitor, the CLEC. Traditionally, the ILEC focused on voice services and by definition was the incumbent in a region. Today, however, the traditional ILEC in a single region may partner, buy, or merge with other competitive or incumbent providers in new

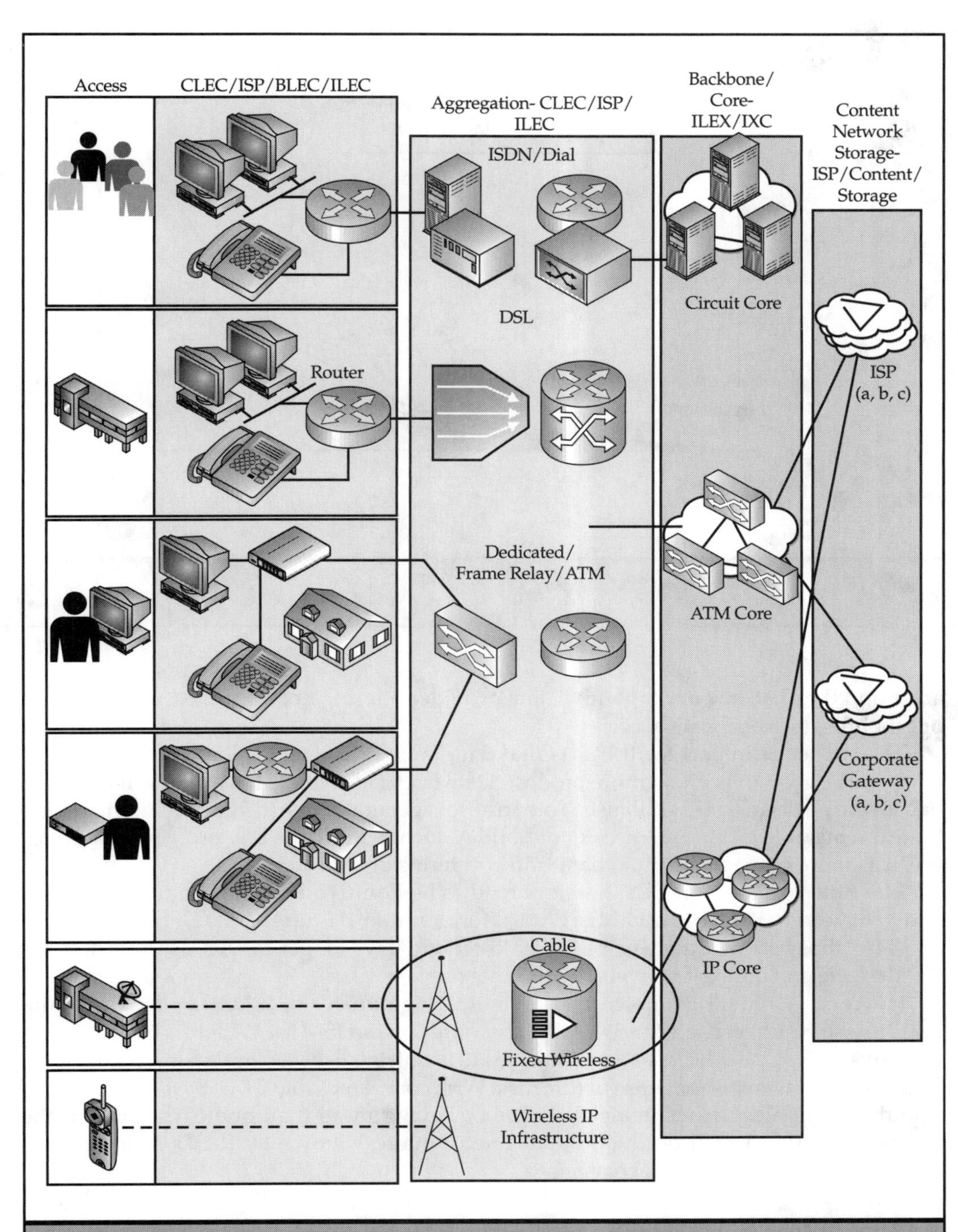

Figure 6-1. New world Internet elements and market segment ownership

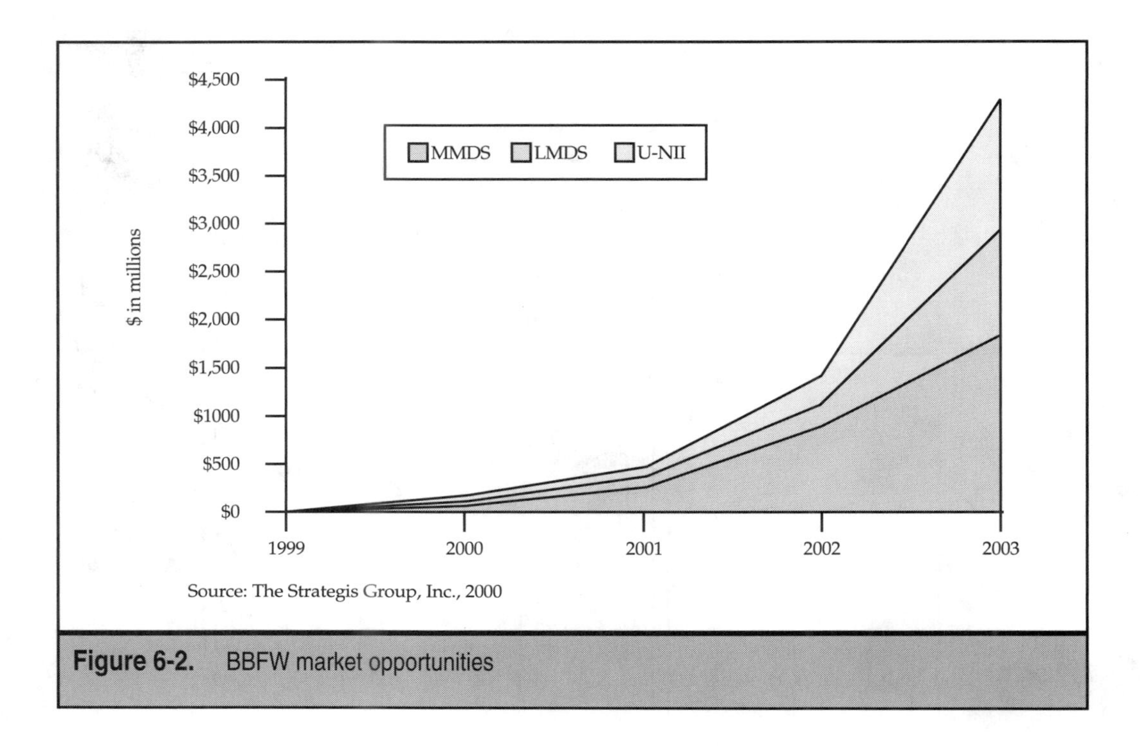

Figure 6-2. BBFW market opportunities

areas and then become a competitive data provider offering not only fixed voice services but mobile voice and data services.

The challenge in part for ILECs is that data is experiencing explosive growth, while voice traffic growth is much more modest. Data traffic also tends to be *bursty,* meaning it has severe peaks in bandwidth use. Voice traffic is generally more predictable, and unless there is a major local natural disaster or holiday such as Mother's Day or Christmas, voice traffic is more manageable than data traffic in terms of load handling.

The BBFW links for ILECs therefore tend to be Point To Point (PTP), to be licensed, and to have very large bandwidths; LMDS, for example, has up to 1300 GHz of spectrum and 144 Mbps bandwidths. The BBFW links tend to have longer ranges, as the beams are tightly focused and exist between two points only.

However, it should be noted that an increasing number of ILECs are looking more carefully at the unlicensed bands—U-NII for outdoor and ISM for WLAN. This is because of delays in acquiring licenses, legal hurdles to acquiring licenses, and in many cases, the ILEC does not want the expense of a license. With companies like Cisco Systems that have superb capabilities for resolving the issue of having more than one competitor in the same area with a U-NII product, the degree of interest shown by ILECs in unlicensed BBFW products is steadily increasing.

ILECs also provide high-bandwidth links between large buildings. A typical ILEC BBFW LMDS deliverable for this scenario is shown in Figure 6-3.

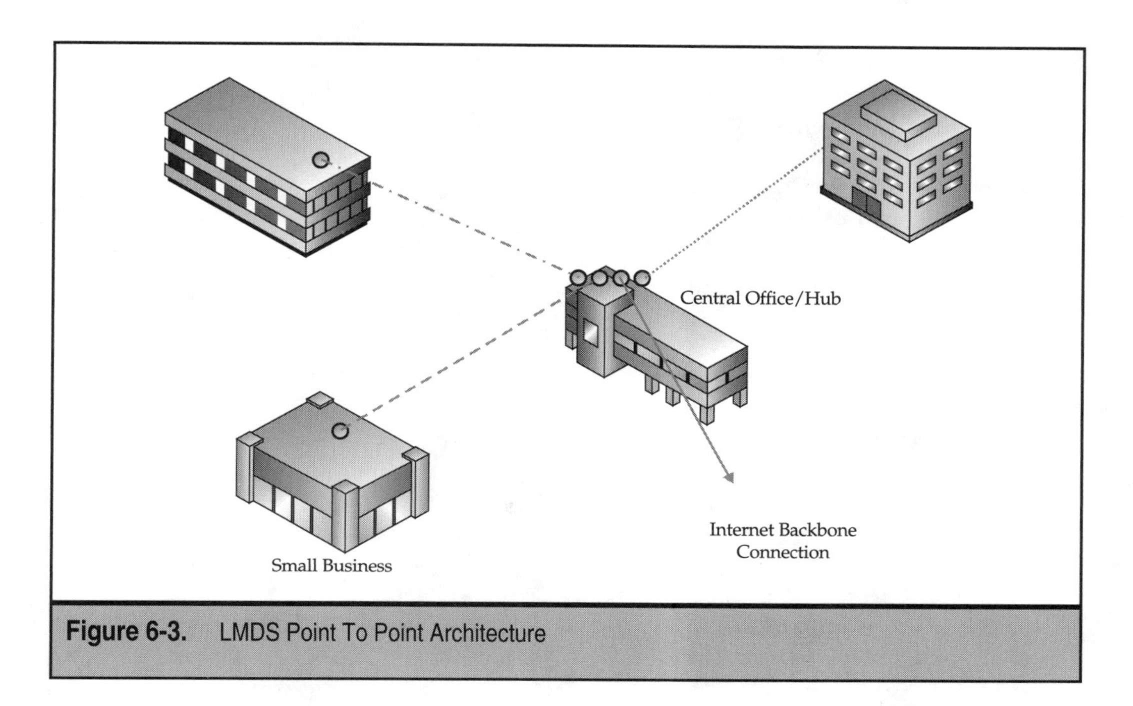

Figure 6-3. LMDS Point To Point Architecture

The competition brought to the ILECs from CLECs has been very good for companies like Cisco, Lucent, Nortel, and 3Com, as ILECs have had to purchase many billions of dollars of network gear in order to effectively compete with the CLECs.

CLECs

CLECs, on the other hand, tend to focus on BBFW and other methods of connectivity to address what is called the *final mile* or, sometimes, the *last mile,* often referred to as the Wireless Local Loop (WLL). CLECs are also commonly the providers of Wireless Local Area Networks (WLANs).

WLL systems typically tee off large trunks and service individual homes or small and medium businesses. This market also includes the Multi-Tenant Unit (MTU), which is a building that houses two or more small businesses, along with the Multi-Dwelling Unit (MDU), which is an apartment or condominium complex. MTUs and MDUs have similar BBFW architectures and are therefore commonly known together as MxUs.

CLECs provide competition to ILECs in the final-mile markets and offer BBFW and DSL as transport mediums. They can also provide toll bypass to businesses and residential customers. *Toll bypass* is where a CLEC can bypass an ILECs local phone network ("local" typically being defined as covering several counties or even half of a state or province).

CLECs came into existence as a result of the breakup of AT&T and other deregulatory changes in the U.S. The federal authorities wanted to create competition, which would theoretically reduce prices to consumers as well as provide better service. The breakup of

AT&T forced it to improve its performance, customer service, pricing, and array of products much as foreign car makers affected General Motors, Chrysler, and Ford. The long-standing incumbents had to improve their products and service, not only to remain profitable, but also in order to remain in business.

Federal law requires that ILECs allow CLECs to connect their gear in ILEC facilities; this concept, called *colocation*, is often a challenging proposition for CLECs, given that ILECs commonly do what they can within the constraints of the law to hamper CLEC efforts. WLL architectures commonly look like Figure 6-4.

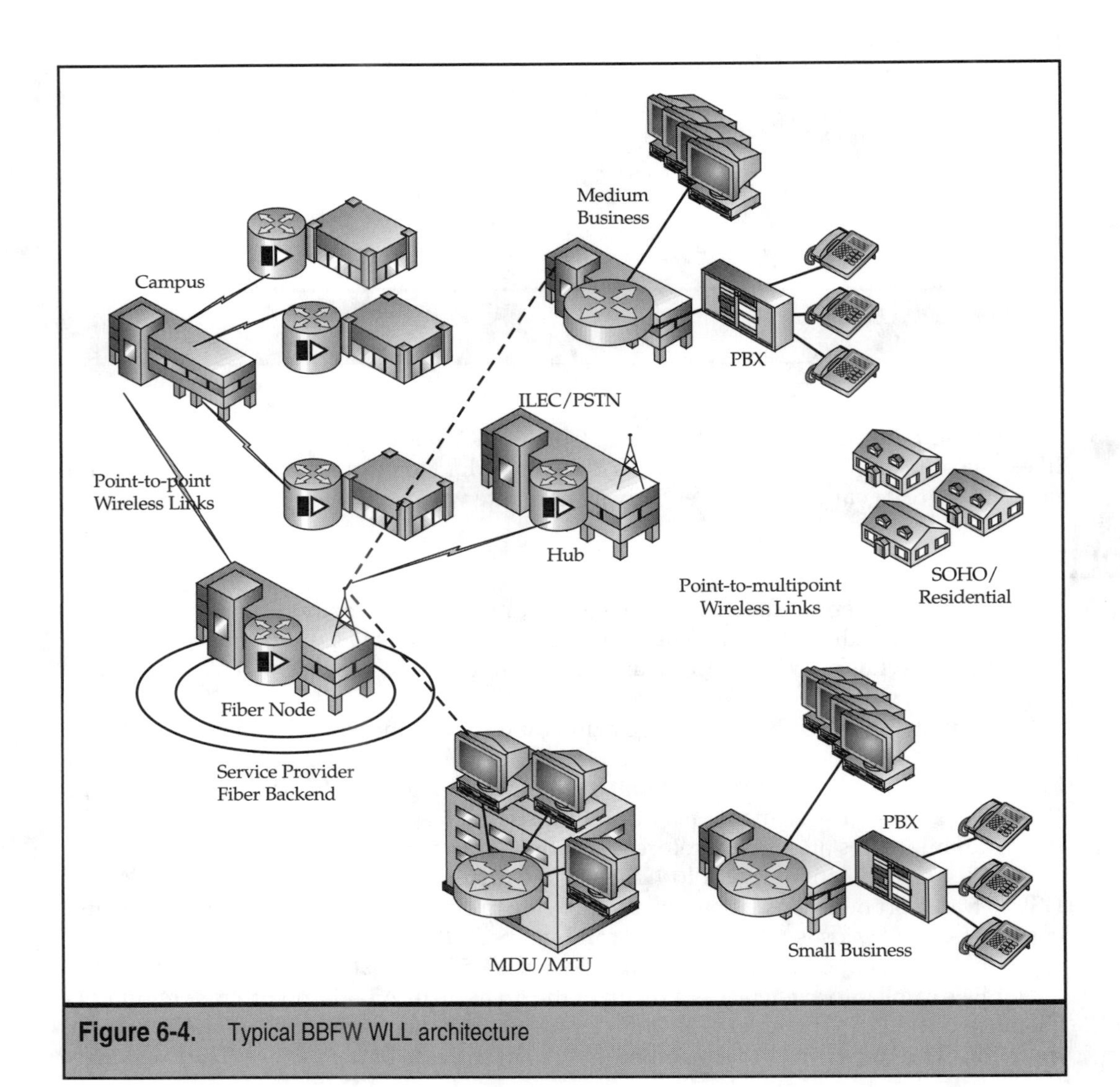

Figure 6-4. Typical BBFW WLL architecture

IXCs

IXCs are companies that connect regional or long distance telephone companies and networks. IXCs have similar network requirements to ILECs in that they generally use very large routers and switches. While ILECs send large trunks of data and voice to and from their networks, IXCs connect the various ILECs. IXCs typically use fiber because of the necessarily immense trunks as well as the absolute requirement for very high reliability. Occasionally, some IXCs use BBFW as a redundant means to protect highly valuable fiber trunks.

Two notable IXCs in the U.S. have extensive plans to use BBFW: Sprint and MCI WorldCom. However, they will use it not in traditional IXC long-haul connections, but as a medium to complete the last-mile connection from the fiber optic cable to the end user for both business and consumers. This is likely a strategy to help fill their own long-haul links with additional traffic. The final-mile conduits could then be used to provide direct services to users or be used by ISPs for actual traffic.

Both of these companies have spent over one billion dollars *each* for MMDS licenses, which are predominantly on the western half of the U.S. for Sprint and predominantly on the eastern side of the U.S. for MCI WorldCom. Both companies have been working extensively with vendors including Lucent, Cisco, and others to provide radios and networks that will provide final-mile services off of their trunks, which will bypass the regional ILECs. Sprint has elected to focus on the final-mile residential market, while MCI WorldCom is focusing on the SMB market. These two companies have exciting business propositions, but the actualization of these objectives has been fraught with delays, immature technology, and deployment bottlenecks.

ISPs

ISPs are dedicated to providing Internet service to large, medium, and small businesses, as well as MxUs and residential areas. ISPs carry virtually no voice traffic, and they very likely won't in any meaningful amount until sometime in 2003, as the market doesn't currently demand it, and Voice over IP (VoIP) technology must mature a bit further before it can be deployed in the residential market. In some geographies, there may also be regulatory implications for ISPs to provide voice traffic; that is, they may need to apply to the local public utilities commissions for the right to deliver voice traffic in addition to data.

Clearly the "granddaddy" of ISPs is America Online (AOL), which, from a modest beginning in 1985, has since grown to be the premier ISP worldwide, having provided more people with an Internet experience than any other group of service providers combined. At this writing, AOL has 29 million users, with specialized Web versions in eight different languages in sixteen countries. They are generally accepted as having the easiest interface to the Internet and are chiefly responsible for the commercial growth of the Internet with a focus on consumers as opposed to technologies.

AOL pioneered technologies such as keywords for simple navigation and the Buddy List to enable instant messaging by showing which of a member's contacts are online. Primarily through these technologies, AOL has created vast online communities with enormously diverse interests. One of the key elements that can be learned from their success is that they have focused on two major areas—ease of use and content—and succeeded in delivering them.

AOL does not, at this writing, have public plans for using BBFW to increase its user base; it instead appears to be adopting a mobile wireless access strategy, similar to that which is pervasive in the European and Asian markets. With plans to provide AOL users access via digital cellular telephones and PDAs like Palm Pilots, the company may well be able to increase the number of its subscribers more rapidly than by waiting until BBFW gains a considerable installed base of users. However, the challenge in part in going to mobile wireless strategy is that the U.S. user is arguably one to two years behind his or her European and Asian counterparts; in other words, U.S. users have a legacy of an Internet experience with large screens, full-sized keyboards, and high speeds—virtually the opposite of what a mobile access experience would be like. There is also, at this writing, considerable controversy as to how much speed the mobile devices will have, in spite of the 3G (Third Generation) promise of speeds up to 384K.

BLECs

BLECs are a new breed of Internet access providers that have taken a clever approach to revenue generation and improving the Internet experience for the building owners and users.

When BBFW, along with cable, fiber, and DSL providers, began seeking to provide Internet access to buildings with commercial or residential tenants, the building owners were besieged by numerous companies, most of whom required unique architectures inside the buildings; for instance, DSL required a different architecture than cable, fiber, or WLAN. After a period of time, the building owners recognized key issues, including these:

- It was cumbersome to accommodate numerous ISPs.
- The building inhabitants faced a confusing number of choices for Internet access.
- Installers occasionally caused damage to rooftops, wiring closets, walls, and structural elements.
- Wiring closets were becoming jammed with various mixes of equipment.
- Building owners recognized they were giving away revenue generation opportunities.

Building owners have begun rapidly taking control of the Internet access between the inhabitants and the Internet at large, and they have reduced the number of vendors from their buildings. An example of the BBFW access model is shown in Figure 6-5.

By controlling access to the building, or group of buildings, BLECs take responsibility for the network elements that reside inside the building—and possibly on the building roof with BBFW. They also charge for the access to the inhabitants and are rapidly moving to broadband to accommodate telecommuters and SOHO users in condominiums and apartment buildings. It is now common in major metro markets to see broadband connectivity touted as one of the prime reasons for selecting one property over others.

Building owners who lease office space to the SMB market have discovered that providing broadband connectivity is as vital as providing space, lighting, heat, and power.

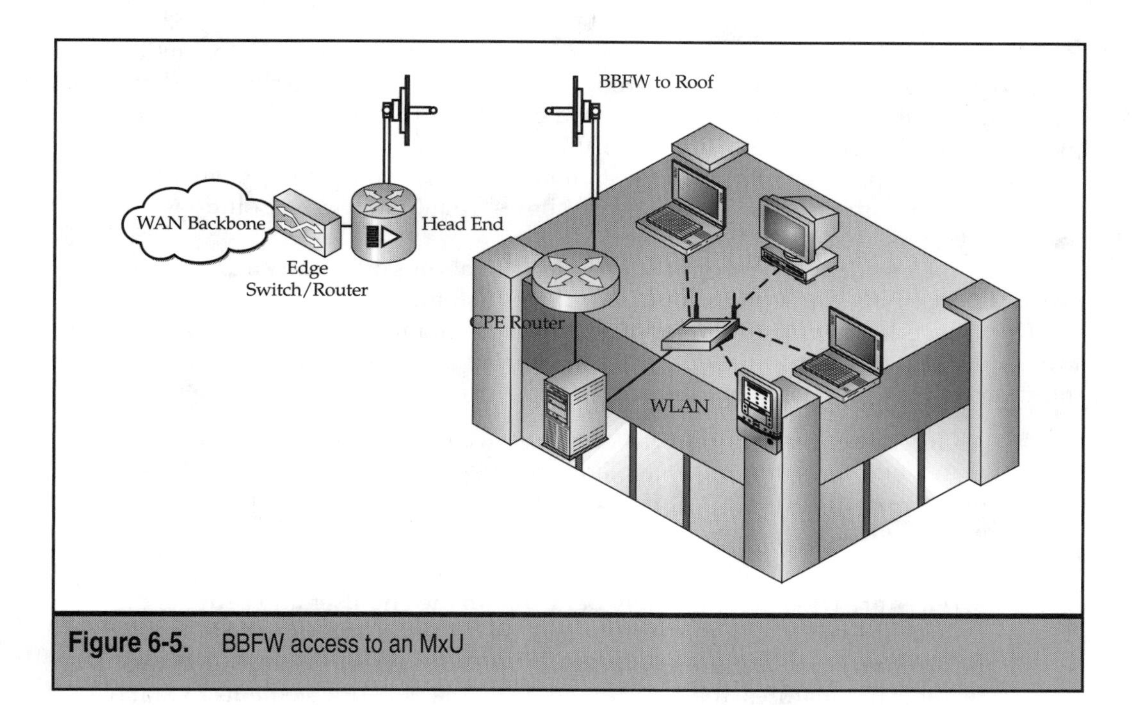

Figure 6-5. BBFW access to an MxU

CUSTOMER ENGAGEMENT QUESTIONS

The dialogue between the various types of BBFW customers outlined previously and the providers of BBFW equipment must be iterative and comprehensive. As BBFW is still a relatively new technology in terms of wide commercial deployment, the average new customer is inexperienced with the array of issues that must be deftly managed. These issues include

- Maximizing revenue and profit generation
- Minimizing capital expenditures
- Reducing system complexity
- Licensing
- Migration from existing network architectures
- Scaling to accommodate rapid user growth and bandwidth requirements
- Network management
- Selection of services and content
- Coordinating the array of ecosystem partners required to install a BBFW system

The iterative process between customer and vendor will enable customers to make certain that they are aware of all the numerous issues related to purchasing, deploying, and maintaining a wireless system. At the same time, it will enable a *solutions* approach to planning, installing, and maintaining a BBFW system.

A solutions approach to a BBFW system includes a number of key elements. One of the best definitions of a broadband solution comes from Matt Irvin, a Solutions Engineer at Cisco Systems, who defines a solution as

"A completely defined and operable service(s) that runs over a given architecture and satisfies a specific business need for one or more customers."

The prospective owner/operator of a BBFW system must ask a number of questions to ensure that a prospective vendor retains a tight focus on the solutions approach. These questions fall into a number of categories, which include

- **Architecture** A comprehensive *end-to-end* collection of network elements that, when combined, allow a platform for services to be delivered. This includes each and every "box," software release, connector, server, antenna, management system, rack, content provider, and more.
- **Service** An application that runs over or is intrinsic to a given architecture. Services are what are ultimately delivered to the end user and provide the true value to connecting to the Internet. The BBFW customer must be briefed in a comprehensive manner on each of the solution elements, but as the relationship matures, the focus moves from the network elements to that of service provision. (See Chapter 5 for further information on services.)
- **Business needs** The *economic factors* surrounding a service offering, including the cost of deployment (often referred to as *capital expenditure,* or *capex*) and operations, as well as the motivation for the customer to offer the service on a forward-going basis (often referred to as *operating expenses* or *opex*).
- **Operation** Provisioning, management, and ongoing maintenance of an architecture and service. This is where the detailed planning for the system on a long-term basis is especially crucial. Issues include interfacing new equipment to older, existing gear (legacy), updating the existing network (migration), and responding to a steadily or rapidly increasing base of customers (scaling).

The Service Providers Target Market and User Benefits

Once again also, if the customer is working with a network company like Cisco, the account team will review the following issues with the customer. If the customer is planning on using an array of vendors, such as one for the radios, one for construction, one for network elements, and so on, it is far less likely that the following issues in their full scope will be detailed. This is another prime reason for selecting a single vendor like Cisco or Lucent to provide an end-to-end solution. Radio, construction, integration, and other subcontractors will have little to no requirement to understand the following issues, much less use an account planning approach to ensuring the customer's success.

One of the first issues to discuss and plan with the prospective vendors, from the network owner's perspective, consists of the service provider's target markets, applications of the solution and service, and user benefits. Again, examples of applications include conferencing, distance training, and telecommuting.

What Are the Customer's Target Markets?

Are the customer's target markets SMB, residential, enterprise, other service providers, or some other markets? The reader will recall that this question is from the perspective of the equipment provider. In this case, the "customer's customer" is the residential and SMB entity. The relevance of this question is that the architectures will vary considerably from residential to SMB, as will the services provided over the BBFW links. The assumptions on returns on investment and other fiscal metrics will vary to an equal degree. The customer expectations, rates charged, and service guarantees will also vary considerably.

What Are the End User Applications Running on This Service, If Applicable?

With applications such as voice or video conferences, which are sensitive to delay and overall quality, the issues of network oversubscription and QoS will come into play. If the customer is planning on focusing on data only, the requirements for network performance can be slightly relaxed.

What Are the User Benefits for the Respective Target Markets?

The service provider must have a laser-like focus on how their customers will be using the network. If the equipment is to be used for telecommuting, the service provider will need to include VPN capabilities and security assurances such as 64 DES, and WEP for 802.11 equipment (indoor WLAN Wireless Local Area Network) if BBFW gear that covers the final 400 feet is also deployed. In summary, the service provider must include service support packages like VPNs to provide a turnkey Internet experience for the end user as well as enhance the revenue stream to the service provider.

Competition

The issue of competition must be considered in any normal business endeavor, but in particular with BBFW, as the service provider will use differing tactics in moving from a position of sole provider of broadband service to a geography to one of competition against other sources of BBFW as well as broadband copper-based competitors.

Because of BBFW's long reach, it can often be deployed as the first broadband medium to a geography. This often enables a BBFW service provider to become the first provider of broadband services to an area. This is in fact the preferred strategy for the service provider, as it enables the provider to charge a premium and capture pent-up demand for broadband services.

The service provider will need to evaluate the competition's service offerings, whether those services are defined, apparent, or even likely, detailing both strengths and weaknesses. Competitive considerations include identifying these factors:

- **The competitor's company** How well funded is the service provider competitor? What is the depth of its deployment and support services? What is the source of its funding—first-round venture capital, or routine institutional financing?
- **Service** If more than one service provider is offering an identical set of services to the same geography, the incumbent (first provider to the area) may wish to broaden its offerings as opposed to engaging in a price war. Competition through reduction of prices is not without its substantial risks. If the incumbent broadband provider can provide good service at a reasonable price, it may lose some customers to churn, but the bulk of them will often remain loyal to the incumbent if only to minimize the hassle of changing service providers.
- **Strengths** Competitive strengths can be in the area of financial reserves, which often buys time for the competitor in that it can attempt a greater array of strategies. They can also be in the area of having a larger established base of customers, which can lead to advances in having a more mature NOC, customer satisfaction programs, and ability to provide gear at a reduced price through volume purchasing.
- **Weaknesses** This is an area the service provider will want to focus on and strategize against. It would be wise to consider that a competitor's weaknesses often change over time as it improves as an operation.
- **Service architecture** Is the competitor's architecture designed for high-performance applications such as voice and video conferencing from the onset, or will it have to migrate to this architecture? This may lead to a more informed decision as to whether the BBFW service provider may want to deploy QoS-enabled gear from the onset.
- **Products** What is the competitor offering in terms of CPE? Is it easy to install, is it stable, and what amount of bandwidth will it provide? Additionally, consider the service provider's competing ability to include QoS.

Service Architectures

This phase of the service provider/vendor engagement sets the foundation to describe the service requirements and will focus on employing network diagrams. These diagrams will detail the service architecture across the core, access/aggregation, and customer premises as appropriate. An example diagram of a Content Distribution Network (CDN) with BBFW service architecture is shown in Figure 6-6.

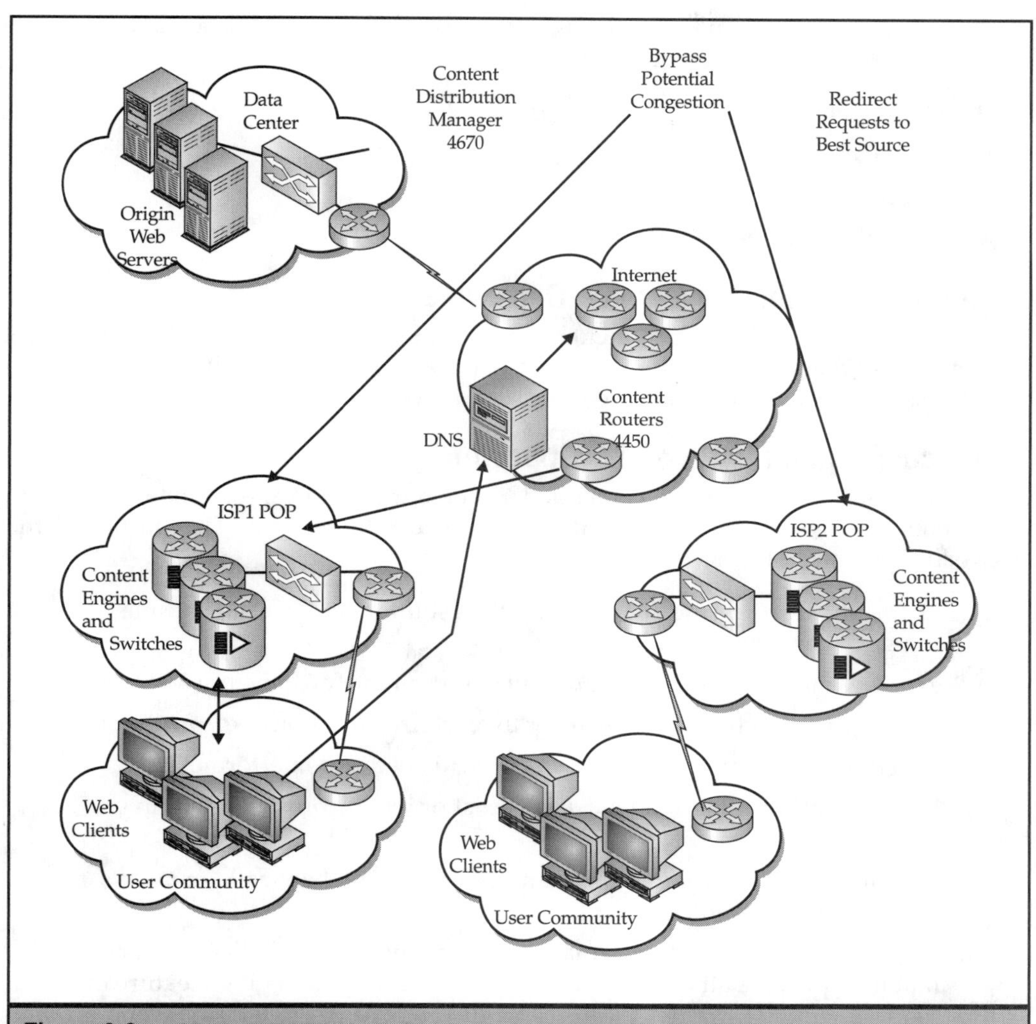

Figure 6-6. CDN with BBFW connectivity sample service architecture

The service architectural drawings, which are sometimes called schematics, are varied and include detailed renderings of:

- Construction
- Traffic modeling
- NOC equipment site maps
- Logical and physical address maps

Customer Timetable for Service Deployment

This portion of the iterative engagement between the service provider and the vendor defines the deployment timetable the customer requires to be successful in its service offering. This section also highlights any contractual delivery commitments and problems.

Technical Requirements of the Opportunity

This portion of the engagement is intended to detail the features, functionality, and requirements for the service needs specified in the previous sections. This portion of the discussion will include these points:

1. Define the features and functions that the solution performs. Examples of things to consider include:
   a) How many times is a feature utilized on a hourly, daily, or weekly basis?
   b) How many users will concurrently perform an action, e.g., provisioning?
   c) How many reports on what objects will need to be performed?
2. Discuss features that don't currently exist or will be provided within six months of deployment.
3. Establish a priority for missing features; categorize them as Mandatory, Essential, or Desirable.
4. Ensure that the product information solicits any known BBFW products that may provide all or a subset of the functionality for a given feature or requirement. For example, if the capability is required to terminate and groom up to 5000 voice lines on a single product chassis, Cisco's MGX8260 may be entered as a candidate product, although it may not support that type of capacity at this given time.

A network provider such as Cisco might use a sample checklist in the form of a table. It should include the appropriate information that should be exchanged between service providers and vendors, as illustrated in Table 6-1.

Customer Engagement Template for Broadband Fixed Wireless

General Business Overview

Review the customer's business objectives for fixed wireless technologies

Questions

Notes

Do you plan to deliver broadband (between 256 Kbps and 2 Mbps) connectivity to small businesses or residences?

Do you plan to deliver broadband (between 2 Mbps and 25 Mbps) connectivity to medium-to large-sized businesses?

Is wireless the primary method of reaching the end user, or is it a secondary method that augments your fiber, cable, or other last-mile (kilometer) strategy?

What services do you plan to offer over your wireless connections?

License Status

Learn about the customer's regulatory status

Questions

Notes

Do you have access to licensed spectrum at this time, or are you planning on acquiring licensed spectrum in the future?

What frequency bands do you or will you have available?

Are considering acquiring an unlicensed band? If so, which one?

Are you considering a frequency band that is not yet allocated, and if so, are you actively involved in the regulatory processes to do this in your market?

Notes

Table 6-1. Vendor Engagement Template for BBFW

Network Use
Inquire into the type of end-user services the customer intends to offer
Questions

What are the end-user services (voice, data, etc.)

What are your competitive advantages with other service providers in your market (voice, video, data mix, bandwidth requirements)

Is your service model based on IP or ATM?

Notes

Radio Use
Determine the type of radio systems the customer intends to use
Questions
Do you plan to exclusively use point-to-point or point-to-multipoint technology, or do you see your services based on a combination of both?

What is the site topology—transmission distances, data rates per link, terrain, obstructions, numbers of hubs and customer premises sites—of the markets that you will be doing business in?

What type of customer premises sites make up your customer mix (single-tenant businesses, multitenant businesses, multitenant residences, etc.)?

Will this be a campus LAN network or will end users be leasing the links from the customer (ISP-type scenario)?

Scaling
Understand the scope and rate of the customer's service rollout
Questions
How large will your initial service rollout be?

Table 6-1. Vendor Engagement Template for BBFW *(continued)*

What is your traffic rate growth profile?

Notes

Current Installed Base
Understand the customer's existing operation
Questions

What network equipment is currently in use?

What radio equipment is currently in use?

What CPE equipment is currently in use?

Notes

Deployment Resources
Understand how the customer will deploy the service
Questions
Are deployment resources captive, or will the customer subcontract?

Do you have a systems integrator selected for all or part of the work: network design (network routing and switching, RF planning), site acquisition and construction?

Notes

Quality of Service (QoS)
Determine the customer's service differentiation
Questions

What levels of QoS do you plan to offer?

Table 6-1. Vendor Engagement Template for BBFW *(continued)*

What oversubscription rates do you plan to use at each level?

What types of service will you deploy? (voice, VPN, etc.)

Notes

Security or VPN Issues

Questions

What level of security will you require?

What types of security challenges do you envision?

Notes

Backbone Requirements

Determine router requirements

Question

Is your backbone ATM or IP?

Notes

Table 6-1. Vendor Engagement Template for BBFW *(continued)*

FIVE KEY WINNING SERVICE PROVIDER STRATEGIES

There are a handful of key strategies that BBFW service providers can employ to assist in building a profitable service provider business. These strategies in summary include

- Secure pent-up demand by deploying in geographies that lack broadband service.
- Know what services to deploy first, second, and third.
- Prioritize BBFW targets.
- Deploy a scalable network that can grow and readily migrate to QoS-sensitive services.
- Make an informed decision on unlicensed frequencies.

Secure Pent-Up Demand by Deploying in Geographies That Lack Broadband Service

There is a reasonable chance that you, the reader of this book, do not have broadband service because it is not available in your area. The tenet of this strategy is—deploy where there is no broadband competition.

With less than three percent of SMBs being serviced by fiber and given the fact that broadband cable providers generally target their resources to the residential as opposed to SMB geographies, this leaves the SMB user with either a DSL offering or BBFW. BBFW offerings to SMBs can command premium prices in exchange for highly reliable services that readily scale for additional users and higher QoS-type services.

On the residential side, most towns with fewer than 25,000 residents lack broadband services of any kind, in spite of a tremendous pent-up demand. BBFW lends itself well to servicing small towns, as it has a twenty-mile reach or more for backhaul and can service a small town from relatively few elevations such as a water, television, or radio tower. If the BBFW provider can begin deploying broadband services with a six-month or greater lead time, that provider has a genuine opportunity to secure a stable and profitable customer base.

Know What Service to Deploy First

There can be little question, given the information gathered by service providers from Sprint and MCI WorldCom down to the smallest ISP, that what most end users want is simply broadband data access to the Internet. Demand for VPN is high where there are telecommuters, but demand for voice and video services lags far behind.

Prioritize BBFW Targets

While it is tempting to go after the residential markets, that is the target that requires the greatest depth for deployment services, both at the onset of the service and going forward by way of customer service. Further, the residential market has the lowest margin of profit, and this distills to the issue of being able to mitigate customer churn due to the expense of the CPE gear. A typical residential user will need to remain with the same service provider for at least twelve to eighteen months in order to amortize the CPE gear. This amortization curve will lessen over time, and this will be a necessity with BBFW gear ultimately costing less than $150 and customer installed.

The primary targets for BBFW service providers are the SMB, MxU, and SOHO markets, for the following reasons:

- For approximately the same deployment effort as in a residential area, a service provider can generate approximately ten times more revenue.
- SMB, MxU, and SOHO customers have lower churn rates than residential customers.

- Migration to more comprehensive services that require greater amounts of bandwidth are more likely with SMB, MxU, and SOHO customers.
- SMB, MxU, and SOHO customers typically require in excess of 3 Mbps of bandwidth, and commonly as much as 5–10 Mbps of bandwidth. This translates to higher generation of revenue and profit through the simple response to the higher bandwidth requirement.

Deploy a Scalable Network That Can Readily Migrate to QoS-Sensitive Services

Deploying CPE devices that are data-enabled only (as opposed also being voice-capable) while including higher QoS services in order to keep deployment costs down is an appropriate tactic. However, at the network edge and core, it's cost effective to use architectures and elements that can readily migrate to high-QoS services such as voice, video, and trunking services. This will not only greatly reduce migration costs but greatly reduce response time to support those QoS-sensitive services.

If an end user determines he or she need voice and/or video, that user can be refitted at the CPE site with an upgrade to the gear. State-of-the-art systems will include the ability for end users to purchase the CPE gear from either an online or brick-and-mortar retail outlet and be able to self-install on a plug-and-play basis. Migrating to gear at the network edge and core that can accommodate unpredicted customer growth or sudden requirements for QoS is far less trivial.

One of the more difficult challenges in building a data network is ensuring that it will scale to accommodate unpredictable growth rates and burst rates. For this, an appropriate modeling tool will provide considerable value. The issue of modeling tools is outside the scope of this book, but major network companies will have an array of them for the customer to use, along with the in-house expertise to generate the relatively accurate projections.

Make an Informed Decision on Unlicensed Radios

As the bulk of the readers of this book will be using radios that do not require broadcast licenses, there remain two predominant choices: ISM (902–928 MHz and 2.4–2.435 GHz) and U-NII (5.725–5.825 GHz). While ISM systems are generally much less expensive than U-NII systems, they generally have far fewer capabilities in terms of scaling, QoS, multipath issues, and available bandwidth. However, to their credit, the lower frequencies have better inherent propagation characteristics; that is, for a given power level, the wavelengths generally travel greater distances.

While companies like 3Com and Lucent are selling a considerable number of 802.11b (2.4 GHz) radios to ISPs at this writing, it is this author's position that users carefully evaluate their needs, bandwidth, and long-term service plans before selecting an 802.11 radio over a U-NII radio. That said, not all U-NII radios are created equal (the same being true for any class of radio, of course). U-NII radios such as those available from Cisco Systems with the VOFDM modulation, or the soon-to-be-released 802.11a format using OFDM modulation, have superior capabilities in terms of rejection of co-channel resolution and multipath. These radios are among the most expensive, if not *the* most expensive, but this is also one of those cases where you get what you pay for, so make a careful evaluation of these radios before making a final element selection, as higher performance can both generate greater amounts of revenue by enabling a greater geographic penetration and reduce capex through having links that remain more reliable.

VALUE PROPOSITION FOR BBFW

Broadband service providers must include BBFW in their portfolio of technology offerings at this juncture in the evolution of the Internet for the following primary reasons:

- BBFW enables service providers to more quickly penetrate geographies that are currently without broadband offerings of any type.
- BBFW can be deployed more quickly than fiber or copper, as it does not require trenching or aerial attachments.
- BBFW does not necessary require colocation, as does cable or DSL.
- BBFW can complement DSL and cable deployments by extending their service ranges.

BBFW allows service providers to more quickly penetrate geographies that are currently without broadband offerings of any type. BBFW has an extensive range that enables the service provider to target the Tier 3 geographies (towns with fewer than 25,000 people). It is generTally cost prohibitive to run fiber out to these geographies, although a fiber node often runs within twenty or so miles of this type of geography.

BBFW can be deployed more quickly than fiber or copper, as it does not require trenching or aerial attachments. After retaining right of way access to either towers or rooftops, a BBFW link can be assembled in as little as half a day. Following the receipt of the network elements and radios, a service provider can begin generating revenue within twenty-four hours. Further, in highly ecologically sensitive states such as California, the service provider is less likely to disrupt ecologically sensitive areas above the ground or culturally sensitive areas below the ground. Environmental studies that can be required prior to trenching can take anywhere from twelve to twenty-four months or more.

BBFW does not require colocation as does cable or DSL. While the law requires that ILECs and cable companies provide service provider access to their facilities to terminate (connect) to and from the PSTN or Internet, the reality is that these colocations are fraught with bureaucratic delays. A BBFW link can terminate on a fiber node or a CLEC T-1 trunk.

BBFW can complement DSL and cable deployments by extending their service ranges. One of the most effective uses for a BBFW link is to attach headend equipment at the DSL or cable node farthest from the CO or MSO respectively and provide backhaul to the beginning of another service area as far as twenty miles away. This eliminates the need for the service provider to lease copper or fiber from a third party, and the link can be installed in a very short period of time.

One of the other uses for BBFW for service providers who already have DSL or copper is to break into a new geography with BBFW, and then follow it up with copper to the areas that cannot be penetrated by BBFW.

SMB ACCESS OFFERINGS

There are really only two things an SMB can do with a network: either generate revenues and profit, or reduce the cost of doing business. BBFW lends itself well to either strategy. In some cases, an SMB can accomplish both at the same time.

First, in generating revenues, an SMB can use BBFW links to broadcast services and content. A content provider can even provide its product to a third party who owns and manages the BBFW link, and indeed it is expected that these types of eco-alliances will proliferate as the specialization envelope is pushed ever further.

Capex will be reduced through the use of VPNs as SMBs access remote campuses and telecommuters without having to install and maintain networks over great distances. Reductions in cost can also occur when an SMB uses BBFW to transport data for storage to either an offsite facility or a third-party storage facility. This same model can be used for medical or legal practices that send large volumes of data or images to remote sites. In other words, using a BBFW link, as opposed to leasing bandwidth from third parties, is a good way to reduce capex.

In summary to this chapter, vendor selection from the BBFW customer perspective or customer selection from the vendor perspective requires a highly iterative engagement where the needs of the customer (service provider) are carefully balanced against the resources and capabilities of the vendor. While providers of radios are often better versed in the actual link performance, the customer should always remember that the deliverable is a network, not simply a set of radios.

PART III

Deployment and Implementation

CHAPTER 7

Deployment: The Make-or-Break Proposition

This chapter will focus on the matrix of competencies required to successfully deploy a BBFW system. We'll include the concept of why the risk for BBFW link performance increases with time, along with a summary description of some the most effective means to resolve this problem.

We will also cover two types of BBFW link deployments, the deployment issues one might incur while deploying links in the MMDS and U-NII frequencies, and also the deployment issues germane to an 802.11 deployment. The latter will focus on indoor issues, as many or most of the deployment issues for 802.11 outdoor links, such as those used on bridges, will be reasonably similar to those negotiated while deploying MMDS or U-NII links.

No BBFW deployment of any appreciable scale is possible without an array of very different skill sets and tools. This is why most BBFW customers rely on ecopartners. We'll also summarize a typical engagement model between a BBFW customer and an equipment provider.

Another issue we'll summarize in this chapter is called *right-of-way,* an important issue given that 85 percent of all antennas are on the tops of buildings as opposed to towers. Landlords know they need to provide broadband as a part of an MxU asset. They are becoming more savvy about rooftops as a source of revenue. What are the differences between antennas mounted on an older building and on a newer one? We will also compare the use of towers versus rooftops.

THE DEPLOYMENT PHASE

Perhaps no phase of a BBFW network is more important than the deployment phase. The number of problems associated with network use can be mitigated by having each element well selected, installed, configured, and managed. BBFW network deployment is entirely nontrivial, and the full array of skill sets and resources necessary to accomplish requires the close coordination of an array of companies. Very few companies in the world today can accomplish this on an end-to-end basis without outside assistance.

There are two types of BBFW deployment within the scope of this effort. The more expensive solution is exemplified by that provided by Cisco Systems through its Clarity acquisition, with links in both the MMDS and U-NII bands and up to 44 Mbps point-to-point capability. The other approach, commonly designated for campus and multi-building business use, is represented by the 802.11b-type deployment with 11 Mbps point-to-point links, as provided through Cisco's Aironet product line in competition with similar products from Lucent, 3Com, and others.

The scope of a deployment can scale all the way from linking a single desk-mounted 802.11 access point to a single PCMCIA card in a laptop, all the way to a nationwide MMDS rollout for multibillion dollar companies. This range of deployment scope is less due to the inherent design of the RF components than a matter of scaling the deployment for dozens to hundreds of sites on a scheduled (and often simultaneous) basis. Very large scope deployments are a unique and rare situation requiring specialized resources, not only in terms of scope, but in terms of sophisticated management tools and experienced personnel in key management positions.

Ecopartners must be conversant in a wide array of issues, from deploying licensed systems on a *Basic Trading Area (BTA)*-by-BTA basis to understanding the most effective tactics for mitigating multipath signals, as well as understanding that a given signal strength for an MMDS system that requires a clear line of site may not satisfy customer requirements but on the other hand may work very well in a U-NII link with OFDM.

There are three major providers of BBFW: Cisco Systems, Nortel Networks, and Lucent Technologies. Each of these companies uses its respective ecosystem partners for deployment and maintenance to such Tier 1 customers as Sprint, MCI, and AT&T. Tier 2 and Tier 3 customers typically use smaller deployment entities and occasionally have captive deployment resources. Most of the Tier 2 and 3 BBFW customers are agile in managing the group of companies required to deploy BBFW and core networks.

Major companies like Cisco Systems that provide BBFW gear have an internal Professional Services group primarily responsible for managing and coordinating deployment partners. Cisco's primary ecosystem partners, for example, include Bechtel Engineering, Price Waterhouse Coopers (PWC), and KPMG, as well as an array of smaller partners like NeTeam.

Bechtel, for example, recognized worldwide for over five decades, is very capable of deploying very large wireless systems. They have extensive captive expertise in site evaluations, construction, logistics, provisioning, finance, and project management. Price Waterhouse Coopers and KPMG are world-class entities with extensive track records in logistics and complex project management. PWC essentially subcontracts the appropriate skill sets while retaining the role as prime contractor and providing or arranging customer financing.

The components and services required to fully install a BBFW network include

- Preliminary site evaluation (paper exercise)
- RF site survey
- Cable routing, customizing for lengths and connectors, and selection for minimal loss
- Mast or tower construction
- Lightning protection systems
- Mechanical assembly
- Construction
- Antenna selection, pointing, and weatherproofing
- Final assembly, test, and documentation
    - Selection, integration, configuration, and documentation of routers and switches
    - Provision of emergency spares at local depots
- Negotiation of right-of-way access (rooftop or tower leases)

Preliminary Site Evaluation

Although only a paper exercise, the preliminary site evaluation is one of the most critical exercises in a successful wireless deployment. It involves an assessment of the size of the network, early calculations for link issues such as distance and level of obstruction, and preliminary determination of frequency issues such as band availability and reuse factors. Tools like the DeLorme™ topography mapping software are used for preliminary evaluation of the regional terrain. This is the stage where profit and loss can be estimated to ensure that deploying a wireless link makes solid financial sense.

RF Site Survey

RF site surveys are as much an art as a science, and it greatly helps to have an experienced team that is fully knowledgeable on the local radiating environment. The reason for that statement is that the radiating characteristics of an area are unique to that area for a given frequency, direction, and transmission power. It is essentially impossible to catalog a certain site and use that data to accurately predict how a different site will propagate even with identical frequencies and power settings. However, a well-seasoned team can acquire a nearly instinctual feel for how a certain radio link might perform in a given area if they have worked that area for some time. The importance cannot be overstressed of having every BBFW link site surveyed in advance of installing the link.

A site survey team usually consists of two individuals with a pair of signal generators and spectrum analyzers who perform the task of evaluating the signal strength over a particular path. They do this by measuring the received signal. The signal generators and spectrum analyzers operate in a given range of frequencies and can cost anywhere from a few thousand U.S. dollars to thirty thousand dollars or more, depending on the frequency measured. A range of frequencies can be analyzed with the same tools. This is an effective technique for evaluating point-to-point links or small point-to-multipoint links where all end points are already known.

On some site survey teams, a production version of the BBFW gear is mounted in a pair of heavy duty vans with pneumatically telescoping masts that have antennas attached at the top. This type of approach is used in point-to-multipoint networks when a number of large geographic areas must be surveyed. Basically, the head-end base station is set up and turned on. The van acts as a movable piece of CPE equipment. This is generally the approach being taken in the U.S. with the MMDS rollout. This approach enables the most accurate measurements and especially provides great amounts of mobility in that the gear does not have to be repeatedly unpacked, set up, and used and then disassembled, repacked, and reloaded. The cost savings alone in assembly and disassembly can justify the cost of these vans, which can easily be in the thirty-five thousand dollar range or more.

The primary task in performing a site survey is to ensure that the amount of energy received at the receiving end of a link will be sufficiently strong enough to operate in all weather conditions, and to ensure that the signal arrives without having been too weakened by distance, obstructions, or fading of the signal.

A good site survey team will take measurements at both ends of a link, as the properties that degrade the signal can have different effects in the two different directions. The relevant part here is that there are receivers at both ends of the link. It is important to remember that even though the radiated power may be the same at each end of the link, the received power may be different due to occlusions in the Fresnel zone; the location; and the *value* of items such as buildings, bodies of water, and rooftops. The reference to value in this case is an indication of how much the physical structure affects the path between the transmitter and receiver.

A site survey team will take the measurements on both ends of a proposed link if they know the exact locations of the buildings or towers, or they may take measurements over a general geographic area that has a cluster of SMB targets. Figure 7-1 indicates the way a site survey team may take measurements over a general area as opposed to incurring the expense of checking for signal strength at every single SMB building site.

It is not uncommon for BBFW network users to prequalify a certain SMB area, such as that which would reside in a business park. This is what the reference *general SMB area* stands for.

This general qualification or prequalification of an area enables service provider sales teams to focus on prospective customers for a given area. Once a specific SMB customer is identified within that area, a site survey and a link budget calculation would still likely be warranted for that specific area because general area qualifications do not take detailed terrain considerations such as trees, small hills, and the location, angle, and composition of nearby buildings.

The two reasons are: 1) The prequalification only certifies that a sufficiently strong signal is available in a given area. The specific radiation characteristics of the actual site may vary from the original, for reasons mentioned in the preceding paragraph. Changes of a even a few feet from the original measurement point can have a dramatic effect on the

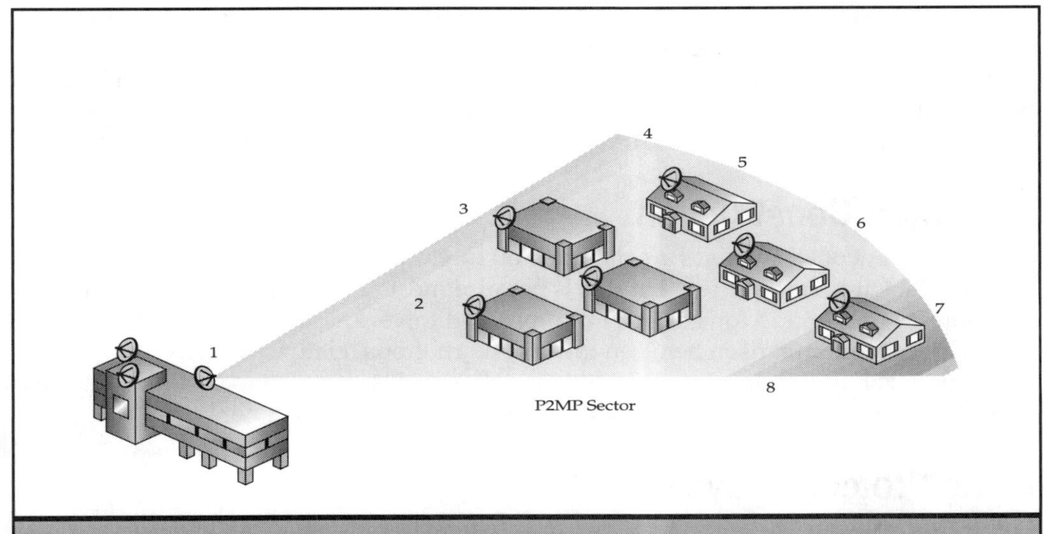

Figure 7-1. A site survey of a general SMB area

fading properties of a signal, and 2) new and prospectively competitive radiating sources may have been introduced to the area since the initial site survey was performed, also altering the RF environment.

Cable Routing, Customizing for Lengths and Connectors, and Selection for Minimal Loss

Surprisingly, it's the installation of cables that can comprise the bulk of the entire time spent installing a BBFW link, once the site is prepared and the gear has arrived. This is because the cables that run between the outdoor gear and antennas down to the routers and switches inside the building must be cut to custom lengths and run through areas of the building that are out of sight, and must have complex connectors soldered on in a fully weatherproofed manner. If even the slightest amount of moisture gets into the cables through the connectors or via a breach in the shielding, it propagates down a considerable amount of the cable and dramatically alters the performance of the cable such that the cable will have to be discarded. As cables can cost several U.S. dollars per foot all the way up to twenty dollars or more per foot, the connectors must be installed carefully and skillfully.

The cables commonly run into the building along with many other cables that are unrelated. The ducts or trays that carry the cables may well already be virtually full before the BBFW installation begins, and as anyone who has run cables through a building can testify, it is hard and time-consuming work.

Signal loss occurs over the cables, as the electrons operate in an environment that is not free from resistance. The general rule of BBFW cabling is to have no less than 3 dB (50 percent) loss per 30 meters of cable. In installations where the cables can run 120 meters or more, larger diameter cable is used to reduce the amount of *cable loss*. Cable loss is one of the calculations used to ensure that radios have the proper power settings and in fact do not exceed federal guidelines for maximum output.

Special attention to cabling must be given in that most electrical system failures are traced to faulty cables, as opposed to a problem with a router/switch component. Further, this type of problem can be difficult and time-consuming to detect and correct.

Mast or Tower Construction

Fully 85 percent of all antennas are mounted on rooftops as opposed to towers. If one ponders this fact, it becomes clear that this is because most tall elements in a metro area are buildings as opposed to towers. Both rooftops and towers have pros and cons. Table 7-1 summarizes the comparison between towers and rooftops relative to BBFW.

Table 7-1 should be used as a general guideline and is intended to provide a snapshot only. There are many exceptions to a number of the issues shown here.

Lightning Protection Systems

Lightning protection systems are required by federal law in the U.S. at distances not to exceed 100 feet of contiguous cabling. Many types of lightning arresters and protection are

Issue	Rooftop	Tower
Access	Generally easy	Requires specialist
Principal liability	Leakage	Lightning strikes
Elevation advantage	If near center of metroplex	Can be designed in
Radio installation difficulty	Generally low	Generally high
General cost	$1,000 - $5,000	$1,000–$2,000 per foot of elevation
Local regulatory issues	Generally none	Can be prohibitive
Space for lease	Not always	Generally yes

Table 7-1. Summary Comparison Between Rooftop and Tower BBFW Installations

available, but in all cases they are intended to minimize the attraction of lightning by ensuring that the outdoor gear does not build up a static charge and, in the case of a lightning strike, to ensure that the energy destroys only a minimum amount of gear. With lightning capable of carrying more than two million volts and approximating the heat of the sun (one million degrees), there can be little doubt that a strike will cause damage. Strike durations can vary so much that damage from lightning can vary from being relatively slight, and limited to the loss of an antenna, or at the other extreme quite high, resulting in the loss of a portion of a tower or building.

Antennas and other outdoor gear such as transverters are far less expensive than the routers or switches that the radios connect to and that are the primary items. In other words, lightning arrestors do not eliminate damage; rather, a lighting arrestor is intented to limit the damage to the least expensive items impacted by a lighting strike. Where possible, the best form of lightning protection is not to put the antenna at the apex of a structure, and instead let something else take the hit. Second, proper grounding is mandatory and required by law. And finally, lightning protectors are cheap insurance as opposed to replacing the high-cost equipment they are intended to protect.

Mechanical Assembly

Mechanical assembly includes mounting the antennas to a short mast of generally four to ten feet on a rooftop. It can also extend to the assembly of a 120-meter tower with all attendant radio gear attached. Mechanical assembly can also include running the cables from the roof or tower to the indoor gear, but this is generally left to the electrical teams.

Mechanical assembly cannot be discounted, as this work includes designs that will withstand extreme weather, including wind, hail, and lightning.

Construction

Construction can include items like assembling a tower or mast, but it generally extends to the installation of concrete pads on rooftops to support the outdoor radio gear and to provide an optimized workplace for the BBFW technicians. Loading calculations must be made to ensure that the rooftops will sustain the weight of an additional structure. Concrete is typically used, but alternatives such as wooden decks can be substituted when other solutions exceed loading requirements.

Newer buildings are more likely to have rooftops that are most suitable for the installation of BBFW gear. Older buildings were, more often than not, constructed with little thought to sustaining much foot traffic. This is why older buildings have roof sealants that are quite adequate for rain and heavy weather but do not withstand the abuse caused by the foot traffic that is a necessary part of installing and maintaining a roof-mounted BBFW system.

Antenna Selection, Pointing, and Weatherproofing

Antenna selection is generally recommended following the site survey, but in some cases the antenna is not selected until the beginning of the final assembly phase. The primary purpose of an antenna is to shape the energy beam, either to focus sharply on a specific building as part of a point-to-point link or to radiate a wide area for a point-to-multipoint service. The most common antennas are

- Semiparabolic or parabolic
- Omnidirectional
- Sector
- Yagi
- Patch
- Reflective panels

Each of these antennas has unique performance characteristics and is commonly deployed for an array of reasons. The best performing antenna in terms of gain is the *parabolic,* sometimes referred to as a *dish* antenna for its shape. A parabolic antenna has high gain, and the radiation pattern is typically only a few degrees. This focused antenna is most often used for point-to-point links.

An *omnidirectional* antenna looks like a straight rod. True to its name, it radiates in a flat circle in a relatively equal manner in all directions. This is a fairly common antenna for wireless LAN environments where broad coverage is needed with no directivity to a specific point on the geography. The drawback to this type of antenna is that it has a relative short range compared to a parabolic antenna.

A *Yagi* antenna has performance characteristics somewhere between a parabolic antenna and an omnidirectional antenna. A good example of a Yagi antenna is the common television antenna. As cross elements are added, the antenna becomes more directive. This antenna is also common in LAN environments where directivity is needed over reasonable distances.

A *patch* antenna is generally shaped in the form of a square or rectangle and for residential use is approximately 12–16 inches square. Patch antennas mount relatively close to a wall or roof. Patch antennas can provide excellent coverage and have a somewhat narrow radiation pattern. Additionally, the low cost, small unobtrusive footprint, and forgiving nature of the patch antenna as the result of suboptimal installation make it very well suited for residential or small business wireless WAN environments.

Reflective panels are used more commonly in areas where a signal needs to be bounced around the side of a building or other solid object or vegetation (such as a tight stand of mature trees). They are passive in that they do not require power to work, and they operate true to their namesake; that is, they merely reflect a signal from one direction to another; they are quite effective both in terms of cost and performance.

Final Assembly, Test, and Documentation

When the outdoor gear has been assembled and mounted, and the cables have been run to the indoor gear (which can reside either in an air-conditioned room on a rooftop or the base of a tower or, in the more likely scenario, in a closet or operations room deep inside the building), it is time for the final assembly and test of the system. State-of-the-art systems like those provided by Cisco Systems have *loopback* capabilities, which means that individual elements such as the transverter, antenna, router blade, and cables can be isolated for test purposes. Generally, each end of the radio link is loopback tested before lighting up the entire link and fine-tuning the performance.

Tests performed include traffic burst, OFDM settings, and link margins. Importantly, the results of these tests should be documented, as they will best approximate the baseline performance of the link when it was first installed. It is the experience of this author that it is a rare BBFW site that has current and comprehensive documentation, and I would submit to the reader that this opportunity should not be passed up, in spite of the pressure to put the link into full service immediately following the stability tests, which typically take approximately 24 to 48 hours.

Selection, Integration, Configuration, and Documentation of Routers and Switches

The selection of either routers or switches or both to complete the network is one worthy of an extensive dissertation, and is mostly beyond the scope of this work. However, a series of criteria exist that must be fully evaluated in order to best design a network that meets the solution requirements of the network owners and users. The criteria have an order that should be followed in order to arrive at a network design based on an empirical evaluation process. The following criteria should be evaluated in order to provide the most optimal network:

- Evaluation of business objectives
  - Capital spending schedules
  - Projected return on investment and other pro forma projections
  - Target market commitment
  - Target market migration, e.g., move to include residential after SMB play

- Target geography commitment
- Staffing level projections

- Vendor evaluation and selection
  - Evaluation of vendor network design and business proposals
  - Equipment evaluation
  - Vendor support commitment evaluation
- Site preparation
  - Head-end and CPE sites
  - NOC site
- Reception of network and radio gear
  - Power
  - Racks, servers
  - Cabling
  - Installation of routers/switches
  - Configuration of routers/switches
  - Installation of radios
  - Network testing
  - Documentation
  - Vendor technical assistance verification
  - Initial traffic flow monitoring

It should be emphasized that, in virtually all network deployments, the phases of final testing, documentation, and link use virtually overlap each other, a fact that calls for a modicum of discipline in order to complete each of the final steps prior to releasing the link for network use.

By way of a final note, it is generally a good idea to test the link and procedure for working with the vendors' technical assistance resources. Better to learn how to work with this valuable resource when the network is operating in a normal mode than to have to learn the engagement process, in addition to attempting to resolve network problems, in real time.

Provision of Emergency Spares at Local Depots

Some vendors, like Cisco Systems, have major spare parts depots within an hour of major accounts and major metropolitan areas. It is not uncommon to have mission-critical spares in place within 60 minutes of failure. Not all vendors can provide this level of service, and not all vendors provide it at no charge. However, networks are complex, and virtually none of the components can be repaired on site. As stated in the earlier "Cable Routing" section, special attention to cabling must be given, in that most electrical system failures are traced to faulty cables, as opposed to a problem with a router/switch component.

Negotiation of Right-of-Way Access (Rooftop or Tower Leases)

Many BBFW networks will require the leasing of rooftop space on buildings that the network owners do not control or even lease. Negotiation of right-of-way goes well beyond the simple renting of space to mount an antenna and includes the following items:

- Power source and payment for same
- Cable runs
- Use of a closet or room for indoor gear
- Air conditioning for indoor gear if a hut is used on the rooftop
- 24 × 7 access to the roof and network gear closet to resolve network emergencies
- Assurances or bonds in case of inadvertent damage to the building
- Assurance of site access should the leasor change
- Exclusivity relative to use (generally requires a premium payment schedule)
- Real estate lease negotiation expertise

In most areas of the U.S., building owners charge approximately $1000 U.S. per month for each antenna installed on their rooftops. Additional monthly charges are often added to this figure when the BBFW network owner requires space internal to the building and/or an air-conditioned hut on the rooftop to house equipment. Construction insurance or bonds are commonly posted to protect the building owner from damage to the building during the construction phase and to further protect the building owner from the risks of fire and lightning strikes that may occur due to the elevation and materials used with antennas.

Unless the BBFW user owns properties on which to locate antennas and/or towers, it would be wise to consider outsourcing this portion of the project. The reason for this is that there is a wide scope of the issues that need to resolved such that they won't disrupt link operation once the network comes online. A simple example of this would be if the owner of the building on which the BBFW antennas were located should change. The BBFW network owners would obviously desire contract clauses that enabled them to retain their antenna locations or be compensated for relocating the antennas to another site equally suitable.

SELECTION OF DEPLOYMENT PARTNER

The scope of issues involved in selecting subcontractors is nontrivial; it includes the following:

- FCC certified (if installing licensed frequencies)
- Financial solvency
- Local environmental experience
- Depth of resources
- Exceptional logistics and project management

- Established relationships with sub-subcontractors
- Local permits
- Finance
- Systems integration
- Forward-going risk profile assessment

FCC Certification (for Installing Licensed Frequencies)

Vendors or contractors who install BBFW gear that requires a site license, such as that required for MMDS or LMDS in the U.S., must be FCC certified if they are going to handle any of the actual radio gear. This means that the technicians actually handling the equipment will need to have been professionally trained, tested, and certified, in addition to the company retaining a certification. This certification does not extend to the handling of the routers or switches at the edge of the network, nor to any of the core gear.

This licensing process has been implemented due to the fairly high radiated power that can be generated by licensed systems, in particular those that have links traveling the maximum distance allowed by the system design. As a word of caution, it's generally always a good idea to walk behind an antenna when on a rooftop if you are not familiar with the gear in place. The FCC, in cooperation with the IEEE, has set forth guidelines for maximum amounts of exposure; these documents are identified as

- ANSI Publication C 95-1
- FCC Dockets 93-62 and 96-8
- FCC Document OET-65

These documents are only indicated for the very few readers who might actually have need of them, and the author has referred to them because they are not very well known. The key point to this is, of course, to use common sense: don't stand in front of parabolic antennas for any appreciable period of time, and don't spend more time in close physical proximity to antennas than you need to. In general, leave this to the professionals.

Financial Solvency

One of the more common mistakes made by BBFW network owners who are new to the business is to fail to verify the financial solvency of the subcontractors they are using. Subcontractors who are not financially solvent can induce a tremendous amount of delay in getting an RF network up and running. While vendors that are not publicly traded will likely be more difficult to acquire financial background information on, a BBFW user can certainly require bank and credit references and verify financial stability with standard credit checking tools. Vendors should also be willing to provide numerous customer references.

In the age of the Internet, when any entity with a sophisticated Web site, can appear much larger than they truly are some old-fashioned leg work will help to ensure that all vendors will be able to complete their portions of the effort on time and with a fully professional level of competence.

Local Environmental Experience

Experience with the local radiating environment cannot be underestimated. Companies with a lot of local experience will likely already have performed site surveys in the spectrum the BBFW customer requires, and will therefore also be familiar with certain areas that are more favorable than others for the purpose at hand.

Companies with this type of experience can also save time and money by making information available on the geographies that are easiest to service for a given radiation pattern and frequency. Deciding on a specific geography goes beyond simply accounting for optimal customer target locations, co-channel resolution, multipath generation, and distance. In determining the most suitable target geographies, you must also consider access to elevations for antennas; such access may or may not include buildings and towers. Towers generally have good access by vehicle, but not always, as there are obviously tremendous differences in how well roads are maintained. Some towers are located on private property where there is virtually no road maintenance. While this may not be a major issue in the summer months, it may become a major issue in the spring or winter months after heavy weather.

Depth of Resources

There is a tremendous amount of demand for BBFW installers by a wide range of BBFW customers, and the number of RF installers does not seem to be increasing proportionally to the demand. Major BBFW providers have recognized this, and occasionally pay in advance for their services to ensure timely support. This leaves open the potential for a bottleneck for the smaller BBFW player in the deployment phase of a BBFW network.

In the customer engagement phase with the various vendors, it is prudent to discuss the company's current and projected backlogs and to determine if a major contract with the vendor is imminent. This won't be as much of an issue with the larger deployment companies, but it can most certainly be an issue with the smaller regional deployment companies.

Exceptional Logistics and Project Management

There are four fundamental elements to truly professional logistics and project management:

- Experienced and empowered logistics and project management personnel
- Sophisticated project management software
- State-of-the-art data-gathering information and procedures
- State-of-the-art communications of project status to project personnel and customers

Few would dispute that the single most important element of successful project and logistics management is that of experienced individuals who have both the responsibility and the authority to make project decisions in real time. A BBFW service provider would be wise to evaluate a deployment partner's key personnel in these positions, evaluating their tools, experience, and reporting and communications tools and methods.

One of the most important assets a deployment vendor can provide is their ability to manage complex projects through the use of sophisticated scheduling and communications tools and state-of-the-art project management business practices. As an example, more than 80 percent of all network gear ordered from Cisco Systems is ordered online. Virtually all of the technical support occurs online, and the payment transactions are also electronic. Even the technical support is Web enabled, but has an audio interface option such that the customer can review network settings with Cisco's technical assistance department and also be able to carry on a phone communication through the Web interface. Not only does this enable Cisco to provide superbly fast service, but customers can be apprised in real time as to the status of their orders.

Established Relationships with Subcontractors

Each vendor will require the support of other vendors. A BBFW customer or service provider should discuss the number and types of subcontractors the vendor typically uses to establish BBFW links. Issues evaluated should include whether any of the subcontractors (or even the prime contractor) has union employees and whether or not the termination date of any union contracts occurs within the schedule of deployment. Due diligence should also involve reviewing the policies and procedures that the subcontractors have regarding immigration laws. This is a hot issue with subcontractors in all areas of the country, but especially in the southern and western parts of the U.S.

The review process should also verify that the subcontractors have bonded and screened employees who will enter secured facilities, as well as verify the employee work experience prior to joining the subcontractor.

Local Permits

A series of local permits may be required for a BBFW link to be established. These permits include but are not limited to construction, antenna location and approval, and environmental impact. Contractors should be well versed with the local permit processes, lead times, and requirements.

Finance

A BBFW customer will require financing in virtually all cases. A combination of financing can be acquired from the major equipment suppliers, such as Cisco and Lucent, but certain deployment entities will also provide financing.

Major equipment providers will provide financing that mostly consists of a line of credit for their BBFW and network gear. The customer will typically need to put up anywhere from 1/4 to 1/3 of the capital necessary for the equipment, and the equipment provider will supply a line of credit for the balance of the gear. Also, in some cases an amount equal to approximately ten percent of the financed amount can be made available for the purchase of noncompetitive gear, as well as an amount equal to approximately five percent of the financed amount to finance customer overhead expenses. The customer in all cases will need to provide substantial credit, a business plan, and other detailed information in order to qualify for this financial support, and of course, these numbers vary from vendor to vendor and deal to deal.

Following the capital put up by the customer, the line of credit and small cash allowances by the major equipment providers, there is typically a shortage of capital on the order of approximately 15 percent, which can be provided in some cases by the deployment partner. This of course will be much more likely with the larger deployment partners such as Getronics Wang, Bechtel Engineering, and KPMG. These entities are not likely to participate in deployment programs under five million U.S. dollars, but the author encourages the reader to contact these companies directly for more detailed information.

Systems Integration

The preferred deployment partner is one that has both RF and network element experience. This limits the field considerably, but also improves the probability of completing the complex task of integrating radios into a state-of-the-art network. Most Systems Integration (SI) partners will have extensive relationships with a number of equipment providers and will have numerous individuals who have been trained and certified on the various types of equipment. SI personnel will also be fluent in construction requirements, configuration of routers and switches, documentation, and training of BBFW NOC personnel.

Forward-Going Risk Profile Assessment

There is a high probability that the local radiating environment will change over time. This is because metropolitan areas are not generally static; that is, they change over time. A building with sides that are predominantly glass that is assembled near BBFW links can have a dramatic impact on link performance. In some cases, it can actually improve the link performance, but in many more cases, it will deteriorate link performance.

Long-range links that cross rural areas and lack sufficient elevation can be affected by crops that ripen, or by metalized buildings commonly assembled in rural areas. In systems with OFDM that operate in the U-NII bands, link degradation can occur if a building near the link path is torn down, as the building may have provided shielding from a competitive unlicensed link or provided a multipath signal that was used as part of the aggregated signal strength.

The key to mitigating the effects of a changing local environment is vigilance in monitoring the link performance. State-of-the-art systems will provide alarms and send pages to NOC personnel if a link degrades quickly, but links that degrade over weeks or days will need to be monitored. The records that will need to be routinely reviewed are called *histograms*, which serve as detailed records on a day-by-day, hour-by-hour, or minute-by-minute basis, depending on how the integrators and NOC personnel configure the monitoring functions of the network.

WLAN (802.11) DEPLOYMENT ISSUES

With an ever-increasing understanding of the importance of pervasive computing and broadband mobility, the subject of deployment would not be fully covered without reviewing the primary principles of what happens in the "final 400 feet" of the BBFW link, that is to say, 802.11 or similar deployment such as HomeRF.

These 802.11 issues are slightly different from conventional outdoor BBFW deployment issues, and while there is an abundance of outdoor 802.11 gear, this section will focus on the issues relative to indoor deployments.

Many of the keys to a successful indoor 802.11 deployment are similar to those for the outdoor version of BBFW, which include a site survey, customer interviews, and careful planning. There are often many more surprises discovered during an indoor deployment, and they come from sources that wouldn't seem apparent. Figure 7-2 indicates some of the more common problem sources relative to an indoor deployment.

Figure 7-2 indicates some of the problem areas, including microwave ovens, metal storage lockers, and file cabinets. While microwave ovens provide a direct source of interference to 802.11 receivers, such as client cards in laptop computers, other items in this illustration are sources of multipath interference.

In-Building Design Considerations

There are a number of in-building design issues to be considered, which include the following:

- Number of potential concurrent clients
- 802.11 shared with Ethernet LAN
- AP utilization increases with the number of associated clients
- Higher speeds from smaller cells
- Autorate negotiation

Number of Potential Concurrent Clients

A state-of-the-art 802.11 system will allow approximately 2048 different clients to associate with a single access point (AP); however, the practical number of client associations is far fewer than that and in fact is on the order of about 25 or so before bandwidth becomes appreciably reduced; as a result, data transfer rates become noticeably slow.

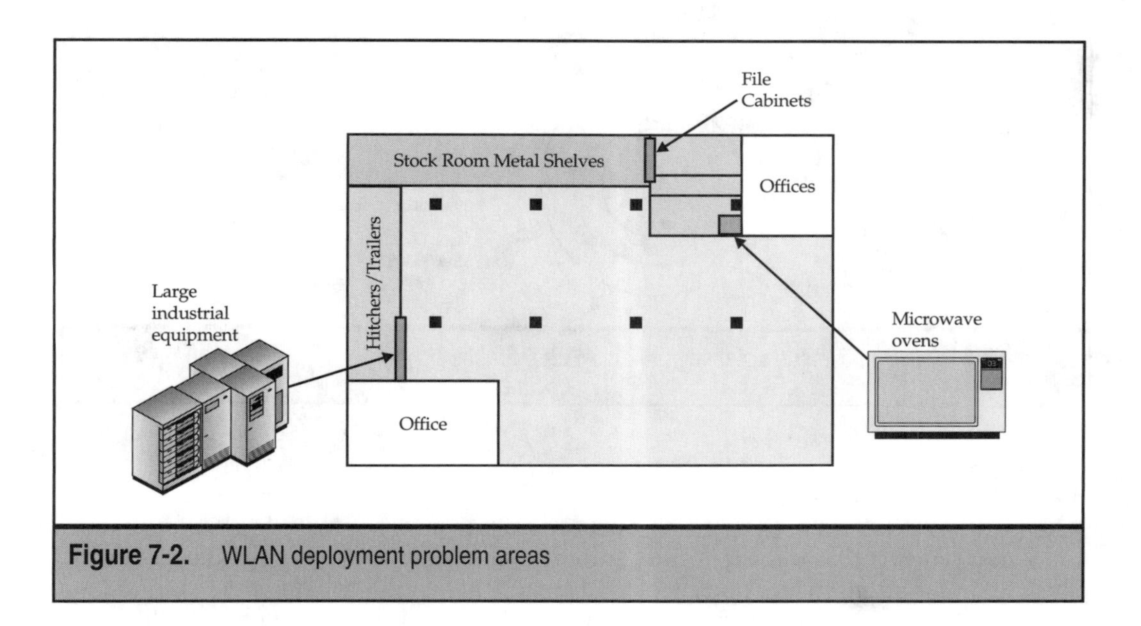

Figure 7-2. WLAN deployment problem areas

802.11 Is Shared with Ethernet LAN

It is important to remember that the WLAN network is merely an extension of the LAN, which is most likely to be Ethernet. This means that the backhaul to and from the servers is performed (generally) by the Ethernet backbone. The traffic management and performance profiles are predominantly set by the total number of Ethernet users, of which 802.11 clients are typically a subset. In other words, in most office applications, the bulk of the users on the Ethernet will be ether-tethered users as opposed to the 802.11 users. The 802.11 system in fact is commonly referred to as *wireless Ethernet*, and operates with predominantly the same protocols and limitations.

AP Utilization Increases with Number of Associated Clients

AP utilization increases with the number of concurrent users, but load balancing can be performed by having more than one AP within the range of the client card. The APs can either be on the same ether port or not. If not, this will enhance throughput and load balancing capabilities.

Higher Speeds from Smaller Cells

Higher downloading and uploading speeds are always a function (in part) of the distance from the AP. Figure 7-3 demonstrates this point.

The approximate range for each of the data rates varies depending on local environments, security settings, traffic load, and antenna selection, but the average range over which one could expect to achieve 11 Mbps rates would be approximately 100 feet LOS from the AP; for 5.5 Mbps, approximately 200 feet LOS from the AP; and for 2 Mbps, from 200 to 400 feet LOS. All these approximations are for indoor use only.

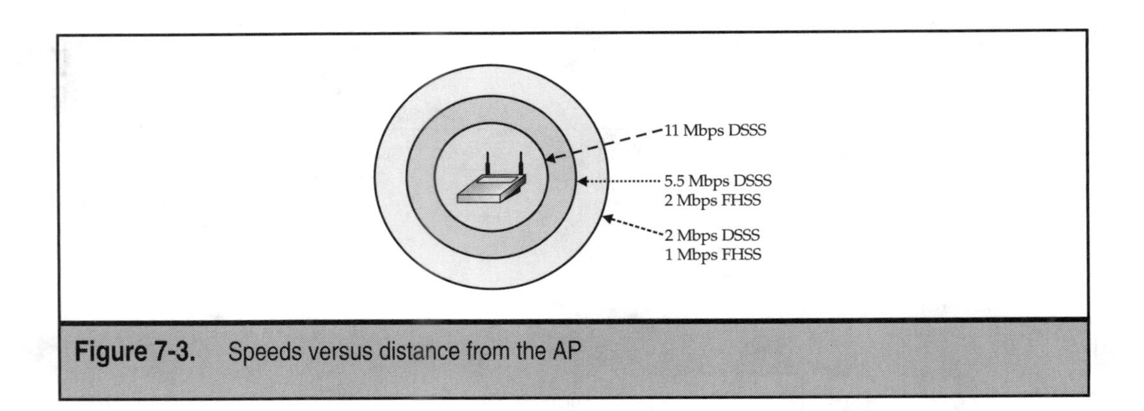

Figure 7-3. Speeds versus distance from the AP

Autorate Negotiation

State-of-the-art systems such as the Aironet products from Cisco include an *autorate* function, which means that the client card and AP will automatically adjust data rates depending on the strength and quality of the signal between the two devices. Autorate is a feature that allows backward compatibility to older 802.11 gear and compatibility with lower-performance gear provided by other 802.11 vendors. (Not all 802.11 gear has the same level of performance.)

It is important to note that a device can have a strong signal that is low in quality due to the amount of noise it carries along with the data. Conversely, one can have a weak signal (low level of power) but fairly high quality. This can occur when the distance is fairly great but the data rates are low and there is little interference from outside sources.

802.11 Deployment Tools

These 802.11 systems are much easier to deploy than typical outdoor BBFW systems. One of the reasons is that state-of-the-art systems like Aironet have client utilities that are easy to use and are integral to the purchased system. While these utilities are not as sophisticated as those found on high-dollar BBFW systems, they are quite serviceable and perform to the limits of their design. The client utilities have a standard browser interface and include

- Site survey
- Configuration
- Usage statistics

The main Aironet Client Utility is shown in Figure 7-4. The client utility used for site survey and forward-going statistics is shown in Figure 7-5.

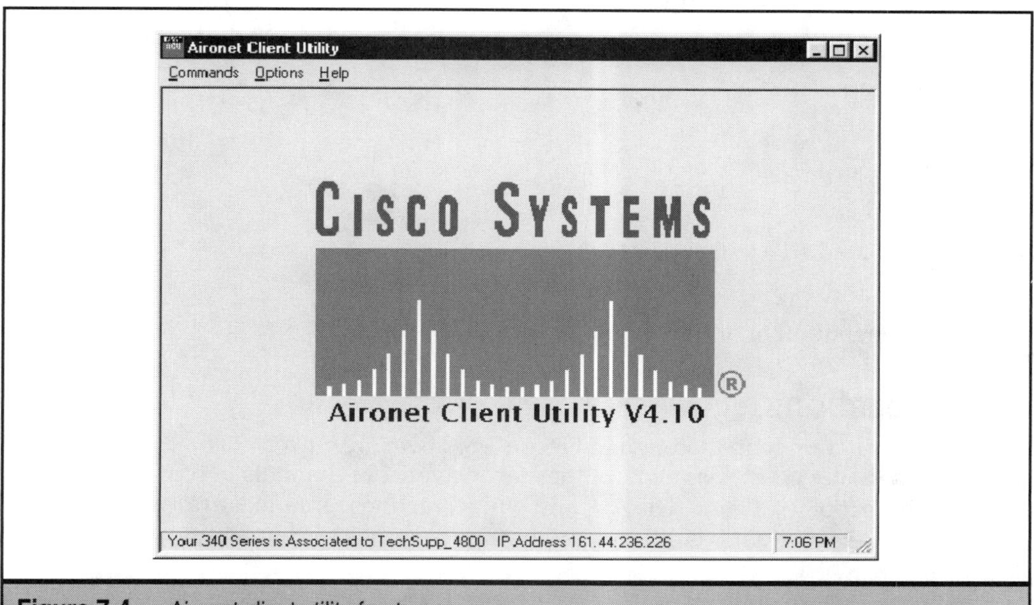

Figure 7-4. Aironet client utility front page

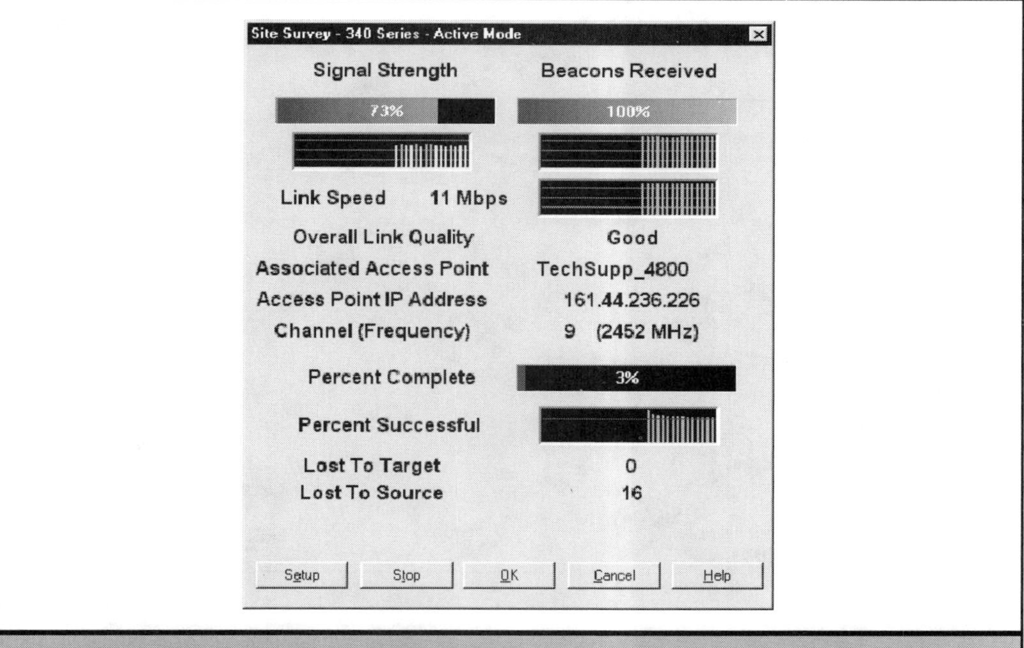

Figure 7-5. Aironet site survey and statistics client

The client utilities are browser-based and easy to use; they include the following information:

- Signal strength
- Link speed
- Overall quality
- Associated AP (which AP)
- AP IP address
- Channel number and frequency

Sample Installations

A sample installation map is shown in Figure 7-6. There are eleven channels in an 802.11 system. This deployment uses eight of the eleven available channels to completely cover a 17,000 square foot retail area. The APs and omnidirectional antennas are mounted in the ceiling, and the retail owner is able to get 11 Mbps coverage to about 60 percent of the total floor space, which is exceptional. Most of the rest of the coverage is at 5.5 Mbps, which is more than adequate for inventory tracking purposes. A telephone system is also incorporated into this deployment using Spectralink phones. While there is no contention resolution for the voice traffic, the voice traffic compared to the data traffic is low and in fact is deployed as a standard item in many hundreds of locations in this retail chain.

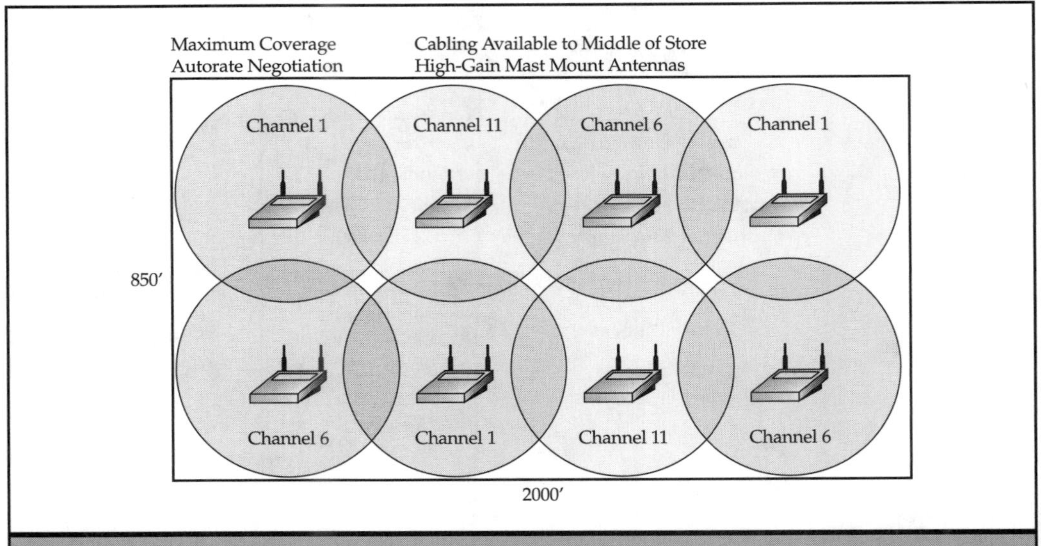

Figure 7-6. Sample installation for a large retail store

The deployment in this national hardware chain has approximately the same number of square feet in the store (17,000) but uses a different approach, which is to mount the APs high on the walls (Figure 7-7). The intent of this architecture in part is to be able to retain a LOS to the client cards, which are occasionally hidden by tall display racks.

Again, this deployment is similar in the amount of area covered, but two types of antennas were used: dipoles to shape the areas for preferred high data rate coverage and patch antennas for outdoor coverage for aesthetic purposes (Figure 7-8).

The more difficult aspect of this deployment that may not be readily apparent to the human eye is that the signal can penetrate floors, which means that the channels used on the floors above and below must be different (Figure 7-9). This is simplified somewhat by the fact that each of the floors is symmetric in layout, making the coverage more predictable (Figure 7-10).

One of the key elements in this deployment was the focus on broadband mobility and pervasive computing, which means that all employees needed to have access to APs at any point in the building. Since the overall bandwidth was not high, a minimum number of APs was used (up to three per area can be used for load balancing).

In summary, 802.11 indoor deployments are generally more simple than outdoor deployments, as the gear itself is easier to use, has lower bandwidth rates, involves shorter ranges, is strictly LOS, and can be installed without the help of professionals.

Outdoor BBFW gear, on the other hand, requires a wide array of eco-partners to accomplish a network deployment that may cover 50 or more square miles. The variables in that amount of geography are substantial in number to say the least, and the links typically have a much higher value than those in an indoor 802.11 architecture.

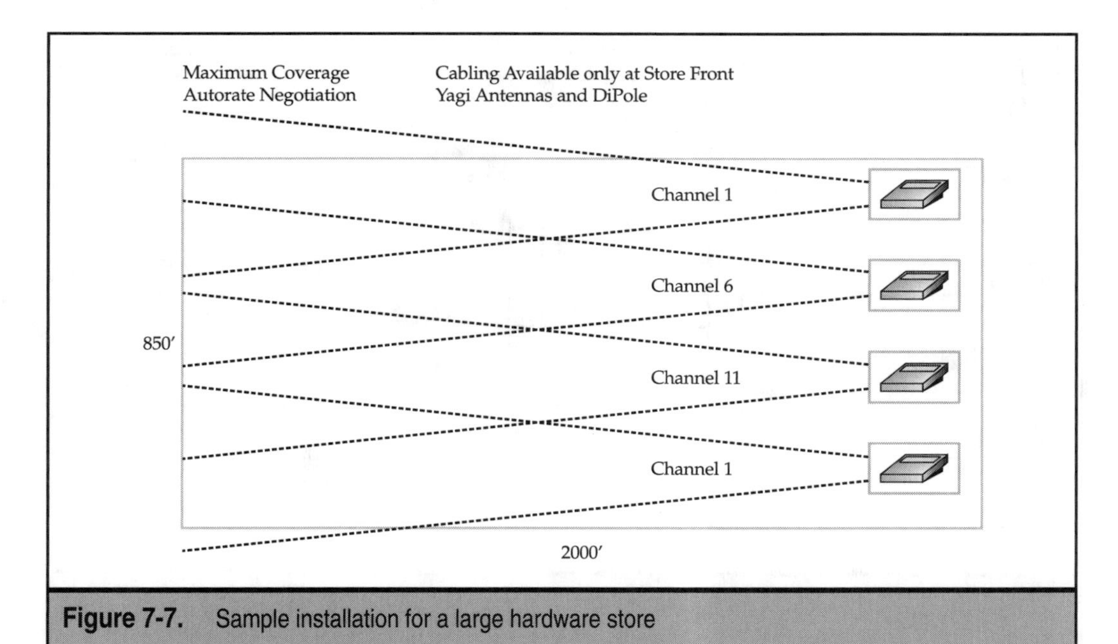

Figure 7-7. Sample installation for a large hardware store

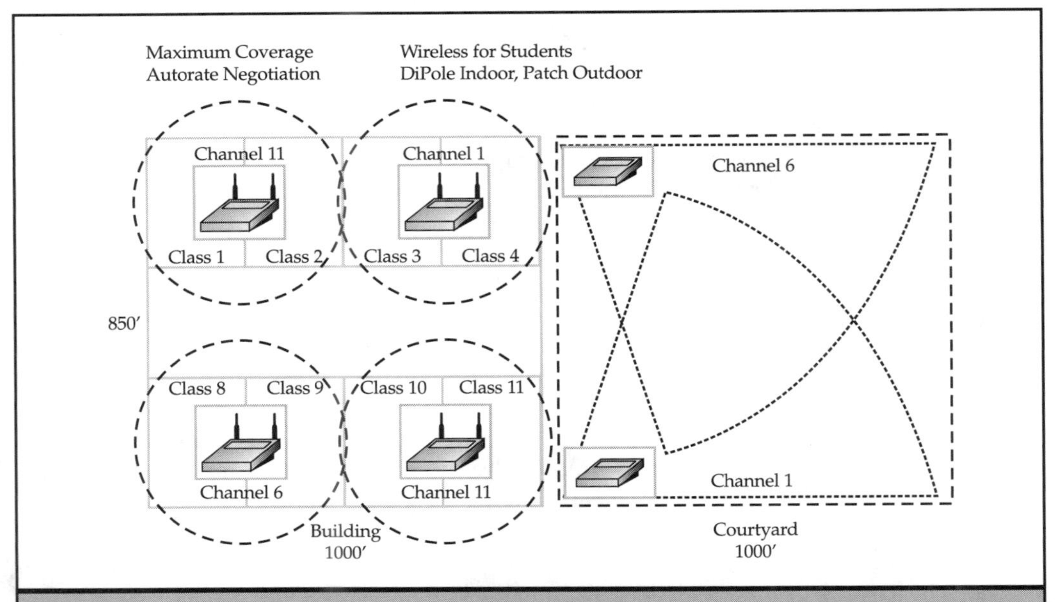

Figure 7-8. Sample installation for a school

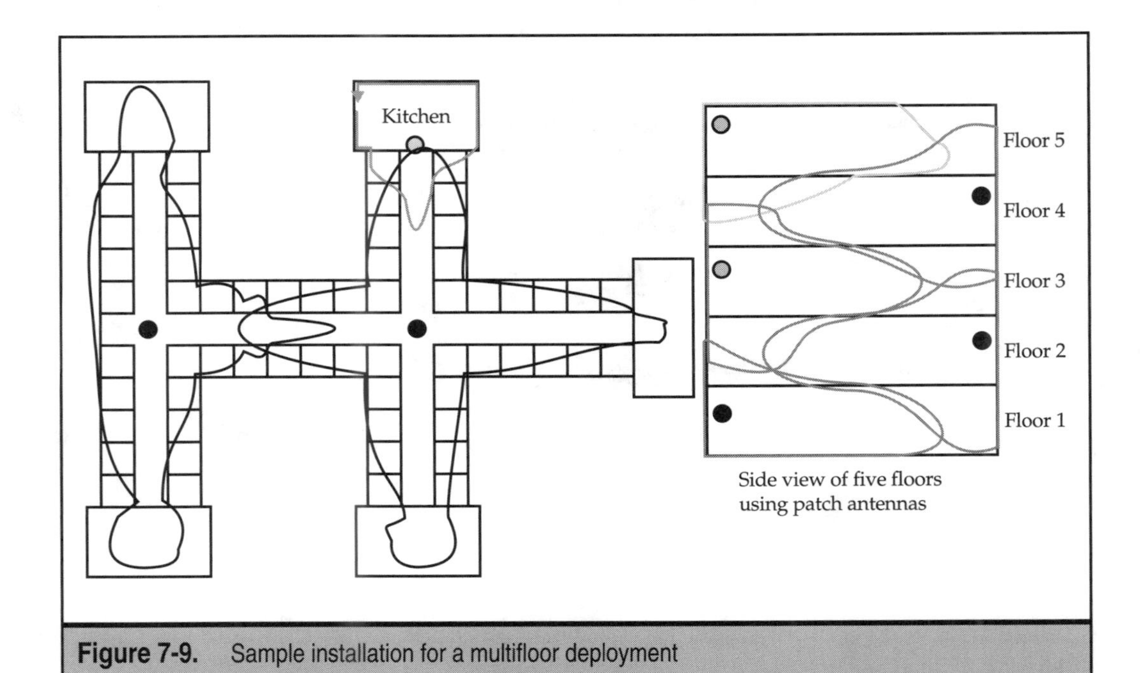

Figure 7-9. Sample installation for a multifloor deployment

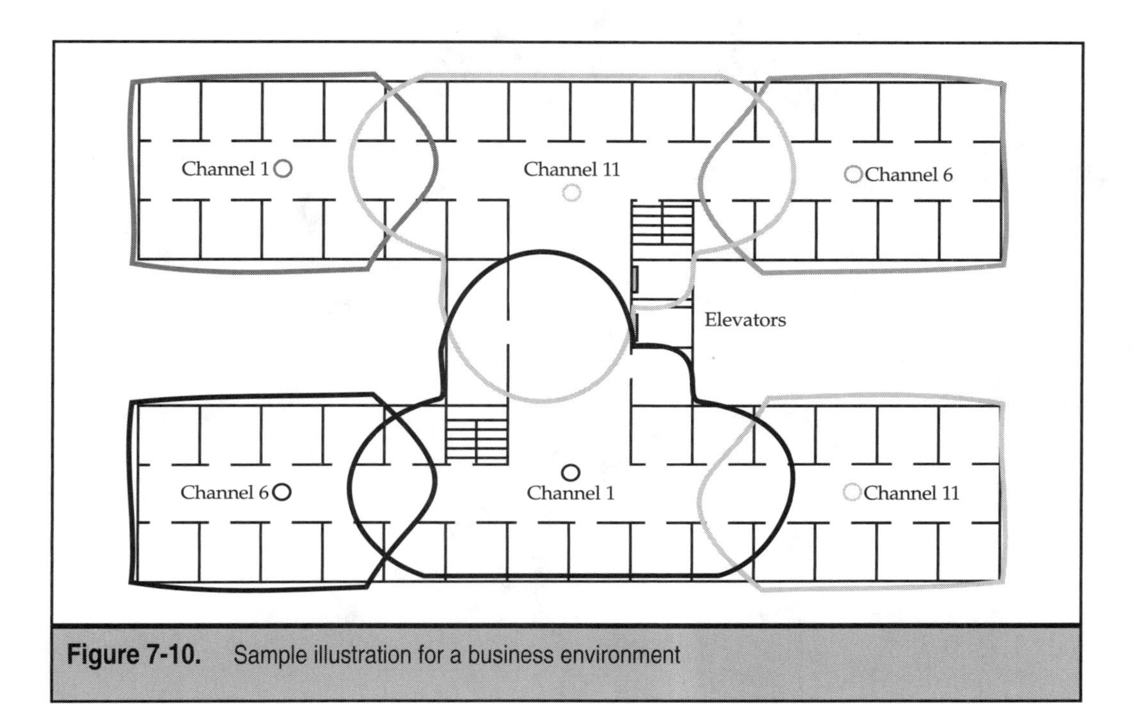

Figure 7-10. Sample illustration for a business environment

PART IV

Wireless Technologies

CHAPTER 8

Modulation Schemes

Numerous BBFW modulation schemes are available, and there is more than one issue to consider when selecting a BBFW system. *Spectral density,* for example, is one of the key metrics for measuring BBFW link performance. This term refers to the number of bits sent out per complete sine wave, or in other words, one complete antenna cycle. To be more specific, if a transmitter were sending out data at 1 Hz, there would be one sine wave per second; if it were transmitting at 1 GHz, there would be one billion complete sine waves in the same interval, and so forth.

Spectral density is largely dependent on the selected modulation scheme. This section will assist the reader in understanding the pros and cons of various modulation schemes and the compromises associated with the use of various schemes.

MODULATION SCHEMES

The objective of a modulation scheme is to transform ones and zeros into waveforms that can be transmitted and received by the carrier frequency of a BBFW link. The term *carrier frequency* has been used throughout this book, and though it is essentially unrelated to the issue of modulation schemes, it may be appropriate to use an analogy to better convey what it means. If you were using a printer, the carrier frequency would be the paper, and the modulated information would be the letters on the paper.

Modulation, therefore, is the technique of turning bits into something that can be carried by the carrier frequency over the air. The carrier frequency has no intelligence; the modulation riding on the carrier frequency carries the intelligence (i.e., data) between two end points.

At lower frequencies (such as 2.5 GHz), the selection of a modulation scheme is very important because there is inherently less bandwidth to work with than at higher frequencies such as LMDS (28 GHz in the U.S.). The proper term with respect to this fundamental is *spectral efficiency,* i.e., how to get the most out of the available bandwidth.

Any signal that can be translated into electrical form, such as audio, video, or data, can be modulated and sent over the air. Data is the easiest to modulate and transmit, while video with voice can be the most problematic, due in part to its reliance on low levels of delay in the transmission and processing phases of moving this type of traffic from one radio to another.

There are many different modulation schemes; Table 8-1 contains a partial list.

All of the schemes shown in Table 8-1 are related to one of three fundamental types of modulation:

- Amplitude modulation, where the output power of the transmitter is controlled by the signal
- Frequency modulation, where the output power is kept constant and the frequency is varied over a small range
- Phase modulation, where the amplitude and frequency remain constant but the phase relationship between the bits of information varies

Symbol	Modulation Scheme
AM	Amplitude Modulation
FM	Frequency Modulation
SSB	Single Sideband
PM	Phase Modulation
CW	Continuous Wave (telegraphy)
PCM	Pulse Code Modulation
VSB	Vestigial Sideband
BMAC	Type B Multiplexed Analog Components
QAM	Quadrature Amplitude Modulation
DSSS	Direct Sequence Spread Spectrum
FHSS	Frequency Hopping Spread Spectrum
BFSK	Binary Frequency Shift Keying
QPSK	Quadrature Phase Shift Keying

Table 8-1. Various BBFW Modulation Schemes

The most common modulation schemes used at the time of this writing for BBFW links are:

- Binary Frequency Shift Keying (BFSK), more commonly referred to as FSK
- Binary Phase Shift Keying (BPSK)
- Quadrature Phase Shift Keying (QPSK)
- Quadrature Amplitude Modulation (QAM)

The reader should note that many other types of modulation exist; with few exceptions, however, they stem from these four base modulation schemes. Each of these modulation schemes represents an increase in complexity over the preceding one, from BFSK to multiple levels of QAM. BFSK will send a "one" with one frequency and a "zero" with another frequency. BPSK will send two states, a "one" with one phase and a "zero" with another phase.

QPSK gets more complex and has four states to represent either a 00, 01, 11, or 10, four phase states, all with the same amplitude. Figure 8-1 indicates the "constellation," which is a collection of permitted maximum phase and amplitude combinations.

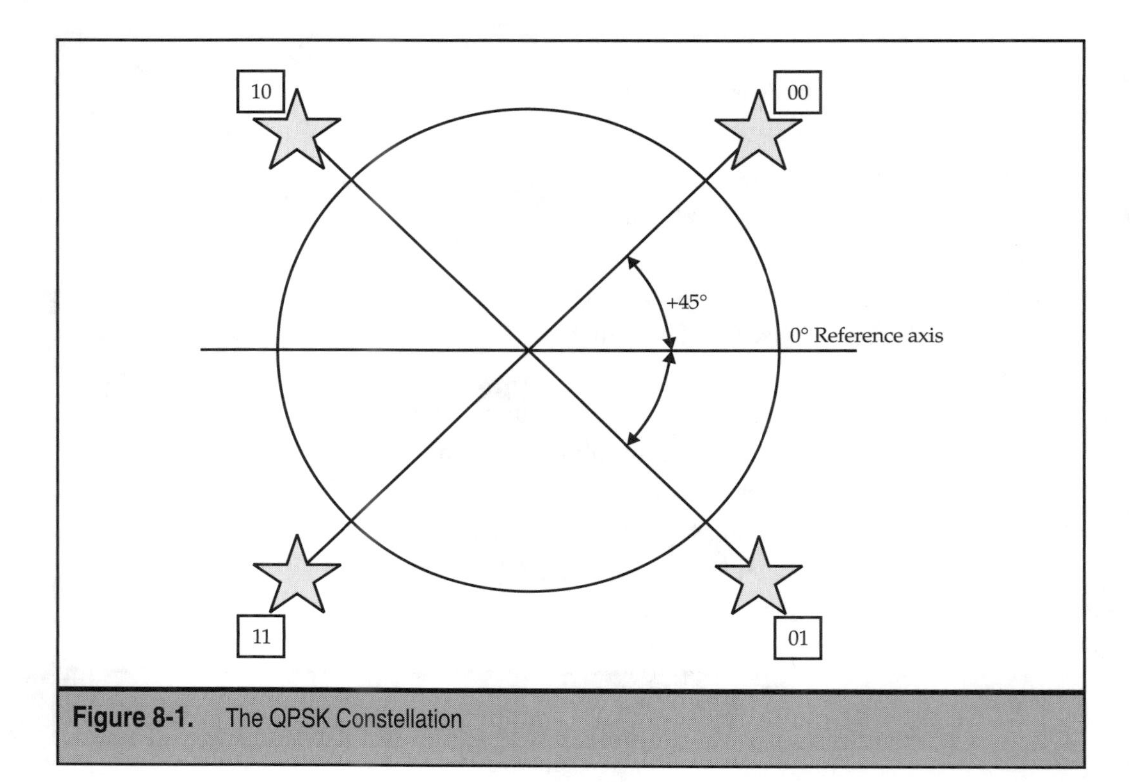

Figure 8-1. The QPSK Constellation

QAM is a technique that modulates two carrier frequencies in both phase and amplitude (using sine and cosine carriers that are 90 degrees apart). Table 8-2 illustrates the fact that as the number of bits increases linearly, the number of phase/amplitude combinations increases exponentially, providing a very high spectral density at even 64 QAM.

This means that with two amplitude values carried by a single carrier frequency, the link can carry two bits as opposed to a single bit, thereby having a higher *spectral density,*

Amplitude/Phase Combinations	Bits per Symbol
16 QAM	4
32 QAM	5
64 QAM	6
128 QAM	7
256 QAM	8

Table 8-2. QAM versus Bits per Symbol

which means that more information is being carried for a given burst of energy from the transmitter. Simply stated, the frequency does not change, but the amount of data transmitted increases as the modulation level increases. This is not without cost. As the modulation complexity increases, so does the probability of an error in the transmission. Errors in transmission (or processing) mean mistaking a one for a zero or not being able to decipher the energy as either as one or a zero. The metric for the relative amount of error is referred to as the Bit Error Ratio (BER).

The quality of the equipment becomes very important when using the more exotic modulation schemes. Quality of equipment refers not only to quality of manufacture but more specifically to inherent quality of design.

There are several trade-offs to be considered when deciding which modulation is most appropriate for a given environment, although the user or network operator will probably be unable to select different types of modulation (but can indeed select different levels of modulation, such as 64 QAM or 16 QAM, for example). Higher frequencies or greater distances tend to favor less complex modulations. Low signal to noise links with simpler modulation schemes generally work better and generally have a lower BER. The trade-off for this simplicity is lower throughput. Some systems provide a single modulation, while others have the capability to change the level of modulation depending on conditions of a specific link.

A number of QAM modulation schemes are common at this writing. In the wireless arena, they range from 16 QAM up to 64 QAM. Today, 64 QAM is more common for MMDS and U-NII band products, while spread spectrum modulation such as FHSS and DSSS are common to WLAN products, particularly IEEE 802.11B–802.11b compliant solutions. QPSK is one of the most common modulation schemes used in LMDS-band solutions because it has such enormous amounts of bandwidth and very high frequency that it does not demand the same levels of modulation complexity.

At this writing, 64 QAM is about as spectrally dense as the transmissions tend to be, as it typically provides the best compromise between the ability to send a higher amount of traffic over a carrier frequency without incurring too much expense, complexity, and time delay for processing. Over time, QAM will of course move well beyond 256 QAM as development teams improve stability and lower the cost of this type of system. Development is under way at this writing on 128 QAM radios and beyond.

In the cabled environment, QAM levels up to 1024 are being tested successfully, but it will be many years before the quality of the components is sufficient to make ultra-high-level QAM cost effective in the BBFW world. Spectral density is especially important in the unlicensed bands like U-NII in the U.S., where there is only 100 MHz of spectrum allocated, and in general, all carrier frequencies below about 26 GHz need more spectrally dense modulation schemes because they generally don't have as much spectrum as those found at 26 GHz and above. The following table illustrates the amount of spectrum allocated in the U.S. for three of the more common bands:

Band	Spectrum Allocation (Kbps)
U-NII	100
MMDS	200
LMDS	1300

It is worthwhile to compare cycles per bit at various frequencies as well. The higher the frequency, the more cycles per bit. The following table provides a snapshot of the difference in the number of cycles per second at 850 MHz and 2 GHz:

Cycles	850 MHz	2 GHz
10 Kbps	85,000	200,000
1.5 Mbps	567	1333

The concept of cycles per bit leads into the concept of a *symbol,* which is a uniquely identifiable signal that contains a certain amount of bits, as determined by modulation complexity. Individual symbols are distinguished by attributes such as duration, amplitude, frequency, or phase. The number of *bits per symbol* is one of the most common methods for determining spectral density. Four or more bits per symbol would normally be considered highly spectrally efficient, while one or two bits per symbol would be less spectrally efficient though still quite serviceable.

In the end, the selection of a modulation scheme is based more or less on the compromise between maximizing bandwidth and building in sufficient resistance to interference. There is a very real distinction between high-end engineering teams whose primary mission is to develop a high-performance BBFW product line and startup companies that happen to have an RF link but tend to be far less sophisticated in their approach. Most experienced electrical engineers can devise some sort of RF link between two devices, but it is an entirely nontrivial effort to produce commercially viable BBFW links, a task that commonly involves teams of 50 to more than 100 engineers.

Multipath and OFDM

RF systems fall into two categories: those using frequencies less than 10 GHz and those with frequencies greater than 10 GHz (referred to as millimeter wave systems). Several bands exist below 10 GHz for high-speed transmissions. These may be licensed bands such as MMDS (2.5 GHz) or unlicensed such as U-NII (5.7 GHz). Bands below 10 GHz can have propagation distances up to 30 miles, while millimeter wave products are generally limited to ranges of approximately 5 miles or less.

RF systems below 10 GHz are only mildly affected if at all by climatic changes such as rain. These frequencies are generally not absorbed by objects in the atmosphere, but they are readily absorbed by vegetation. Products in these frequencies tend to reflect off various objects such as buildings, concrete, and water and thus result in a high amount of multipath.

Since the beginning of development of microwave wireless transmission equipment, manufacturers and operators have tried to mitigate the effects of reflected signals associated with RF. These reflections are called multipath. State-of-the-art BBFW products, currently in the MMDS and U-NII bands, not only tolerate multipath signals but can actually take advantage of them.

Multipath signalling tends not to be an issue in millimeter wave frequencies, since most of the multipath energy is absorbed by the physical environment. However, when these frequencies are used in highly dense urban areas, the signals tend to bounce off objects like metal buildings or metalized windows. The use of "repeaters" can add to the multipath propagation by delaying the received signal.

Multipath is the composition of a primary signal plus duplicate or echoed images caused by reflections of signals off objects between the transmitter and receiver. In Figure 8-2, the receiver "hears" the primary signal sent directly from the transmission facility, but it also sees secondary signals that are bounced off nearby objects. These bounced signals will arrive at the receiver later than the incident signal. Because of this misalignment, the "out-of-phase" signals will cause intersymbol interference or distortion of the received signal. Although most of the multipath is caused by bounces off tall objects, multipath can also occur from bounces on low objects such as lakes and pavements.

The actual received signal is a combination of a primary signal and several echoed signals. Because the distance traveled by the original signal is shorter than the bounced signal, the time differential causes two signals to be received. These signals are overlapped and combined into a single one. In real life, the time between the first received signal and the last echoed signal is called the Delay Spread, which can be as high as 4 μsec.

In the example shown in Figure 8-2, the echoed signal is delayed in time and reduced in power. Both effects are caused by the additional distance the bounced signal traveled relative to the primary signal. The greater the distance, the longer the delay and the lower the power of the echoed signal. One might think that the longer the delay, the better the reception would be. However, if the delay is too long, the reception of an echoed symbol can actually interfere with the primary signal by presenting a second burst of information. In some cases, where there may be no direct path for the incident signal in non-line-of-sight

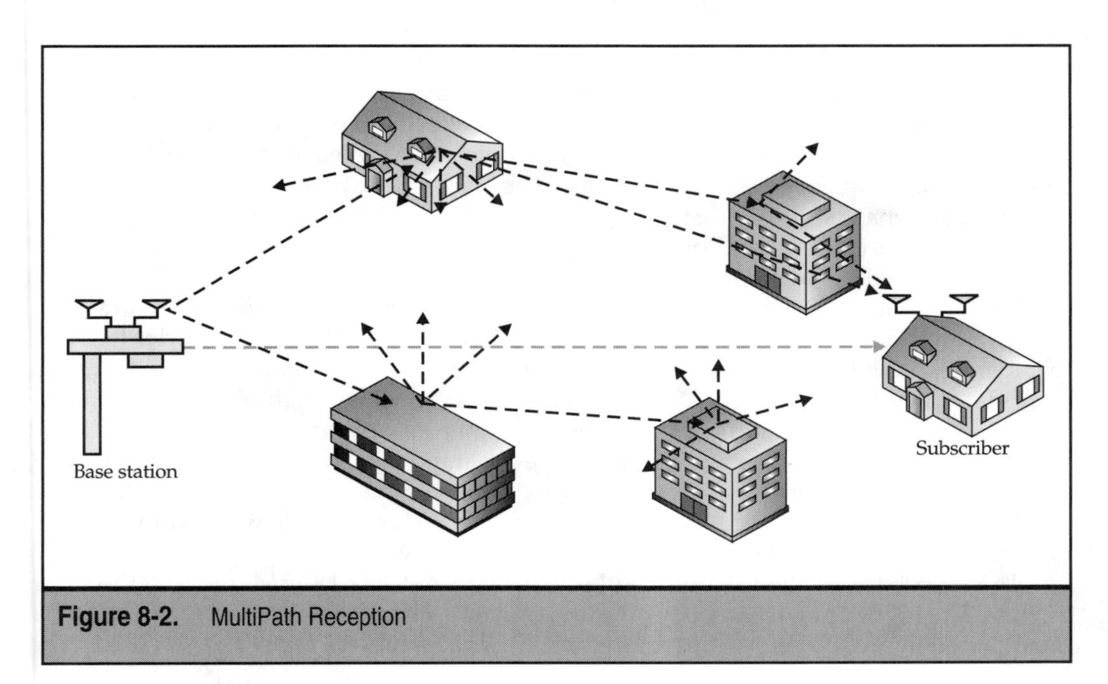

Figure 8-2. MultiPath Reception

(LOS) environments, the "primary" signal may be small (lower in power) in comparison to other, secondary signals.

In analog systems, this multipath situation can be physically sensed, for instance, as a "ghost" image on your television. Occasionally, no matter how much you adjust the set, the image does not go away. In these analog systems, this is an annoyance. In digital systems, it usually corrupts the data stream and causes loss of data or lower performance. Correction algorithms must be put in place to compensate for the multipath, resulting in a lower available data rate.

In Figure 8-2, the echoed signal actually interferes with the reception of the second symbol, thus causing intersymbol interference (ISI). This ISI is the main result of multipath, and digital systems not designed to deal with it give up substantial performance and reliability.

In LOS environments with very short range, multipath can be minor and can be overcome easily. The amplitudes of the echoed signals are much smaller than that of the primary one and can be effectively filtered out using standard equalization techniques. In non-LOS environments, however, the echoed signals may have higher power levels or combine to have much stronger signals than the original signal because the primary signal may be partially or totally obstructed. In this scenario, resolving with equalization is more difficult.

In the previous discussions, the multipath occurs from stationary objects. However, multipath conditions can originate from mobile objects that initiate varying degrees of multipath from one sample period to the next. This effect is called *time variation.* Superior digital systems withstand rapid changes in the multipath conditions, referred to as "fast fading." In order to deal with this condition, digital systems need fast automatic gain control (AGC) circuits.

Many modern fixed microwave communication systems are based on quadrature amplitude modulation (QAM). These systems have various levels of complexity. Simpler systems such as phase shift keying (PSK) are very robust and easy to implement because they have low data rates. In PSK modulation, the shape of the wave is modified in neither amplitude nor frequency, but rather in phase. The phase can be thought of as a shift in time.

In binary phase shift keying (BPSK), only one bit is transmitted per cycle (a complete cycle with bits is called a symbol). In more complex modulation schemes, more than one bit is transmitted per symbol.

The modulation scheme called QPSK (Quadature Phase Shift Keying) is similar to BPSK. However, instead of having only two separate phase states, QPSK carries two bits per symbol. Like BPSK, QPSK is used because of its robustness. However, since it only modulates two bits per symbol, it still is not very efficient for high-speed communications. Hence, higher bit rates require the use of significant bandwidth.

Even though QPSK does not make changes in amplitude, it is sometimes referred to as 4 QAM. When four levels of amplitude are combined with the four levels of phase, we get 16 QAM. In 16 QAM, two bits are encoded on phase changes and two bits are encoded on amplitude changes, yielding a total of four bits per symbol.

This approach to modulation can be expanded out to 64 QAM and 256 QAM or higher. The higher the density in QAM, the higher a signal-to-noise (s/n) ratio must be maintained in order to meet the required bit error rates (BERs).

How the data is encoded also plays an important part in the equation. The data is usually scrambled, and a significant amount of forward error correction (FEC) data is also transmitted. Therefore, the system can recover those bits that are lost because of noise, multipath, and interference. A significant improvement in BER is achieved using FEC for a given SNR at the receiver.

Several techniques have been used to make digital modulation schemes more robust, including: QAM with decision feedback equalization (DFE), direct sequence spread spectrum (DSSS), frequency division multiplexing (FDM), and orthogonal frequency division multiplexing (OFDM).

In wireless QAM systems, DFE is used to mitigate the effects of the InterSymbol Interference (ISI) caused by multipath. Delay spread is a condition when a replicated signal arrives later than the original signal. When this occurs, the echoes of previous symbols corrupt the current symbol.

The DFE filter samples the incoming signal and filters out the echoed carriers. The complexity of DFE schemes causes them not to scale with increases in bandwidth. The complexity of the DFE filter (number of taps) is proportional to the size of the delay spread. The number of required taps is proportional to the delay spread (in seconds) times the symbol rate. For a QAM-based wireless system transmitting in the MMDS band (6 MHz wide channel) to survive a 4 μsec delay spread, the number of taps required would equal 24.

Direct sequence spread spectrum (DSSS) is a signaling method that avoids the complexity and the need for equalization. Generally, a narrowband QPSK signal is used. This narrowband signal is then multiplied (or spread) across a much wider bandwidth.

Therefore, if an SNR of 20 dB is required to achieve the appropriate BER (bit error rate), the total spread bandwidth needed to transmit a digital signal of 6 Mbps equals 600 MHz. This is not very bandwidth efficient. In addition, the sampling rate for the receiver needs to be about 100 times the data rate. Therefore, for this hypothetical system, the sampling rate would also need to be 600 million samples per second.

Although predominately used in the mobile (cellular) arena, Code Division Multiple Access (CDMA) is used on advanced IP RF systems, and it is used to allow several simultaneous transmissions to occur. The value of this is that each data stream can be either voice, video, or data, and broadcast within the same transmission. In practical terms, it gives the ability to provide different services within a given transmission area. A single service provider customer could send video broadcasts; perform CLEC services; and extend or compete with DSL, cable, or ISDN.

With CDMA, each data stream is multiplied with a pseudorandom noise code (PN code), which is another way of stating that each data stream is supplied with its own identification number. All users in a CDMA system use the same frequency band. The intended signal is recovered by using the PN code. Data transmitted by other users or a even a native transmitter is filtered out during the reception phase. It's important to note that all receivers within transmission range will receive the signal because it's on the same frequency, but only the intended signal gets past the filters.

Another advantage of CDMA is that the amount of bandwidth required is now shared over several users, giving more users per transmitter. In systems where there are multiple transmitters and receivers, however, proper power management is needed to ensure that one transmitter does not overpower other users in the same spectrum. These power management issues are mainly confined to CMDA architectures. Frequency division multiple access (FDMA) and time division multiple access (TDMA) systems are more tolerant of fluctuations in power.

As an example, assume that the system is perfect and contains no interference. Each user transmits a signal at 30 dBm before the CDMA encoding, and each signal is reduced to 10 dBm after spreading. Each additional user's wideband energy is added to the composite signal. Each time the number of users is doubled, that signal will rise by 3 dB. When each receiver despreads a particular signal, it is going to get back the 30 dBm signal and see a noise floor. Therefore, the SNR is lowered as additional users are added to the system. At some point, the total wideband "noise" floor will overwhelm the narrowband signal and, therefore, make the despread signal unusable.

In a Frequency Division Multiplexing (FDM) system, the available bandwidth is divided into multiple data carriers. The data to be transmitted is then divided between these subcarriers. Since each carrier is treated independently of the others, a frequency guard band must be placed around it, which is another way of saying that no other data will be carried on an adjacent frequency. This guard band lowers the bandwidth efficiency.

In some FDM systems, up to 50 percent of the available bandwidth is used for guard bands, which prevents them from carrying data. In most FDM systems, individual users are segmented to a particular subcarrier; therefore, their burst rate cannot exceed the capacity of that subcarrier. If some subcarriers are idle, their bandwidth cannot be shared with other subcarriers.

In Orthogonal Frequency Division Multiplexing (OFDM), multiple carrier frequencies (or tones) are used to divide the data across the available spectrum, similar to FDM. In an OFDM system, however, each tone is considered to be orthogonal (independent or unrelated) to the adjacent tones. As shown in Figure 8-3, each tone is a frequency integer (one whole number) apart from the adjacent frequency and therefore no guard band is required around each tone.

Since OFDM requires guard bands only around a set of tones, it is more efficient spectrally than FDM. You can see how this works in Figure 8-4.

Because OFDM is made up of many narrowband tones, narrowband interference will degrade only a small portion of the signal and have no or little effect on the remainder of the frequency components.

OFDM systems use bursts of data to minimize intersymbol interference (ISI) caused by delay spread. Data is transmitted in bursts, and each burst consists of a cyclic prefix followed by data symbols. For example, an OFDM signal occupying 6 MHz is made up of 512 individual carriers (or tones), each carrying a single QAM symbol per burst. As Figure 8-5 shows, the cyclic prefix is used to absorb late arriving signals due to multipath (transients) from previous bursts. An additional 64 symbols is transmitted for the cyclic prefix. For each symbol period, a total of 576 symbols are transmitted, by only 512 unique QAM symbols per burst.

In general, by the time the cyclic prefix is over, the resulting waveform created by the combining multipath signals is not a function of any samples from the previous burst.

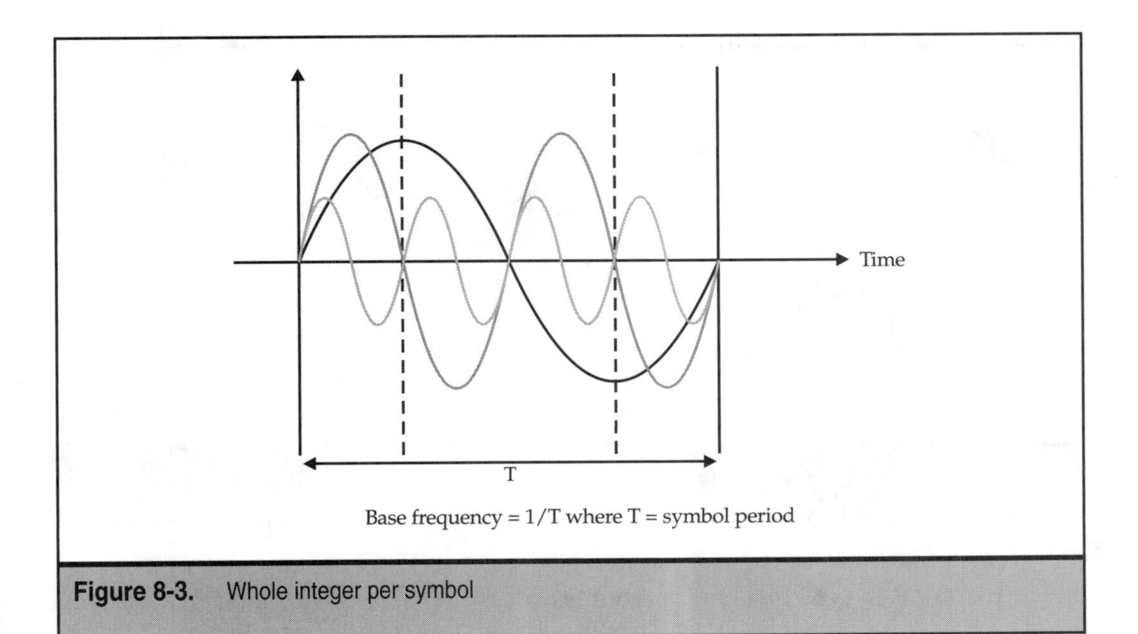

Figure 8-3. Whole integer per symbol

Hence, there is no ISI. The cyclic prefix must be greater than the delay spread of the multipath signals. In a 6 MHz system, the individual sample rate is 0.16 μsecs. Therefore, the total time for the cyclic prefix is 10.24 μsecs, greater than the anticipated 4 μsec delay spread.

Some OFDM systems only use QPSK for the modulation scheme. If 16 QAM or 64 QAM is used, then the amount of data transmitted significantly increases.

In addition to the standard OFDM principles, the use of spatial diversity can increase the system's tolerance to noise, interference, and multipath. Another way to visualize this is to think of a second antenna. You stop your car at a red light and the radio broadcast

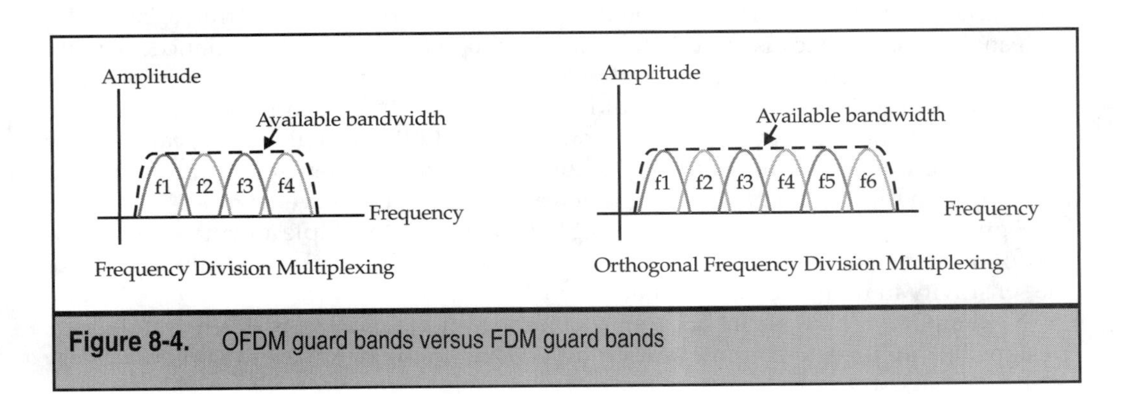

Figure 8-4. OFDM guard bands versus FDM guard bands

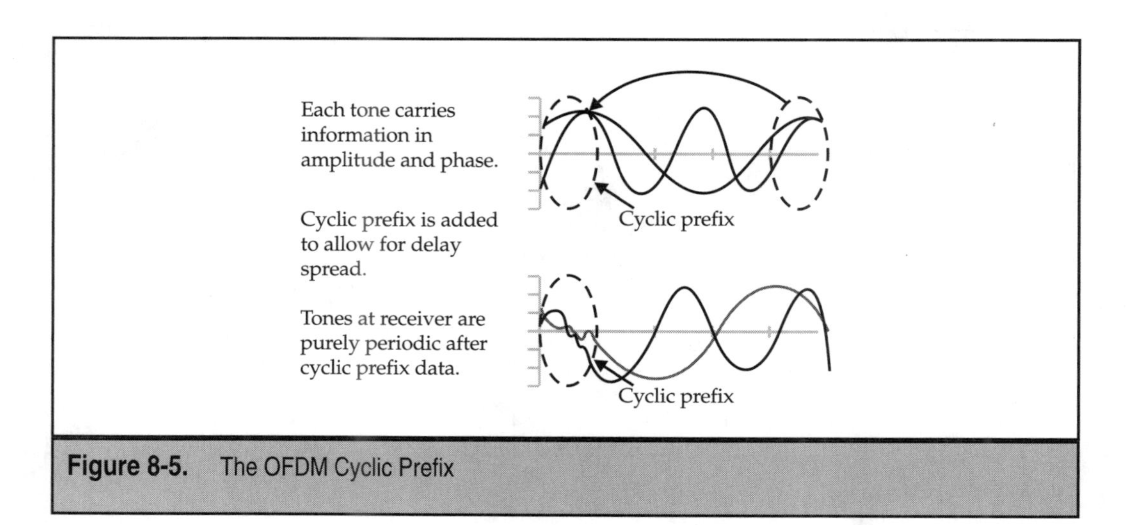

Figure 8-5. The OFDM Cyclic Prefix

turns to static. What you do to regain your broadcast signal is move your car ahead a few inches to a few feet. What you've accomplished with this is to relocate your antenna outside of a void or null in the local reception area.

This is referred to as vectored OFDM or VOFDM. The "V" in VOFDM stands for "vector," and simply means having a second antenna online along with the primary antenna; it does not refer to a dynamic movement of either the primary or secondary antenna. Spatial diversity means two antennas a given distance apart for the same receiver. On virtually every system sold at this writing, spatial diversity is used only on receivers, not transmitters, as illustrated in Figure 8-6. It is expected, however, that spatial diversity will become standard at the transmitter end of each link before too long because tremendous performance gains and increased geographic coverage will more than offset the additional cost for the extra transmit antenna. The technology surrounding the concept of multiple transmit and receive antennas is also referred to as "multiple-in, multiple-out," or MIMO.

Spatial diversity is a widely accepted technique for improving performance in multipath environments. Because multipath is a function of the collection of bounced signals, that collection is dependent on the location of the receiver antenna. If two or more antennas are placed in the system, each could have a different set of multipath signals. The effects of each channel would vary from one antenna to the next; therefore, carriers that may be unusable on one antenna may become usable on another. A generally accepted rule of thumb for the antenna spacing is at least ten times the wavelength.

Significant gains in the SNR are often obtained by using multiple antennas. Typically, a second antenna doubles the sensitivity in LOS deployments and affords up to 10 times the sensitivity in non-LOS environments.

Neither QAM/DFE nor DSSS systems elegantly scale with increases in bandwidth. Designs for high-speed communication links using these techniques yield either very

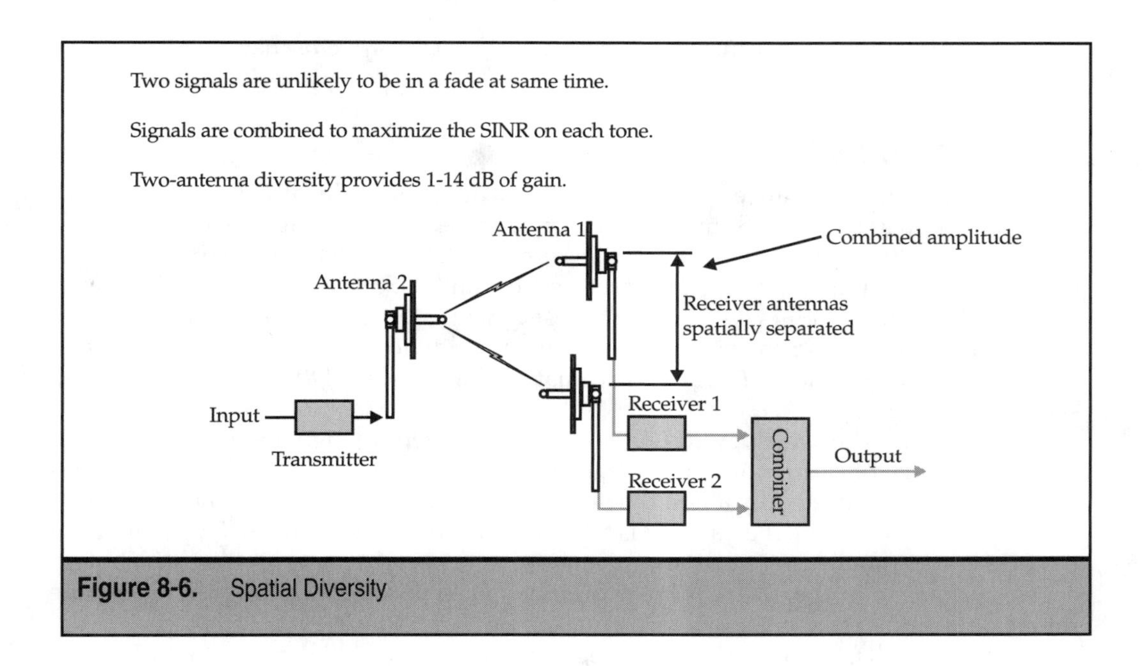

Figure 8-6. Spatial Diversity

expensive and complex systems, or systems that cannot guarantee the BER rates needed to consistently deliver BER in the realm of fiber optics in non-LOS environments.

In comparing today's modulation techniques, OFDM and VOFDM are compelling techniques to resolve for multipath issues existing in highly obstructed environments, although the trade-off is at least a 20 percent rise in cost of goods sold for equipment, which can readily translate to a 60 percent increase in retail price. Having said that, discounts on the order of 20 percent to 40 percent are reasonably common when purchasing an end-to-end network, and the extra cost of goods sold can be offset by the increased number of customers available through greater *harvest rate.*

OFDM products used in obstructed environments provide key elements to enhance the harvest rate over systems that require a clear line of sight to operate. BBFW customers can expect upward of 30 percent to 40 percent additional coverage area for a given transmitter, which translates into a proportional number of customers that can sign up for BBFW service. This reference to additional coverage area is not manifested in greater distance from the antenna but in accessing those areas closer in that are hidden to the traditional line-of-sight-only networks.

BBFW systems can also therefore be deployed in dense urban environments where multipath is pervasive. No longer are point-to-multipoint systems limited to the macrocell models. Microcells similar to those for cell phones can be created. Instead of tall microwave towers, simple rooftop head ends can be used, and instead of macrocells with radii of approximately ten to twenty miles, microcells with radii of three to eight miles. In general, VOFDM allows the user to create simple, spectrally efficient, multipath-resistant, robust data communications links in areas where non-VOFDM systems have had marginal performance.

WLAN Spread Spectrum

Most WLAN products utilize spread spectrum techniques. To legally operate in the 2.4 GHz ISM spectrum in the U.S. and many other countries, a type of signal spreading must be used. There are two kinds of spread spectrum available, direct sequence spread spectrum (DSSS) and frequency hopping spread spectrum (FHSS). DHSS typically has better performance, while FHSS is typically more resistant to interference. Although OFDM is a technique for spreading the signal over a given bandwidth, it is not, by definition, a spread spectrum technique. The FCC is evaluating the specifics of the definition of spread spectrum.

A commonly used analogy to understand the concept of signal spreading (spread spectrum) is that of a series of trains departing a station at the same time. The payload is distributed relatively equally among the trains, which all depart at the same time. Upon arrival at the destination, the payload is taken off each train and collated. Duplications of payload are common to spread spectrum so that when data arrives excessively corrupted, or fails to arrive, the redundancies inherent to this architecture provide a more robust data link.

With DSSS, all trains leave in order, beginning with Train 1 and ending with Train N, depending on how many channels the spread spectrum system allocates. In the DSSS architecture, the trains always leave in the same order, though the number of railroad tracks can be in the hundreds or even thousands.

With the FHSS architecture, the trains leave in a different order, that is, not sequentially from Train 1 to Train N. In the best of FHSS systems, trains that run into interference are not sent out again until the interference abates. In FHSS systems, certain frequencies (channels) are avoided until the interference abates.

Interference tends to cover more than one channel at a time. Therefore, DSSS systems tend to lose more data from interference as the data is sent out over sequential channels. FHSS systems "hop" between channels in nonsequential order. The best of FHSS systems adjust channel selection such that highly interfered channels are avoided as measured by excessively low bit error rates. Either approach is appropriate, and the choice depends on customer requirements, the selection criteria primarily involving a severe multipath or interfering RF environment.

DUPLEXING TECHNIQUES

As you may recall from earlier in this work, there are several fundamental ways in which a link has each end communicate to the other. They are represented in Table 8-3.

Type	Communication	Example
Simplex	One person at a time can use the link.	Walkie-talkies
Half-Duplex	The speaker cannot be interrupted by the other party.	Speakerphones
Full-Duplex	Both parties can transmit and receive in parallel; that is, the speaker can be interrupted by the other party.	Broadband radio/telephone

Table 8-3. Different Duplexing Techniques

In order to accommodate these different duplexing techniques, various *access techniques* must be used. To understand how various access techniques operate, the reader must understand the following key terms:

- **Frequency Division Duplex (FDD)** FDD is a method of transferring two halves of a full-duplex communication by using two or more different frequencies.
- **Frequency Division Multiple Access (FDMA)** FDMA is a method where the data is split up over multiple carrier frequencies.
- **Time Division Duplex (TDD)** TDD is a duplex transmission that uses two different time slots for transmitting, commonly, but not always, found on systems that use the same frequency from the head end to the CPE.
- **Time Division Multiple Access (TDMA)** TDMA is a technique for multiple common users at the CPE side of a transmission. This enables multiple users to transmit to a common head end. This is a very common access technique in the cellular phone industry.
- **Code Division Multiple Access (CDMA)** CDMA is a spread spectrum technique that allows multiple CPE users to access a common head end because each user has a unique code. It is also very common in the cellular phone industry.

For simplex communications, the most rudimentary access techniques can be applied; they can be as simple as not allowing a receiver to receive information or voice while the same radio is transmitting. Both transmitters will be on the same frequency and time domain; and type of link is not typically used in point-to-multipoint systems. This is the least expensive and complex radio access technique.

When designers begin to incorporate a link with half-duplex architecture, they most commonly use a time domain access technique such as TDD or TDMA.

The most complex types of BBFW links are full-duplex. They commonly use an array of access techniques. For example, in some complex BBFW products, the architecture calls for FDM downstream and TDM upstream. This enables both radios to remain on the same frequency but likewise enables broadcast to all CPEs, which can return information upstream on a prioritized basis in order to ensure that the most delay-sensitive traffic, such as voice and video, is not queued behind routine data traffic.

The most efficient systems are full-duplex, as they pass traffic in both directions of the link at all times the system is in use, utilizing different transmit and receive frequencies.

ERROR CONTROL SCHEMES

There is a considerable difference between *error detection* and *error correction.* Error detection is critical for high-end BBFW systems, as it enables the network managers to determine how well the system is performing at any given moment, and equally important, how it is performing over a given time frame such as a day, week, month, or year.

Error detection is commonly referred to in terms of *bit error rate,* or BER, and is typically expressed in the form of a ratio; for instance, a value of 1×10^{-7} would mean that one bit in 10^{-7} bits transmitted would be in error. State-of-the-art systems have bit error rates of 10^{-8} for voice and 10^{-11} for data. These BERs are equivalent to those found in fiber optics systems and are therefore extraordinary.

There are a number of common error correction techniques; in all of them, the data is corrected on the receiving end of the transmission link. The most common forms of error correction are applied such that a block of bits is sent along with the data in order to corroborate that all the data was sent satisfactorily.

Errors are introduced into the data from outside interfering sources such as other radios, multipath phenomena, other electronic equipment in the area, and even solar flares from the sun. Error correction schemes enable the receiving end of a link to correct data to a very high degree. One way to accomplish this is with *parity,* bits which is a method such that the transmitted data includes a parity bit. The parity bit ensures that the number of received bits is either odd or even. If a single bit is corrupted such that it can't be used, or the bit does not arrive at the receiver, the parity bit will notify other processing regimes that further correction, or retransmission of the data must occur.

Of course, the system wouldn't request a retransmission if only a single bit out of thousands or millions was sent in error, but if the overall bit error rate was too high, the transmission could be resent—not for voice or video necessarily, although that is indeed doable with buffers, but fairly routinely for data transmissions that either go over a very long distance or involve very high data rates. Parity bit values depend on the types and quantity of the information sent over the link, and it should be stated that the receiving end of the link can either repair the errors, live with the errors, or have the originating transmitter resend the information.

CRC (Cyclic Redundancy Check) is a common error detection scheme that typically generates a value based on the result of a polynomial algorithm performed on the packet or frame header and data payload. The polynomial result is performed at the transmitter and then again at the receiver, at which point the CRC values are compared and should be identical.

OFDM includes an error correction technique of *interpolation*, which is to say that the data is corrected by interpolation based on the corrected values of training tones, which are equally spaced along the numbers of data carriers. In the U.S., for example, the U-NII band has 1056 equally spaced training tones along the 100 MHz of U-NII spectrum.

This number of training tones is not a frequency-dependent issue but rather, the number of training tones is set by the operator. In an environment where there is a very high noise level, and consequently a low SNR, the number of training tones can be increased to improve the correction. However, as the number of training tones goes up, the number of carrier tones available to handle the actual information goes down. Therefore a balance needs to be established between and acceptable amount of information throughput, an acceptable number of errors, and the number of training tones.

The receiver corrects the training tones to predetermined phase and amplitude settings. The data, which is evenly interspersed between the training tones, is also corrected for amplitude and phase. There is a relatively large number of training tones in the U-NII band (or any other band using modulated by OFDM) in order to correct the narrowest amount of spectrum carrying data. Figure 8-7 shows the training tones evenly interspersed with the frequencies carrying data.

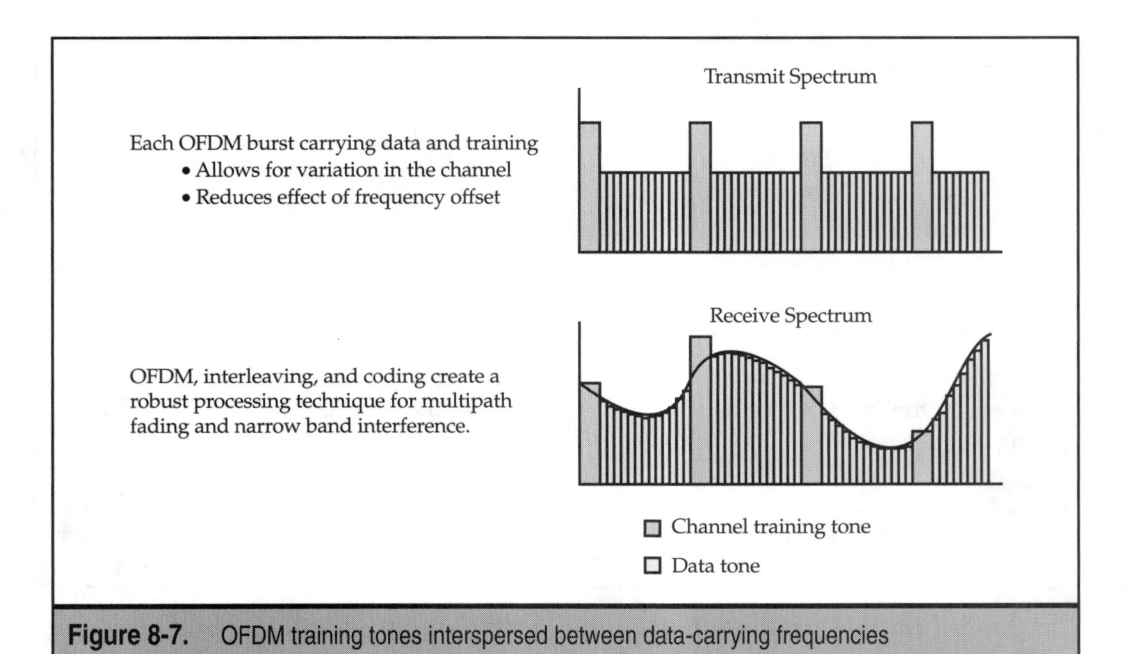

Figure 8-7. OFDM training tones interspersed between data-carrying frequencies

Bits	1	0	1	1	0
Transmitted channel-coded bits	111	000	111	111	000
Received bits (with errors)	101	100	011	101	001
Recreated original bit stream (error free)	1	0	1	1	0

Table 8-4. Forward Error Correction

Forward error correction (FEC) is a correction scheme based on the concept of receiving multiple copies of the same bit and then having the system "vote" on the most likely bit. This scheme is indicated in Table 8-4 above.

Block interleaving is an error correction scheme that separates the most "important" bits and populates them among the "less important" bits. More important bits might be for a CPE address, for example, while a less important bit might represent some of the words in the middle of an e-mail message. Table 8-5 illustrates the concept of interleaving.

Other error correction and detection schemes are commonly used, but these examples should be sufficient to demonstrate the concepts without requiring some degree of prior

Data before entering transmitting radio	1	2	3	4	5	6	7	8	9	10	11	12	13
After interleaving but before transmission	13	6	8	1	11	3	7	12	4	9	5	2	10
Burst errors during transmission	13	6	8	1	11	E	E	E	E	E	5	2	10
Result after deinterleaving at the receiver	13	6	8	1	11	3	7	12	4	9	5	2	10

E = bit error.

Table 8-5. Example of Block Interleaving

expertise in the subject matter and the use of mathematical formulas to adequately explain these other methods.

What the reader should have learned from this chapter is that there is an array of modulation schemes available to the designer; all of which require compromises. A commercially viable BBFW link also requires both error detection and correction given the density of information required in a competitive system. The more data is packed into a BBFW link, the more error correction will be required. This is also true for the very high performance VOFDM system, which relies heavily on processing gain in order to substantiate its performance claims.

CHAPTER 9

System Performance Metrics

The purpose of having BBFW system performance metrics is to be able to measure how available and how reliable a BBFW link is and to determine when to make changes in power, gain, antenna, OFDM setting (if the link has no line of sight capability), or diversity. Link availability has the following two key criteria and is defined as the period of time measured in seconds or minutes per a twelve-month period:

- How long the digital signal remains uninterrupted from interference
- How long the timing between the radios remains in synch

This is most commonly expressed in a percentage or "number of nines"—for example, 99.99 percent availability (see Table 9-1).

Prior to detailing system performance metrics, it is important to understand the various elements that can affect link performance. While some individuals believe that BBFW link performance is solely based on the performance specifications set by the radio manufacturers, the actual scope of how well a link performs is proportional to how well the following items are executed and completed:

- Preliminary site survey
- Site survey
- Link installation
- RF equipment selection
- Link margin management
- Equipment settings
- Site stability
- Traffic load
- Traffic type
- Non-radio equipment performance

While the issues of preliminary site survey, site survey, and link installation were covered in Chapter 7, we shall detail in this chapter some of the issues involved in the other items listed above.

RF EQUIPMENT SELECTION

While there is a wide range of RF equipment available on the market today, this array of equipment can essentially be divided into two types: equipment with NLOS capabilities and equipment lacking NLOS capabilities. The reason the author divides the equipment in this manner is that no other radio aspect affects performance and link metrics to the same degree, with the possible exception of whether the radio is microwave or millimeter wave. NLOS radios will work where LOS radios won't, and NLOS radios will generally perform much better in the same geographies at similar power levels while carrying similar traffic types and loads. Links with NLOS capabilities have a much greater capability in terms of maximizing the quality of the received signal.

At the time of this writing, there are no NLOS radios available in the millimeter range (> 10 GHz), and this is primarily due to the fact that multipath signals on this band tend to be random and, therefore, not repeatable. Additionally, the energy from millimeter wave radios is much more readily absorbed into the radiating environment, or dispersed so badly that the levels of energy that reach the receivers are too small to be of benefit.

LINK MARGIN MANAGEMENT

Link margin is calculated in dB at the receiver and is directly related to link availability. An adequate link margin is essential to ensure that the link performs at the same level of reliability in all conditions, including both indoor phenomena, such as suboptimal link management and installation techniques, and outdoor phenomena, such as changes in the signal properties associated with the multipath environment or increased interference from co-channel and adjacent channel sources. One may think of link margin in terms of an engine throttle that is not fully opened. The unused portion of the throttle potential would be analogous to link margin.

Table 9-1 shows a sample used for a state-of-the-art system for a NLOS link.

Virtually no customer would tolerate a link that was down for 87 hours over a period of 12 months, and this is shown for reference only. In fact, most customers demand that the BBFW links have either a *3 nines* or *4 nines* reliability, which is a reference to the total number of nines. Large telco's (telephone companies) generally insist on 4 nines or greater reliability.

The number of nines required is proportional to the amount of power received. On a full duplex system, this can be nontrivial, as each end of a BBFW link commonly performs differently from the other. This should be expected, as the link ends can be anywhere from 100 meters to 20 miles or more apart.

Link Availability Requirement	Amount of Downtime per Year	NLOS Link Margin
99%	87 hrs 35 mins	8 dB
99.9%	8 hrs 46 mins	12 dB
99.99%	53 mins	18 dB
99.999%	5 mins	22 dB
99.9999%	30 secs	26 dB (approx)

Table 9-1. Link Availability As Related to Link Margin

Further, recalling the premise that power drops by the square of the distance in a free space and LOS environment, interference is predominantly a local effect. As interference sources are virtually always closer to one end of the link or another, they typically have a significantly greater effect on one end of the link. Reflective sources causing multipath signals in a microwave system are virtually always direction dependent; in other words, the multipath taken by the carrier frequency in one direction may not be the same when the carrier wave is traveling in the opposite direction.

Link Metrics

Link metrics predominantly focus on the issue of *errors*. This reference is in regard to what percentage of the time a link is online, and also the percentage of errors within a traffic stream.

Errors are measured in seconds and are referred to as *errored seconds* (ES), *severely errored seconds* (SES), and *consecutively severely errored seconds* (CSES). Errors are also measured in *downtime minutes* (DM).

ES occur when there are data errors within the period, and are generally not consecutive seconds. This type of error represents a condition not generally deemed to have serious performance impact on a link, even for high QoS traffic, because the information is often recovered during the DSP portion of the data transmission. In essence, an occasional bit here or there is lost, but the loss is not so significant that the intelligence cannot be recovered.

The more critical of the two measurements is SES, which is defined by the International Telecommunications Union (ITU) as a period where the timing between the radios is lost for greater than one second and the *block error rate* is greater than 30 percent. Block error will be explained a bit later in this chapter.

Radios must remain in synch (synchronized) with each other for the purpose of communicating information that is germane to the performance and settings of the radios. This information is not the same kind of information that the users at one or both ends of the link are interested in; that is, this information is not part of a standard application package such as Word or PowerPoint. This information can be sent between the radios in the same carrier frequency as the radios use, and this is accomplished by using time slots that are reserved for system maintenance and monitoring. On some systems, this information is sent outside the frequencies used for carrying the traffic that end users require. This constant dialog between the two end-point radios is critical to the proper operation of the system.

High-speed clocks are used to keep both radios in synch with each other, and there is usually a *master* and *slave* relationship between the two radios, regardless of whether the architecture is point-to-point or point-to-multipoint. In both architectures, the headend radio is set as the master, while the other end of the point-to-point link, or CPE radio in a point-to-multipoint architecture, is set as the slave. The master radio maintains the primary clock; the non-headend radios are slaved to that clock.

When the radios lose *lock* or timing from each other, the end-user information ceases to be transmitted. The headend radio must go through a reboot routine, which can vary greatly from manufacturer to manufacturer. As an example, some manufacturers, but not all, require that all elements of the headend go through individual loopback tests to ensure that each primary physical element of the radio is operating within normal parameters. The headend radio must then be reconfigured to take the state it was in prior to losing lock on the other end(s) of the BBFW link. Recovery lock loss can take from a few seconds to several minutes to no time at all, depending on the cause of the loss of synchronization.

Following this procedure, the radios at the other end of the links are usually *pinged* by the head end, which means they receive a signal to transmit their condition, MAC addresses, and so forth. In other systems, the radios that are not at the head end do not await a ping, but rather go into a ping mode following loss of lock to the head end. This is a state where the non-headend radios ping the head end and await instructions for synchronizing their clocks to the headend clocks, as well as setting configurations for power, gain, addresses, and so forth.

The pinging state is typically a signal that does two primary things: it alerts the headend radio that a non-headend radio is in the local radiating environment, and it informs the headend that it is awaiting instructions for reconfiguration. Pings can occur at a nominal interval but are typically done on the order of one per second.

In point-to-multipoint systems, there is an issue of what happens when, for example, several hundred to 500 CPE radios lose lock with the head end. This would almost always be due to a headend failure. In this case, there are at least two common ways to bring that many radios back online. If the system is using a Frequency Domain Modulation (FDM) in the downstream, the radios can be pinged and reconfigured based on frequency, which can be done typically on the order of a few seconds. If the CPE radios all operate on the same frequency, as is the case with an FDM downstream and Time Domain Modulation (TDM) upstream architecture, then the headend radio can ping a certain number of CPE radios at a time, based on their MAC addresses. This will enable the CPE radios to come back online fairly quickly and in an orderly manner.

The other way to get a large number of CPE radios back in lock with a headend is to have them ping at certain time slots. This can be done a number of different ways: by setting their time slots at the original configuration of the cell, by splitting the time slots into even/odd sessions, or by dividing the total time slot available into manageable-sized groups of CPE radios.

Block error rate is a reference to the information that cannot be recovered by the radio. Some radio manufacturers and designers believe that a block error loss of 30 percent, as defined by the ITU, is too generous. Current state-of-the-art systems can maintain a block error rate of less than 5 percent of the total time a link is measured. It is noteworthy that radio manufacturers are not bound to comply with ITU specifications; rather, they are more commonly used to provide a starting point for standards, but the ITU also encourages compliance to its standards in order to reduce the number of standards radio manufacturers build their radios to. As mentioned earlier, the block error rate is part of the equation for calculating the Severely Errored Seconds (SES) metric.

Consecutive Severely Errored Seconds (CSES) is the most critical type of error in a link, because it features the greatest period of disruption in the link and the worst kind of disruption: lengthy periods of time where radio lock is lost and data transmission has either ceased or is lost. If the CSES count is long enough, it can cause one or both radios or routers to reboot, which further adds to the data delay. Data can be re-sent, but in many cases with state-of-the-art systems, a resending of data does not happen if the CSES is too long in duration. What is deemed "too long" in terms of duration is unique from manufacturer to manufacturer.

Links are measured for performance and reliability over a set period of time, and in most BBFW systems, a great deal of performance data collection occurs when the links first come up. The measurement period often changes as a user has an increased level of confidence in the performance stability of the link. Initial periods of measurement of ten days or more are often recommended once the link has become stable, to ensure that the link margin and other link features are set correctly.

This length of time will vary from customer to customer. The period of ten days generally allows for the full cycle of any changes to the physical environment that exists between both ends of a link to change—for example, parking garages filling and emptying, planes flying in and out of the area, large trucks moving down a freeway over which a link resides, and so forth. In millimeter wave systems, the link should be measured in heavy weather as well as in conditions where there is no precipitation or high amounts of particulate in the air.

Both microwave and millimeter wave links could also be affected by variations in *atmospheric refractive conditions*, but this is difficult if not virtually impossible to predict and is generally an issue with links that have ends that are over the horizon. Refractive conditions in the atmosphere occur in two primary types, the first being where the signal does not follow the curve of the earth over long-distance links (approximately 20 miles or more). In this case, the radio signals continue out into space.

The other kind of refractive condition is when the signal is captured by a thermal or high humidity layer in the atmosphere and is reflected back down to earth. This can provide a multipath effect in certain links where the antennas are generally over the horizon or at distances approaching this condition. This type of refraction is called *K* factor and is actually harnessed for super-long links and is quite predictive in nature although it typically requires very high power antennas and low frequencies. This type of system is not typical to the BBFW market at the time of this writing.

Regarding commercial BBFW links, however, atmospheric refractive conditions are not typically a challenge and are very dynamic in terms of how often they can occur, their typical maximum effect on a BBFW, and how long they remain in effect. It should be duly noted that link performance is measured on an ongoing basis as a best practices approach for as long as the link remains in existence. A final quick note on atmospheric refractive effect is that it occurs in certain geographies more than others. A reputable site survey and deployment entity will be more than able to quantify this effect from a historical basis.

Regarding the mapping and archiving of BBFW link performance, the greater the period of time over which link performance is gathered, the better the network managers and technicians will be able to determine how well the link is performing. It will also guide them as they bring up greater numbers of links in a common geography with common equipment.

A sample link analysis table assembled by Greg DesBrisay of Cisco Systems is shown in Table 9-2 and provides an indication of the types of information gathered to determine how well a link will perform.

	Longitude	Latitude	
Location of Site 1	77.22861	38.9216667	
Location of Site 2	77.105	38.9636111	
Climate Factor (0.1 to 0.5)	0.3		From hot and humid to cold and dry
Topology Factor (0.25 to 4.0)	1		From very smooth to mountainous
Number of Antennas (1 or 2)	1		
Payload (L=1,M=2,H=3)	1		Data throughput, low, medium, high
Center Frequency	2.500	GHz	
Bandwidth	6.0	MHz	
SNR Requirements	17.0	dB	
Thermal Noise Floor	-174.0	dBm/Hz	
Noise Figure	4.0	dB	
Min. Required Signal	-85.2	dB	
Transmit Power	30.0	dBm	
Transmit Antenna Gain	28.3	dBi	
Receive Antenna Gain	24.0	dBi	
Transmit Cable Loss	1.0	dB	
Receive Cable Loss	1.0	dB	
Total Gain	80.3	dB	

Actual Link

Distance (mi.)	Free Space Loss (dB)	Received Power (dBm)	Fade Margin (dB)	Availability (%)
7.2428683	121.8	-41.5	43.8	100.0000

Typical Link

Distance (mi.)	Free Space Loss (dB)	Received Power (dBm)	Fade Margin (dB)	Availability (%)
1	104.6	-24.3	61.0	100.0000
2	110.6	-30.3	54.9	100.0000
4	116.6	-36.3	48.9	100.0000
6	120.1	-39.8	45.4	100.0000
8	122.6	-42.3	42.9	100.0000
10	124.6	-44.3	41.0	100.0000
30	134.1	-53.8	31.4	99.9985
50	138.5	-58.2	27.0	99.9812

Table 9-2. Sample Link Margin Analysis for a LOS Link

The preceding type of table quickly enables the user to determine the likely availability of a given link under conditions including range, power, antenna selection, and bandwidth. It is important to remember, however, that link availability is not system availability. The microwave link is only one portion of the complete system. Total system availability is a calculation based on all parts of a network.

It is also important to note that the link may not perform at its peak levels of efficiency until both ends of the link are fully debugged and tuned. When the link is first brought up, the initial measurements are typically taken during the first ten minutes of operation, then again one hour later, and then for a consecutive ten-hour period once the network managers are satisfied the link is working at an optimal rate of efficiency and reliability. The first ten-hour periods generally carry test traffic primarily consisting of code words provided commercially by firms that specialize in providing test traffic for BBFW links.

Table 9-3 is typical for that used in a state-of-the-art BBFW system to log-error rates. It indicates time since last error, error free seconds (EFS), ES, SES, CSES, and downtime minutes (DM). This table is archived and compared over time with newer data. Statistical analysis of this information is critical to long-term success. Histograms can be generated from this information to enable the user to quickly determine whether key performance profiles are showing gradual degradation in link performance; early detection of degrading performance can allow for scheduled maintenance, as opposed to emergency repairs.

Min Ago	Time	EFS	ES	SES	CSES	DM
10	2d04h	60	0	0	0	0
9	2d04h	60	0	0	0	0
8	2d04h	60	0	0	0	0
7	2d04h	60	0	0	0	0
6	2d04h	60	0	0	0	0
5	2d04h	60	0	0	0	0
4	2d04h	60	0	0	0	0
3	2d04h	60	0	0	0	0
2	2d04h	60	0	0	0	0
1	2d04h	60	0	0	0	0
0	2d04h	60	0	0	0	0

Table 9-3. Typical Error Statistic Table

State-of-the-art systems have several techniques for monitoring link performance and determining fault areas. A critical tool is the *loopback* test. This test enables the network technician to isolate key BBFW components such as the antenna, ODU, cables, router blades, routers, and DSP and RF portions of the radios. In other words, tests can be performed on each of these components one at a time. This is an exceptionally useful capability, because there is no more difficult challenge for a network operator than determining *where* a problem is. *Correcting* a problem is relatively easy compared to locating a problem.

The standard document used to set limits for maximum rates of error is ITU G.826. A sample of the upper limits is shown in Table 9-4.

Link margins are generally dynamic in response to interference and do not remain constant. This means link margins must be both monitored and managed to ensure that the link is maintained at levels set by the customer. This service can be performed by the installer or service provider by way of a *service level agreement* (SLA). The SLA sets forth the terms of the guarantee and the roles of both parties in terms of who does what when link margins go below a pre-agreed setting.

To a point, the link margin is managed autonomously in state-of-the-art systems, in that the BBFW element management software has the capability, within certain limits, to increase or decrease the radio output power, gain, diversity settings, and traffic rates. This does not eliminate the need for monitoring and analysis but can adapt for short-term random issues associated with a dynamic environment.

Equipment Settings

One of the most important aspects of determining link metrics is how well the network technicians monitor the links and make adjustments accordingly to the radios. As often as not, network technicians are kept very busy simply managing the entire network and don't invest a lot of mental energy in the BBFW links until an alarm is raised. These alarms can be sent to network technicians by page or phone call. The first thing the technician will do is examine the histograms of the BBFW link in terms of performance. From there, the technician may make manual adjustments to the link or call the technical assistance center of the equipment provider or installer.

Error Limits (following ITU Specification G.826)	Maximum Allowable Errors
Errored Seconds (ES)	Eight per ten-minute period
Severely Errored Seconds (SES)	One per ten-minute period
Consecutive Severely Errored Seconds (CSES)	Zero per ten-minute period (the threshold for this counter is 10)
Degraded Minutes	Zero per ten-minute period

Table 9-4. Maximum Error Limits in ITU Specification G.826

One of the advantages of a BBFW system that is provided by a network company such as Cisco is that the radio is managed on the same type of software that the balance of the network is managed on. This reduces the amount of time spent training network operators and enables those operators to better understand the radios and how they perform. In the case of the Cisco radios, they appear simply as a standard routed link to the network technician.

State-of-the-art systems have automated controls to reset both gain and power, and these settings are adjusted on a packet by packet basis. The number of these adjustments is typically set by the network managers.

If the installers have performed their tasks in compliance with best practices, they will have a series of settings that they will provide the network managers. These settings are nothing more than a simple document that instructs the network technicians to set various settings to an indicated point for a reasonably predictable result. If the links are showing excessive degradation, the network technicians can refer to several preconfigured settings before calling the technical assistance center for support.

Site Stability

Changes in the environment that resides between the antennas can have a very great effect on the performance of the link. One of the most common examples of this is when a building is constructed in between the two antennas. This new construction doesn't have to reside directly between the antennas; in fact, it can reside 15 degrees or more off the direct path between the two antennas and still have a detrimental effect on link performance.

The same is also true if a building is removed from the radiating area, as this may also dramatically alter the multipath effect. A renovated building can also affect link performance and require an adjustment if the exterior of the building is dramatically changed; for example, a large of amount of highly reflective material is either added or removed from the building, or the building becomes the site for broadcast headend in a similar frequency band, which can produce either adjacent or co-channel interference.

Rainy conditions can affect even microwave multipath elements, as the water can effectively provide a smoother surface from which the energy can bounce off. Even the addition of artwork to the front of a building, or larger reflecting pools, can produce changes in the multipath effect. This is also true if a large concrete plaza or parking lot is added near the path between the two antennas. In summary, a wide range of environmental changes can affect the link performance of a BBFW network. This does not mean that the link provider is completely at the mercy of the local physical environment. Rather, this is where the additional link margin calculations taken when the link was first powered up become valuable. If the link was set up correctly with sufficient margin calculated into the performance, then only small network adjustments are generally required to maintain a high-performance BBFW link.

Site stability can also refer to the cycle of normal traffic within a period of a time, which is typically one work week. If the two antennas reside across a municipal airport that has a lot of air traffic on weekdays but not much on weekends, link performance can be affected, although higher-end systems will make changes in real time to accommodate this delta.

The same is true for vehicular traffic should it crest a hill near the area between two antennas or be at the right distance and angle incident to the path between the antennas.

References to Link Performance

One of the most commonly asked questions when dealing with prospective BBFW link owners is, "What is the range of this link?" Most BBFW prospective users that ask this question disregard the other essential elements that contribute to link performance. This is from a general lack of BBFW experience, and those new to BBFW aren't typically aware of the myriad issues that combine to take a fully comprehensive view of a BBFW link.

The correct way to characterize a link is to respond to the preceding question along the lines of, "This point-to-point link will provide you with a 45 Mbps full duplex signal at a range of 20 miles LOS with a BER of 10^{-8} for voice and video traffic and 10^{-11} for data traffic. We calculate approximately 99.99 percent availability over a period of 12 months based on these specific pieces of equipment under these conditions." This response, while only to be taken as an example, is sufficiently comprehensive that it will enable the prospective owner of a BBFW link to know a lot about their link without having to delve into the details, which is generally left in the domain of the network technicians and engineers.

The following key issues must be considered when determining the full scope of how a link should be evaluated, all of which are inseparably interrelated:

- Range
- Bandwidth
- BER
- QoS capability
- Other items

Range

Range is a function of an array of items including frequency, power, modulation, bandwidth, antenna elevation, and BER. With enough elevation and antenna gain, a state-of-the-art point-to-point link can have a range in the UNII band in excess of 20 miles with a BER of 10^{-8} for voice and 10^{-11} for data at 45 Mbps of full duplex (making the link a 90 Mbps link by the standards used by some pure radio vendors).

Range is always a critical issue because customers rarely have the option of antenna placement; in other words, the antenna needs to be mounted on or very close to their facilities. Therefore, reductions in range are generally not possible.

Range is also very different when it is LOS vs. NLOS. As a general (and very rough) rule of thumb, an NLOS link for a given frequency, bandwidth, and BER is approximately one-fourth that of a LOS link. This depends to a great degree on the *value of the occlusion* or, in other words, how much the objects between the antennas affect the transmissions. As stated earlier in a previous chapter, the proximity to one or the other antenna, angle, elevation, and composition all affect the range of a prospective link.

Two other items can dramatically affect the range of a link: whether or not the occlusion in the path moves (like trees or airplanes), and whether the vegetation is deciduous or coniferous. Vegetation that matures over time will also have the same detrimental effect as vegetation that changes from a dormant to leafy state.

A well-circulated article appeared in the press during the time of this writing that detailed how a multibillion provider of BBFW links installed a high-speed link to a residence in the winter only to discover that the link quit working in the late spring. Upon investigation, it turned out the installers made site survey measurements through trees that were barren of leaves. When spring arrived and the trees produced their foliage, the BBFW network would not work.

Elevation is one of the best cure-alls for range, but when the system uses VOFDM in an unlicensed frequency, some consideration must be given to using more elevation than absolutely necessary when the carrier frequency is in the unlicensed band. The reason, as stated previously in this work, is that buildings, terrain, and other occlusions can be used to greatly reduce the issues of co-channel and adjacent channel interference.

One other issue to consider, in addition to range, is how well the geography within the range is covered. In many cases, it may be better to have less range but better geographic coverage. Having the lowest number and smallest size nulls (areas where the signal is not strong enough to be of service to a client) is a practical concern to service providers for a given area of coverage. If the BBFW owner can accommodate this issue—as opposed to simply attempting to radiate the maximum number of square miles from a headend, or using a more expensive solution such as radiating an area from more than one angle—they will reduce the overall challenge of providing BBFW service to a specific geography.

A BBFW cell planner would probably be wise to ensure that high target areas such as office parks, downtown metroplexes, and concentrations of condominiums and/or MTUs do not suffer from weak signal strength as a result of attempting to maximize the range from the headend. To summarize this point, target selection is mission-critical, as opposed to simply going for maximum range or maximum square miles.

Bandwidth

Bandwidth is a function not only of modulation but also of range, because as the power drops, so does bandwidth. Inversely, range can be increased if bandwidth is reduced. Figure 9-1 summarizes bandwidth vs. range.

This is an important consideration for BBFW users because they must take into consideration the issues of scaling as they grow their customers and their customers continually demand increasing amounts of bandwidth. Over time, bandwidth must be of a sufficiently high quality that it can carry voice and video, which means there must be enough bandwidth to set aside a significant portion of it for the higher QoS settings dictated by time- and delay-sensitive services such as voice. Figure 9-2 compares the more common BBFW links, and for good measure includes xDSL and even next-generation mobile capabilities.

The more important issue relative to bandwidth is BER because it's possible to receive so much corrupted data that the system cannot deliver useful data into the receiving

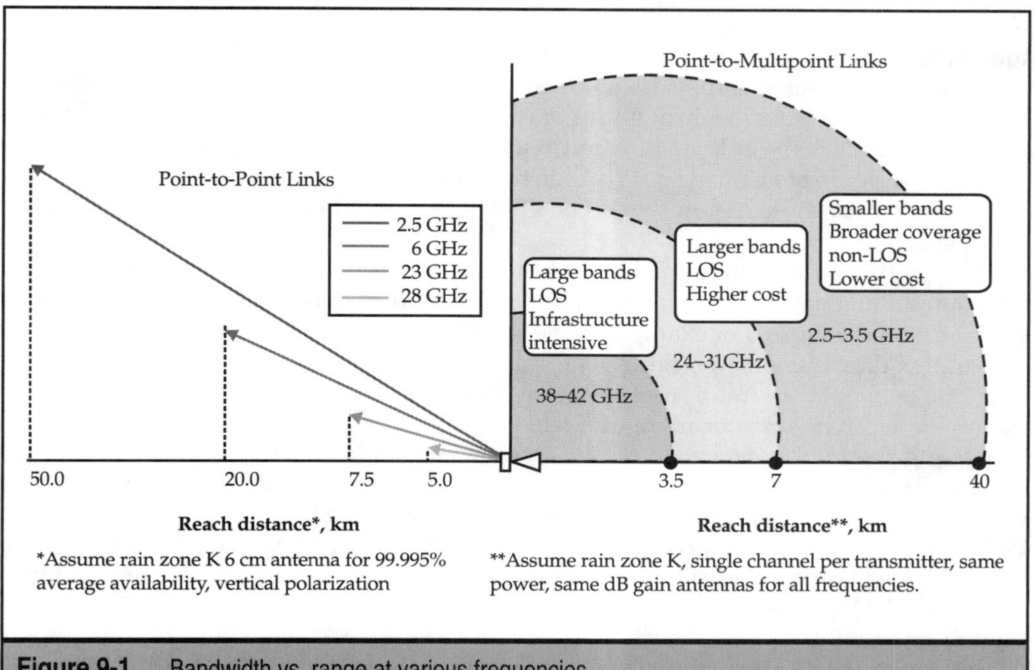

Figure 9-1. Bandwidth vs. range at various frequencies

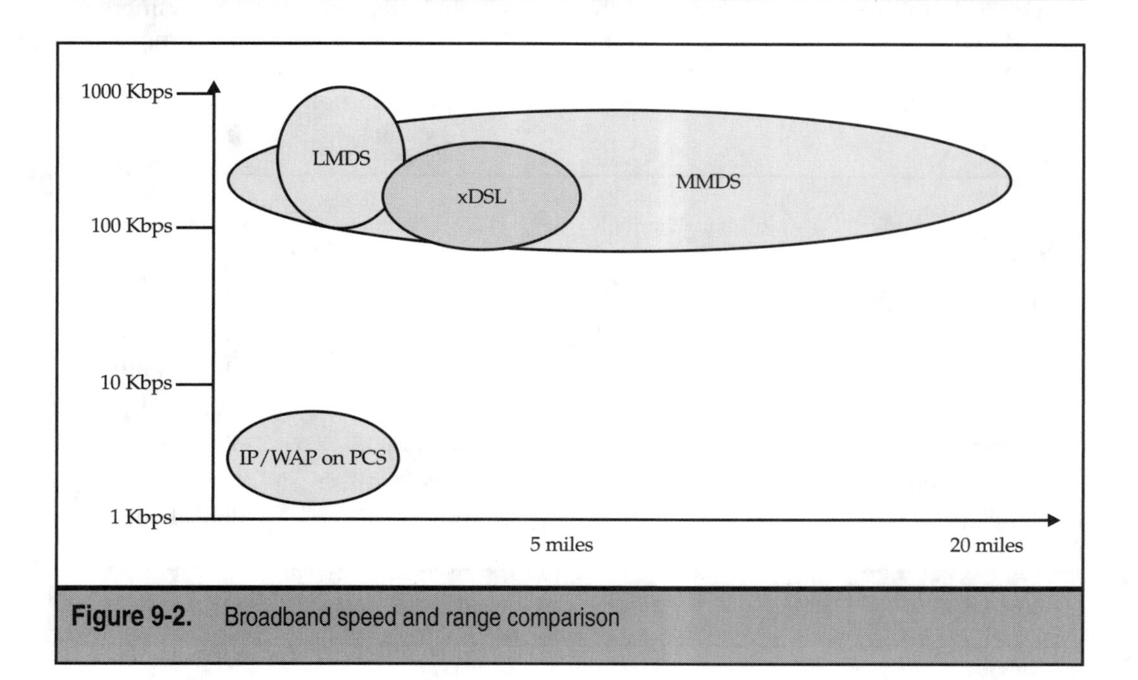

Figure 9-2. Broadband speed and range comparison

network. In general, it is better to have lower amounts of bandwidth with high quality than to have high amounts of bandwidth with high amounts of noise. State-of-the-art point-to-point systems should be able to have data rates on the order of 45 Mbps of *goodput* (also referred to as throughput or net throughput) on a full duplex basis, which means both ends of the link can transmit at the same time. State-of-the-art multipoint systems should be able to provide data rate on the order of 19 Mbps downstream with generally a slightly lower upstream rate. In an OFDM system, the data rates are a function of the bandwidth occupied.

Bandwidth is also determined by how much processing occurs to recover and verify information that has arrived via multipath or from either a co-channel or adjacent channel interference source. For example, some state-of-the-art systems will enable a 45 Mbps full duplex data rate on a point-to-point link and approximately a 19 Mbps full duplex data rate on a point-to-multipoint system with very little OFDM processing.

As there can be three or more different settings of OFDM, depending on how much processing the information requires on a post-transmission basis, the data rates can be reduced by approximately one-third on a middle OFDM setting, and the data throughput can be as little as one half of the full rate. In fairness, though, if the link required the highest OFDM setting, getting 22 Mbps for a point-to-point link obviously would be far better than having a link that simply would not work at all, have poor reliability, or have throughputs far below 22 Mbps. Table 9-5 illustrates how a typical BBFW system with OFDM settings might perform.

Information can also be sent over various amounts of available spectrum. The more spectrum used to transmit data, the greater amount of data that can be sent. Although range tends to decrease as bandwidth increases, reliability and availability increase. This is why Table 9-5 indicates 6, 3, and 1.5 MHz channels. Subscriber units (CPE) almost always have lower power levels and smaller antennas than those found at the head end in point-to-multipoint systems. These two items definitely have an effect on

VOFDM Setting	6 MHz Channel	3 MHz Channel	1.5 MHz Channel
Downstream			
High throughput	22.4 Mbps, 21 dB	10.1 Mbps, 21 dB	4.2 Mbps, 21 dB
Medium throughput	17.0 Mbps, 18 dB	7.7 Mbps, 18 dB	3.2 Mbps, 18 dB
Low throughput	12.8 Mbps, 14 dB	5.1 Mbps, 14 dB	1.6 Mbps, 14 dB
Upstream			
High throughput	19.3 Mbps, 24 dB	**8.1 Mbps, 24 dB**	4.2 Mbps, 26 dB
Medium throughput	15.2 Mbps, 19 dB	**6.2 Mbps, 20 dB**	3.2 Mbps, 20 dB
Low throughput	11.4 Mbps, 15 dB	**4.4 Mbps, 15 dB**	1.4 Mbps, 15 dB

Table 9-5. Sample Throughputs for Given OFDM Settings (MMDS)

the upstream data rates, which are commonly reduced by 10 percent or more from the downstream rates.

QoS Capability

Bandwidth in and of itself does not guarantee high levels of QoS; rather, bandwidth is apportioned on a QoS basis with transmission priority given to the higher QoS-rated traffic, and radios do not necessarily perform at either Layer 2 or 3 of the OSI stack. This is a function of how the radio is integrated into the router or switch at each end of the link. Currently, QoS requirements are not in high demand with customers, but most of the industry believes it will become a vital issue as users migrate from the relatively simple transmission of data to VPN, voice, and then video.

Having a BBFW link managed at Layer 3 of the OSI model ensures that as the type of traffic increases in complexity, the customer does not have to replace the existing system. As stated elsewhere in this work, a system should ideally be designed and deployed for how it will be used approximately 18 to 24 months down the road.

The BBFW portion of the network is the access or edge portion of the network. QoS must be implemented not only throughout the LAN on each end of the link, but also through every network the traffic travels through if high QoS levels are required.

Other Issues

There are other items that contribute to overall throughput, such as how well the antennas are pointed (or whether they become misaligned due to extreme weather, or become detuned due to ice buildup if a radome is not used). LAN traffic congestion can cause bandwidth fluctuations, and bandwidth variations can even be ascribed to how well the end user's PC is working. (Although it's granted that PC performance is not closely related to how well the link operates, it can give the appearance of a slow connection.) In summary, there are a number of places where information throughput can be throttled on either side of a BBFW link in addition to phenomena occurring directly on the link itself.

CHAPTER 10

Differences Between Headend and CPE Gear

While both ends of a BBFW link require radios and other equipment to perform the task of connecting two sites wirelessly, there are considerable differences in the radios themselves and the BBFW architecture at each end of a point-to-multipoint link.

As stated earlier in the book, the two primary BBFW architectures are point-to-point (P2P) and point-to-multipoint (P2MP), although there are some variants on the point-to-multipoint architecture. The preponderance of this chapter will address point-to-multipoint systems, because there is much less commonality at each end of a point-to-multipoint link than a point-to-point link.

POINT-TO-POINT ARCHITECTURE

Point-to-point systems require essentially identical radios at each end of the link; however, as the last chapter described, there is typically a master radio and a slave radio to ensure operability through the necessary sets of commands and to ensure the fastest configuration or reconfiguration in the event of a link being brought back online after failure.

Also, as stated in the last chapter, this difference between the two radios in a point-to-multipoint link is primarily software as opposed to hardware, but there can be no doubt that most CPE side P2MP radios are nearly always much smaller and less expensive to build than its headend counterpart. The issues of clock synchronization, configuration, security settings, frequency lock, and so on are fundamentally identical to the point-to-multipoint architecture, except that in a point-to-point link, they are optimized for a pair of radios, which are usually identical, as opposed to a single headend interfacing up to 500 CPE radios of varying configuration in a point-to-multipoint architecture.

As the master radio can be managed and monitored from virtually any browser location, the NOC can be at either end of a point-to-point link or at a remote site. Point-to-point links are not typically used for generating revenue, although there are examples of companies providing backhaul for other entities (a very common model in the copper and fiber industries). Backhaul may be considered as the main pipe that brings all of the data to and from a BBFW headend or wireless LAN. Figure 10-1 indicates a typical backhaul scenario, which is shown as a pair of fiber rings, but can be fiber, copper, or BBFW.

Point-to-point links are more commonly used to connect remote campuses, to backhaul cellular or BBFW cell sites, or to extend telco copper services such as DSL or cable services to remote sites. The users of each end of a point-to-point site are generally the same.

Backhaul is one of the most important issues when deploying a BBFW system over a metropolitan area. Headends (HEs) can have a P2P link to a second HE, which then has a series of sectors that are serviced. One of the key challenges in this type of architecture is that the size of the backhaul pipe becomes very large very quickly; for example, when you have BBFW cells with an OC-3 aggregate information rate, the backhaul pipe can go to OC-12 with little difficulty. In unlicensed systems, such as U-NII in the U.S., the most

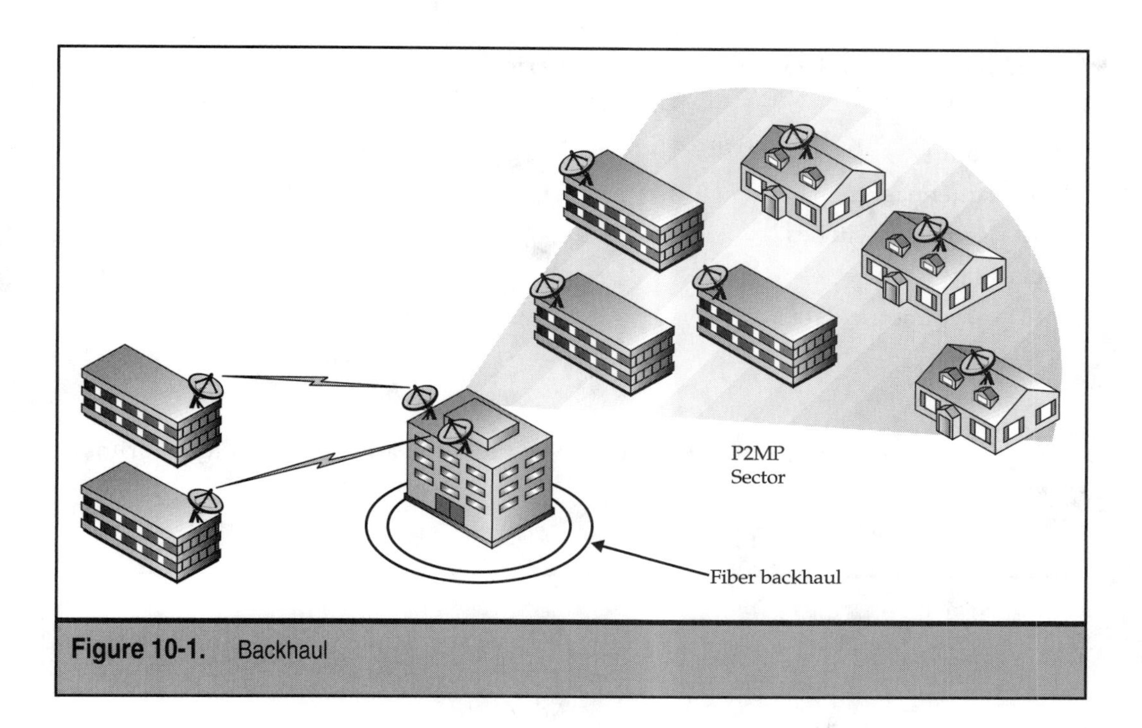

Figure 10-1. Backhaul

optimal manner to accommodate backhaul is with fiber, assuming a fiber node is close enough to the HE to be serviceable.

There are many hardware vendors that provide P2P links in a variety of frequency bands, licensed, unlicensed, and registered. These products provide various levels of throughput, up to 100 Mbps and more in some situations. Currently, most of these vendors use proprietary technologies. Some are using and proposing open architectures, but as of this writing, there are no standards-based platforms available.

Point-to-point and point-to-multipoint HEs generally aggregate to larger routers or switches, as BBFW is commonly only one of the types of access used to connect end users.

POINT-TO-MULTIPOINT ARCHITECTURE

P2MP radios far outnumber P2P radios. While a P2P link will be limited to two radios (therein the term point-to-point), a point-to-multipoint sector can have to 1000 client radios or more interfacing a single HE radio. The typical range of CPE P2MP radios is between 200 and 500, with the average being closer to 200. This is primarily a function of the shared bandwidth per sector. The greater the number of CPE radios, the less bandwidth per CPE and greater the number and length of delays by the client users who access the HE with a BEST EFFORT policy.

The headend of a P2MP link generally varies greatly from the CPE radios in the following key areas:

- Radios
- Antennas
- Routers/switches
- Slave status
- Asymmetric data flowsP
- Cost
- Use

Figure 10-2 and Figure 10-3 show a state-of-the-art headend radio and its primary elements.

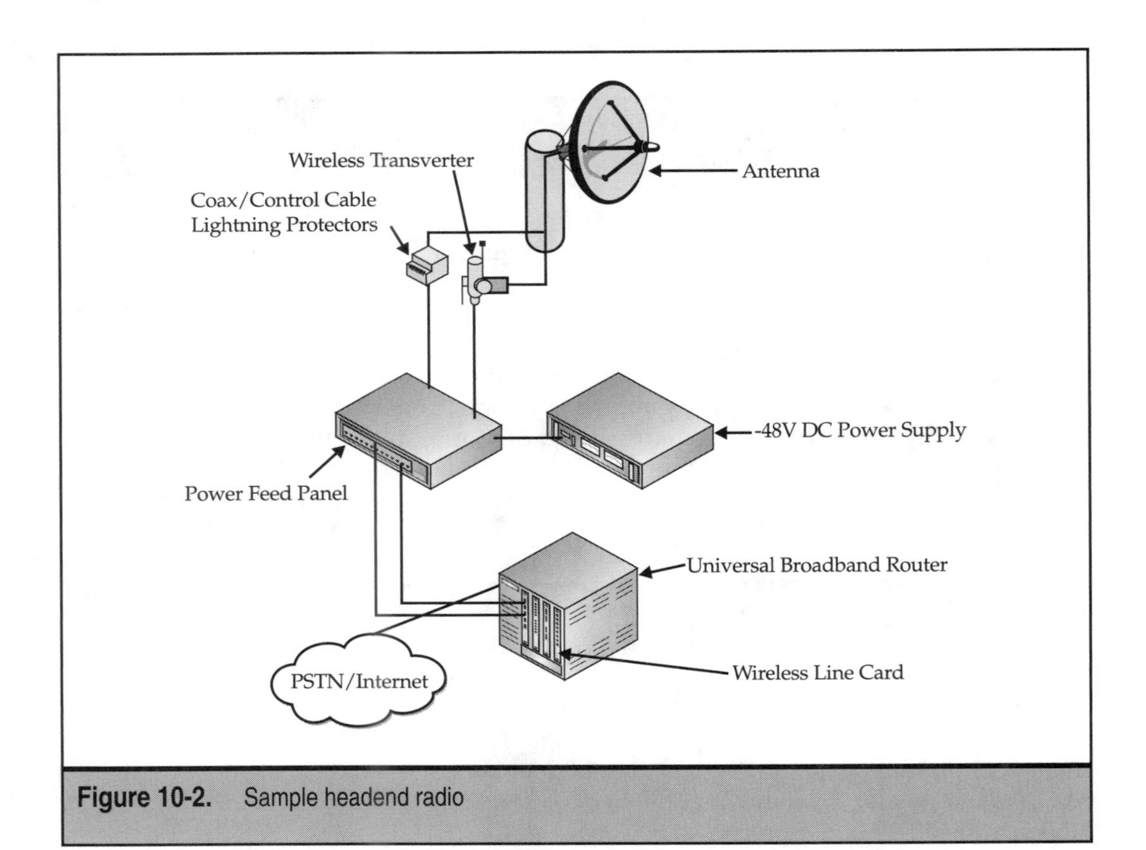

Figure 10-2. Sample headend radio

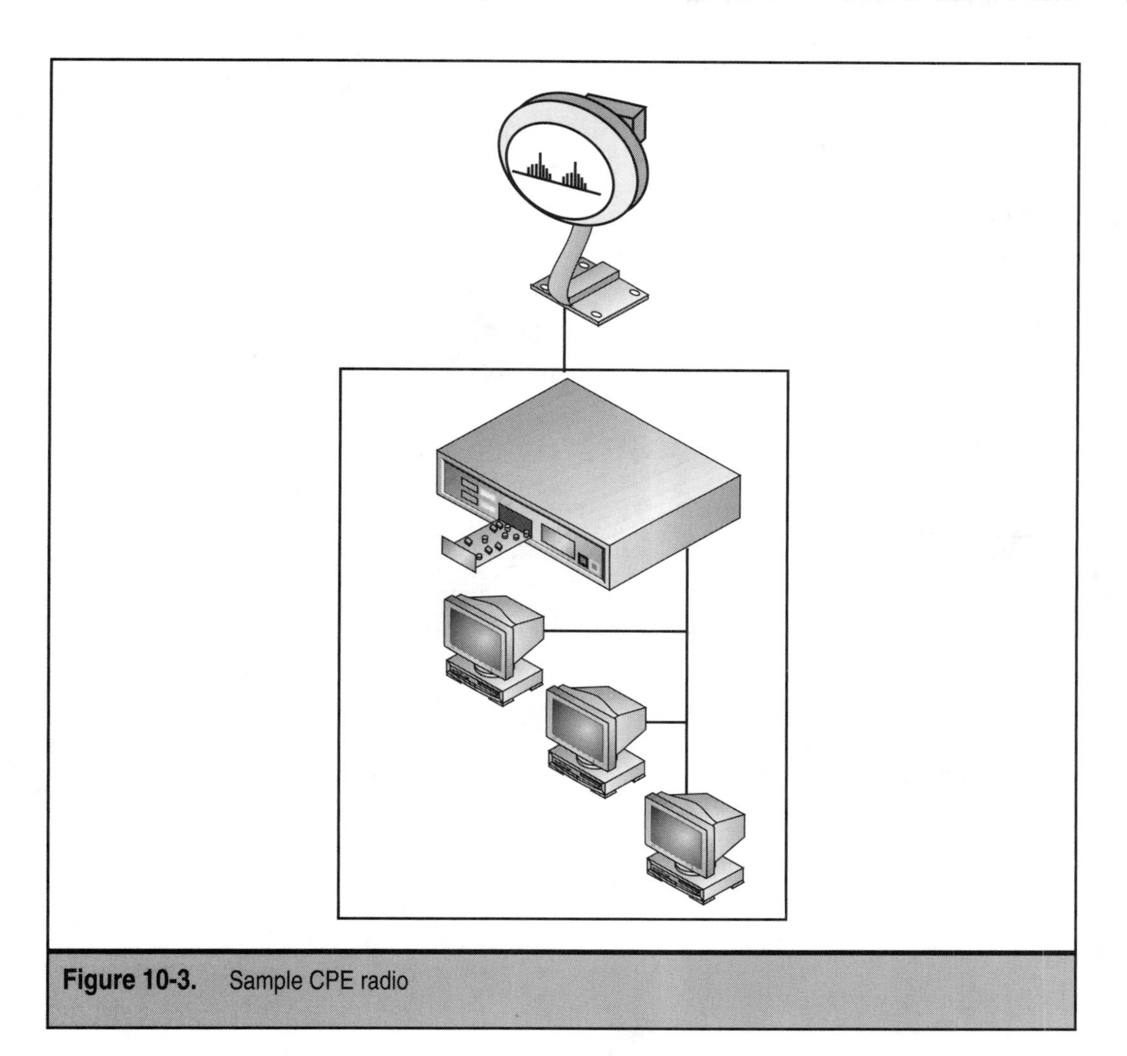

Figure 10-3. Sample CPE radio

Radios

Tables 10-2 and 10-3 provide a snapshot of the differences between HE and CPE radios. As there is a fairly wide range of specifications, this table is not intended to be comprehensive, but representative only.

Antennas

Antennas tend to be much larger at the HE than at the CPE sites. There are several practical reasons for this. Primarily, however, the HE radio typically communicates with many

	HE	CPE
Radios per sector	1	Up to 1000
Size	Up to 1 cubic foot	Generally 9″ × 9″ × 4″
Weight	Up to 40 pounds	Generally 10 pounds
Output (U-NII)	75mW*	75mW*
Output (MMDS)	2.0W*	150mW*
Typical antenna	Parabolic dish, Yagi	Patch, Yagi, small parabolic dish
Radio cost	$5K – $30K	$300 – $3K

* Power output of the radio from the antenna depends on the antenna gain and can be limited by regulatory agencies.

Table 10-1. Snapshot Comparison of HE versus CPE Radios

hundreds of CPE radios at varying distances. By way of analogy, if you have a flashlight with a beam that is adjustable, you could light up an area with either a long, narrow beam or a wider beam with shorter range. Because HE antennas typically service an area from 60 to 90 to 120 degrees, they are typically rectangular sector antennas (looking somewhat like a 2×4 piece of wood).

The size of the sector antenna is important, not so much for transmission purposes, but for reception purposes. Typically, the larger the antenna, the higher the receiver gain, and so in many U-NII or MMDS applications, the headend antenna tends to be approximately four feet in length.

Installation in high-wind conditions may have the installation and system-design individuals opting for a parabola with a smaller diameter, but a radome can be fitted on the front of the dish to help reduce wind shear load as well as to eliminate the buildup of ice on the antenna. Ice buildup on an antenna can, under certain conditions, detune the antenna.

Yagi antennas are often also used at HEs because they have a wider area that they can radiate, and consequently are easier to install. The offset for this improvement is a slightly lower level of performance. Most television antennas mounted on residential homes are Yagis. The Yagi antenna is very common in the wireless LAN buildng-to-building environment because it is typically inexpensive, easy to install, and somewhat forgiving in its alignment. Yagis come in many sizes, with varying numbers of elements depending on the beam width needed for the specific application. These antennas usually are not recommended for BBFW systems, as their gain is somewhat limited.

On the CPE side, parabolic antennas or patch antennas are optimal for performance, and when used are commonly on the order of 18 inches in diameter. Think of the CPE side as a point-to-point network. The CPE can only talk to the headend, so it is important to focus all of the transmitted power directly to the headend. Not all homeowner associations allow externally mounted parabolic antennas, and in many of these cases, a patch antenna is used, which is a small, square-shaped antenna that is also approximately 18 inches across. In some situations, Yagis can also be used, but not commonly in CPE deployments to residences.

The trade-off is, of course, antenna focus. Parabolic dishes can have a very small beam width and very high gain. As you move to smaller patch-style antennas, the gain drops and the beam width grows, thus reducing the distance between the headend and the client. Trade-off assumptions such as these need to be factored in at the initial site survey and business case analysis because they can have a dramatic effect on the number for CPE units deployed, type of client and area covered and the cost/ROI of a given network.

The ODU and antennas tend to be integrated into most residential CPE devices. This is done to minimize the area required to attached both the ODU and the antenna to the outside of a structure, and to simplify the installation. In some high-performance residential CPE units, the ODU can be located approximately a meter from the antenna. This is done to enable the deployment teams to more easily tune the CPE device, but more specifically, it allows the use of higher-power outdoor units and larger antennas such as the dish mentioned earlier.

For SMB deployments, parabolic antennas are generally the devices most commonly selected, and they range from 18 inches in diameter to greater than four feet. Obviously SMB sites generally have far fewer restrictions than those found in residential deployments, although residential BBFW deployments will eventually outnumber SMB deployments by two orders of magnitude or more.

Routers/Switches

There is a considerable difference in the routers and switches used at the HE as opposed to the CPE sites.

Table 10-2 provides a snapshot of the different router/switch families used at the HE vs. the CPE site. Cisco part numbers for routers are used because Cisco equipment greatly outnumbers competitive equipment in the field today. While switch part numbers are not indicated, corresponding capabilities may be assumed. Instead of focusing on the actual router part numbers in Table 10-2, focus primarily on the device capabilities. Cisco part numbers for routers are used because Cisco equipment greatly outnumbers competitive equipment in the field today.

Table 10-2 indicates that the HE uses larger routers with far more capabilities, which is intuitive because the HE is an aggregation site for the sector or group of sectors.

900 Family (CPE Only)	2600 Family (CPE Only)	3600 Family (CPE or HE)	7200 Family (HE or PTP)
10BaseT	10BaseT	10BaseT	10BaseT
Ethernet	Ethernet	Ethernet	Ethernet
	100BaseT	100BaseT	100BaseT
	POTS Voice	POTS Voice	POTS Voice
		ATM 25	ATM 25
		Token Ring	Token Ring
		E & M Voice	E & M Voice
		Multiport modem	Multiport modem
		Serial synch/asynch	Serial synch/asynch
		BRI	BRI
			FDDI

Table 10-2. Snapshot Comparison of HE vs. CPE Router/Switch Capabilities

Slave Status

CPE radios in P2MP deployments are the slave radio without exception (as far as this author knows). However, there are mesh networks in which numerous radios, or indeed any radio, can be the master radio, which relegates the other radios to slave status. A *mesh* network is a network where radios communicate to a number of peers, and, in general, there is no master/slave architecture. Mesh RF networks are rare and typically more common to the ISM applications, so their coverage is considered outside the scope of this work.

Asymmetric Data Flows

Data flow is commonly not symmetric in terms of the direction its traveling in BBFW networks. The generally accepted premise is that CPE users tend to download larger files to their location than those that are uploaded to network and eventually to another end user. This premise is not without controversy, however, and QoS mechanisms combined with adept network management can have symmetric data flows, a state in which the information being downloaded to a user occurs at the same speed at which data can be uploaded from the end user to the network and subsequently to another user (who would be downloading the information). BER and most CIR QoS levels commonly have higher downstream than upstream data rates. Downstream is generally referred to as the direction of information from the HE to the CPE; upstream is the obvious counterpart of this reference.

Market	Average	Burst	Lines	Usage
Large business	1 Mbps	8 Mbps	1 – 4 T1	Day
MxU	512 Kbps	2 Mbps	1 – 2 T1	24 Hours
Medium business	384 Kbps	1 Mbps	1 T1	Day
Small business	128 Kbps	1 Mbps	4 – 16A	Day
SOHO	64 Kbps	512 Kbps	2 – 4 A	Day
Residential	24 Kbps	384 Kbps	2A	Night

Table 10-3. Current Information Rates

Table 10-3 indicates the typical rates of information flow at the time of this writing.

Note that the rates indicated above are typical at the time of this writing and do not reflect the data rates desired by these markets. It is a reasonable assumption that most if not all of these markets would accept higher average and burst data rates if the rates charged were within the range of that which the market would bear. This table is intended to indicate the nominal data rates at present in order to show the major differences between these market spaces.

Cost

HE equipment is, with very limited exception, far more expensive than CPE equipment. This cost of CPE equipment may be one of the most notable reasons that BBFW has not been deployed far more than it has at the time of this writing. In addition to cost, ISP entities will need to eventually recognize and adopt the cellular telephone model along with the DSL and cable modem business models, which are established upon the premise that CPE devices do not have an up-front charge.

The cost of the CPE gear is paid for over time and guaranteed with service term agreements of one to two years. These service agreements stipulate that the CPE user must retain a contract with the service provider for a minimum period of time or pay a one-time charge. This one-time charge is intended to cover the cost of the CPE gear and its installation.

The issue of equipment cost and how it is recouped goes beyond the service provider and extends to the equipment providers such as Cisco, Lucent, and so forth. At the time of this writing, the major manufacturers rely on wide profit margins that are relatively consistent from different types of equipment placed across the network. In other words, business units that develop and sell CPE are under pressure from executive management teams to maintain wide profit margins similar to the larger products such as high end routers.

CPE equipment drives HE equipment sales, which drives core sales. The key to this is to have CPE gear with reduced margins that are recouped by high-volume sales. If the manufacturers of CPE and other network gear altered their pricing structures, BBFW links would become more common. This challenge is an issue with the SMB market, but is an especially sensitive issue in the residential market, which is considered the Holy Grail for service providers. However, price points, line-of-sight issues, and competition from cable and DSL are currently limiting deployment to the residential market.

Use

Usage of CPE equipment varies greatly from user to user and market to market. Most residential users want to use broadband for simple e-mail and Web browsing. Because the fastest-growing segment of the Internet population in the U.S. consists of individuals over 60 years of age, the transmission of digital photographs is becoming one of the most common uses, following the use of simple e-mail.

When a user is new to the network, a number of registration-type events need to occur in order for the CPE user to have access to the network. In state-of-the-art systems, the following are the primary registration events:

- User Registrar
- Network Registrar
- Modem Registrar
- Access Registrar

User Registrar

After a new user has their equipment installed and aligned, they must register with the service provider using the equipment. This is often accomplished by the deployment teams in order to speed up the completion of the link and to minimize the chance of the CPE user contacting customer service to walk them through the registration process.

When the new user has completed the user registrar process, they are registered on the network owned by the service provider, which enables the end user to have access privileges along with billing information to a credit card, for automatic bank account withdrawal, or for traditional mailed monthly billing.

Network Registrar

When the new user has completed the network registration process, they will have DHCP and DNS services in order to have multiple PCs at the CPE location. Network registration also enables the CPE user to have access to the network from virtually any telephone connection in the world via DHCP.

Modem Registrar

Following the modem registrar completion, the user will have TFTP (trivial file transfer protocol) and TOD (time of day) service for the various PCs they may have on at the CPE site. TOD is necessary to synch the CPE LAN with the HE radios and router clocks.

Access Registrar

The access registrar adds services that enable a secure type of authentication to verify that the user attempting to access the network is the approved CPE user.

Detailed CPE Registration Process

The preceding registrar processes are the key registration events that must take place between a CPE and HE prior to the transmission of traffic between the two link ends.

Assuming the HE has a number of CPEs, which is the more common architecture, the full menu of radio functions prior to the link transmitting information in a typical state-of-the-art radio is as follows:

- Power up
- Scan allowable frequencies and synchronize to HE
- Select upstream and obtain transmission profiles
- Adjust power and gain
- Time synchronization
- Frequency adjustments
- Establish IP connectivity
- Connect to DHCP server
- Establish TOD
- Establish link security and authentication process
- TFTP download service configuration for CPE router
- Begin transmission of traffic

Power Up and Scan Allowable Frequencies

When the CPE radio is powered up, it automatically begins to scan for frequencies the HE will be transmitting. The CPE device will only search a subset of the total frequencies sent out by the HE because not all the CPE devices will be tuned to the same frequencies. Having said that, there are fairly common architectures that use a TDM approach, which means that all of the CPE devices will use all of the allowable frequencies, but will only transmit and receive at certain time intervals. Once the CPE radio has locked onto the HE, it will synchronize its internal clocks with those at the HE.

Select Upstream and Obtain Transmission Profiles

The network managers will predetermine a minimum allowable upstream and downstream transmission rate for the CPE device. The CPE will have a MAC address that will be mapped to a preset transmission profile at the NOC. Once online, the CPE users can opt for more bandwidth or a higher QoS setting depending on the service offered by the service provider and the level of service currently purchased by the user.

Adjust Power and Gain

Power and gain must be adjusted based on range and multipath effects of the link. CPE radios that are closer to the HE are generally set to lower power and gain settings. In some state-of-the-art systems, the closest a CPE radio can be to the HE without using special attenuators between the antenna and ODU is on the order of 200 feet.

Time Synchronization

One of the most important events to occur between the CPE and HE is clock synchronization. This is especially important for TDM-based architectures.

Frequency Adjustments

Detailed frequency adjustments are commonly the next event. This event may recur when there is a substantial temperature difference between the HE and CPE gear, because on some systems, the printed circuit boards will have discrete components that change frequency based on temperature. This is generally not much of an issue unless the HE radio is at a substantially different temperature; for example, more than 20 degrees Fahrenheit. Equipment sensitivity on this issue varies greatly and this event is not common on most well designed and fabricated systems.

In extreme weather deployments, the major RF elements will reside in temperature-controlled environments.

Establish IP Connectivity

This is an event that only takes place on systems that transmit using IP. While most network owners agree that IP is the protocol to which they should commit their systems, the preponderance of BBFW equipment uses the ATM transmission protocol.

Connect to DHCP Server

Connectivity to a DHCP server enables the CPE user to have multiple PCs on their LAN or to retain the ability for their personnel to connect to the network via laptops from virtually any telephone connection in the world.

Establish TOD

TOD enables the CPE to be synchronized to the HE and is necessary to provide accurate time stamps for network management, troubleshooting, upgrading, and billing.

Establish Link Security and Authentication Process

At this juncture, the BBFW link is nearly complete and is the point at which the security algorithms and authentication protocols are deployed. These algorithms are rarely native to the radio and are not, in fact, options that are integrated to the routers or switches attendant to the HE radio.

TFTP Download Service Configuration for CPE Router

This is information passed on to the CPE LAN edge router, which can then be propagated to the other routers or switches at the CPE LAN site.

Begin Transmission of Traffic

Following the successful completion of these events, the CPE network technicians typically test the link with off-the-shelf software that sends a series of code words and link loads to ensure link integrity and to enable them to provide a baseline link configuration. These tests can be anywhere from ten minutes to ten hours in duration, as stated in the previous chapter. During these tests, the link is optimized, OFDM settings are fine-tuned, and so forth prior to the link being used for standard CPE LAN traffic.

CHAPTER 11

BBFW Security

Security is a subject with enormous scope and depth that requires years of training before a true measure of expertise can be claimed. In addition to that, security issues are rapidly evolving, so genuine fluency in the subject requires ongoing industry monitoring and participation. Accordingly, this section is not intended to be comprehensive; a single chapter in a book cannot even remotely address the scope and depth of security issues.

This chapter will focus on two primary issues: the invasion of a BBFW link, and the hacking of a wireline LAN access. To combat such security breaches, many encryption schemes are available to encrypt data before transmission. This chapter reviews four of the most common security applications. It also discusses the *Data Encryption Standard* (DES), with a specific look at 64 DES, as well as RSA (named after its creators, Rivest, Shamir, and Adleman), the standard that grew from 40- and 56-bit versions of DES.

Finally, the chapter will review the *Wired Equivalent Privacy* (WEP) protocol, which is primarily used on 802.11 unlicensed links used to protect from wireless eavesdropping, and RC4, a variable key size stream cipher mechanism with byte-oriented operations (designed by Ron Rivest for RSA Data Security). This chapter focuses on the WEP version provided by Cisco Systems for its Aironet product line, because, at the time of this writing, Aironet represents the state of the art in 802.11 radios. Chapter 13 will focus directly on 802.11 radios.

THREATS TO SECURITY EVER INCREASING

Security is an ever-burgeoning issue on the Internet, with the number of hackers increasing rapidly, in the U.S. and in many other countries. In addition, the proliferation of hacking tools made available over the Internet increases the prospect of losing valuable data or suffering even more disruptive consequences, such as denial of service. Research information released in the late 1990s indicates various ranges in the cost of security breaches. The U.S. Federal Bureau of Investigation estimates the annual losses due to security breaches to be between $1 billion to $4 billion; the accounting firm Ernst & Young estimates the losses at $5 billion.

One example of the necessity of implementing data security measures is the fact that in 1998 there were approximately 330,000 laptop computers stolen in the U.S. If these computers were not properly protected with security features such as even a rudimentary password protection program, the probability of security breaches is almost certain.

During the research portion of this chapter, the author was amazed to discover that common browser search engines during a single query will provide nearly as much information and access to hacking tools and methods as references to resources to prevent such activity. Very little effort is required to find hacking software, and tutorials for beginner hackers are common.

Another great concern with this trend is that as hacking software proliferates, the degree of expertise required to utilize hacking software has inversely diminished. Figure 11-1, sourced from the U.S. General Accounting Office, illustrates this point.

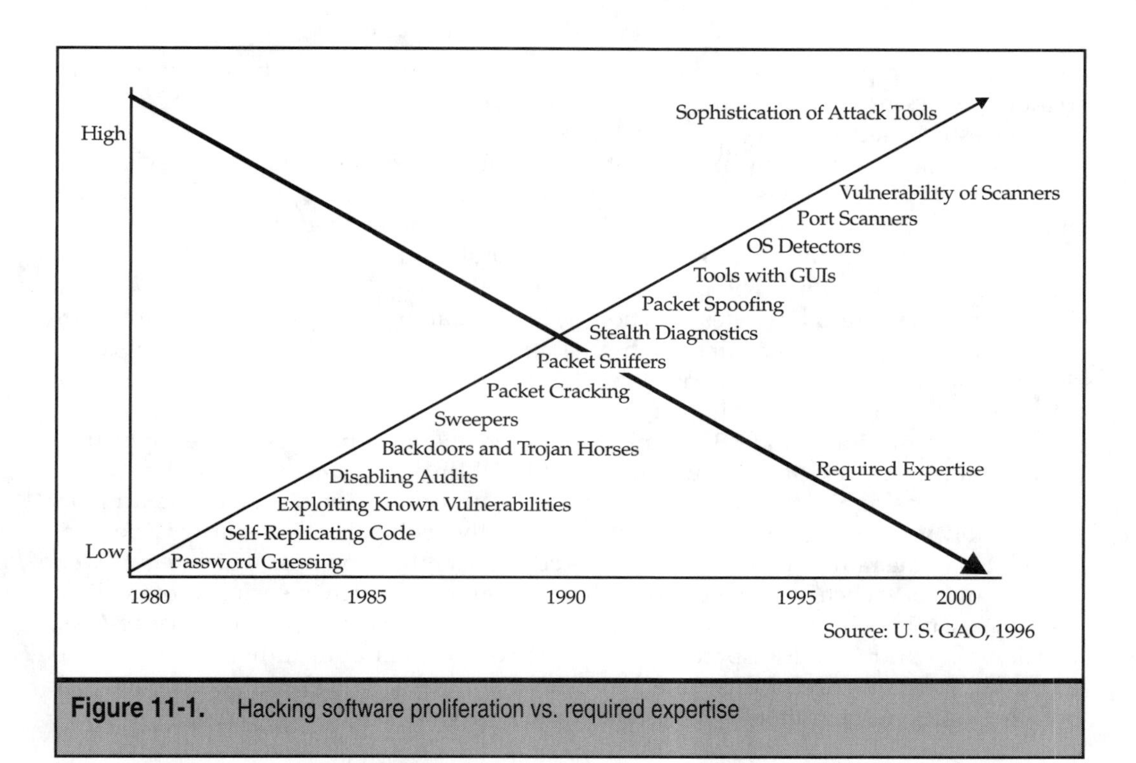

Figure 11-1. Hacking software proliferation vs. required expertise

A *USA Today* newspaper article dated May 30, 2001, declared that law enforcement is woefully underfunded to provide surveillance, tracking, and apprehension of business to business hackers. The article further stated that law enforcement officials could only muster a prosecution rate of an estimated 0.1 percent of all hacking activity. This very low number of reported and prosecuted incidents may be because the majority of activities go undetected and hence unreported. Additionally, there is a significant percentage that do get detected, but still go unreported, for fear of bad publicity to the company that may have stock price implications or other undesirable fiscal effects.

Another piece of the security puzzle is that many security breaches are not done deliberately, but through an innocuous action that was meant to assist in achieving a certain work goal but ended up creating a security problem. An example of this is when personnel at a company purchase their own 802.11 wireless client cards and use them at work without the proper security settings.

The combination of a worldwide group of ill-intentioned individuals with high-performance hacking software (along with a moderate degree of determination and a reasonable amount of time) and a general inability of law enforcement to mitigate the situation (and, until recently, few laws to deter such activity) means that today's service providers must be literate in the many issues of security. Further, they must deploy dynamic and comprehensive elements to minimize the probability of disruption and loss to their businesses.

The issue of security is yet another reason why BBFW links must be considered as a network element, not merely an array of radios relatively isolated from the service provider's or CPE LAN. However, the key premise of this chapter is that hackers will far more likely attack the user's network through means other than the BBFW links, although accessing the network through the RF links should not be discounted, especially on 802.11 networks, which will be covered later in this chapter.

Failure to secure the wireless portion of the link along with all other network elements increases the threat against high-reliability networks and the protection of confidential information. It is also true that the larger the service provider is, the more attractive it becomes as a hacking target.

As service providers and BBFW link owners and operators progress in the sophistication of their networks, distinguishing the BBFW link from the rest of the network becomes increasingly difficult. In addressing the issue of security in state-of-the-art links that are at Layer 3 of the OSI stack, one has to carefully distinguish the security issues that are related to the radios themselves as opposed to the end-to-end link, which requires either a router or switch at each end. By placing a router or switch at each end of the link, the link essentially becomes a routed connection between two customer premises LANs, which makes the radios transparent to hackers or other users, both authorized and unauthorized. The relevance of this statement is that security becomes an issue as network breaches occur at the routers and switches on the network edge.

BBFW LINK ACCESS AS HACKING MEDIUM

There are four primary objectives for hackers:

- Denial of service
- Theft of information
- Theft of service
- Use of networks as stooges

Denial of Service

The most common form of attack, which occurs most frequently with BBFW links, is denial of service, because it's easier to accomplish, and hackers often find more value in disrupting the network or making it inoperable than actually stealing information or engaging in theft of service for their own purposes. As networks often have more than one method of getting data on and off various LANs, a DOS attack can cripple one of the access mediums, multiple access mediums, the entire LAN, or, in the worst case scenarios, an entire network. DOS does not specifically require the ability for the hacker to breach any portion of the network, because networks with BBFW links can be relatively easily jammed, which has the same effect as DOS.

Theft of Information

While the media has portrayed the most common hacker objective as theft of information, it occurs less frequently than the results of other security breaches. Theft of information requires the hacker to place a value on the information and to be able to convert that value to an asset in their favor. Examples of this would be the theft of credit card numbers or bank account information, such as personal identification numbers (PINs). Hackers either use this information directly for their own gain or, more commonly, sell the information to third parties, who then use the information.

Theft of Service

Theft of service is where the hacker simply wishes to use the network either to transmit their own traffic or to store applications and other software, which may or may not be illegal for them to own. In a market where broadband is expensive, it is easy to understand this temptation.

Use of Networks As Stooges

One other common objective for a hacker is to use a network as a *stooge* to attack other networks. This will be covered in detail later in the chapter as part of the Denial of Service detail.

Networks are hacked through the access or edge portions of the networks. To be more specific, what hackers engage are LANs and groups of LANs. This is generally accomplished via wireline access, but with the incorporation of BBFW into a network, a completely different set of tools and tactics is added to the possible methods of attack. While tools for hacking via wireline access are prolific and easy to use, this is not necessarily the case for hackers who wish to invade networks via one or more radios.

The way hackers enter a LAN via radio gear is to *sniff* out a radiated signal through the use of sophisticated receivers located a BBFW link. To elaborate, this means the hackers must first locate a receiver and then receive transmissions. At that point, the hackers work at decoding the transmissions and then *spoof*, or in other words, imitate an authorized user. One other way to do this is to spoof the HE gear and hack into the CPE units. Target selection depends on the value of the targets, which in some cases may be the CPEs and in other cases the HEs.

In addition to actual penetration of the network via spoofing, the wireless network is also vulnerable to *jamming*. That is, the intentional increase in noise or signals at a targeted receiver in order to swamp the receiver and its buffers such that it can no longer accept authorized traffic. This attack is called *denial of service* (DoS) and is shown in Figure 11-2. In any event, these tactics are decidedly nontrivial, and most hackers will find it easier to access the network via non-RF means.

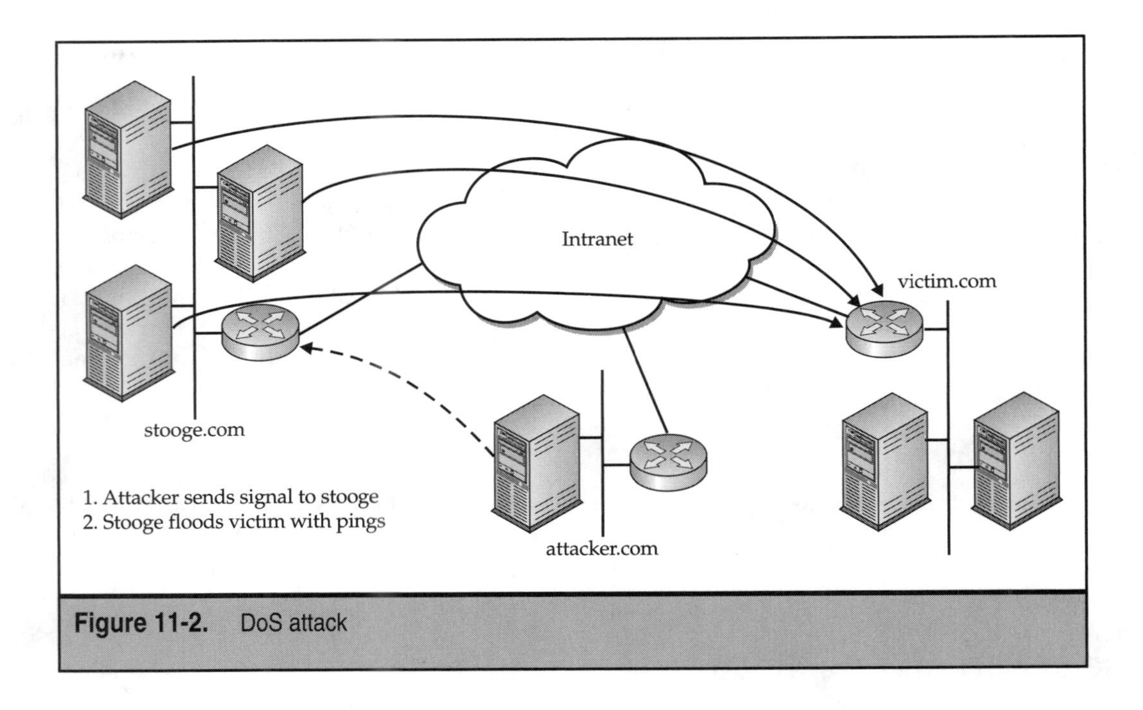

Figure 11-2. DoS attack

Challenges of Penetrating a BBFW Link

The challenge of penetrating a BBFW link for nonauthorized activities is not a trivial task. It requires an understanding of not only the specific modulation techniques used but also some very specific physical issues. Examples of these challenges to the hacker include:

- Relative physical proximity to the target
- Extensive RF hardware and software mounted on mobile platform
- Extensive RF expertise
- Cracking state-of-the-art spreading and modulation schemes such as OFDM and 64 QAM

Although the combination of the preceding elements does not eliminate the possibility of attack, it certainly raises the cost of the attack and limits the number of criminals that have this type of access and availability. The concept of making a hacker work very hard to disrupt the use of a network is at the heart of most security mechanisms. With a virtually infinite number of targets, along with the fact that a significant number of them are relatively easy to penetrate, hackers will generally not attempt to invade well-secured networks; it's easier to simply breach a network that is relatively unprotected.

Relative Physical Proximity to the Target

In order for the perpetrator to access the networks via the BBFW radios, they would have to be relatively close to the actual link. A hacker could be as far away as five or more miles from a transmitter, although if accessing the network, they would have to radiate signals that are identical to those from any normal CPE site or near the HE. This would still pose a substantial challenge in finding a perpetrator in this scenario, as the term *proximity* is relative. Chapter 10 goes into some detail regarding the synchronization and other necessary dialogue that must occur between the two radios.

Once the hacker began radiating from their mobile platform, they could be detected and found with directional antennas by law enforcement officials, provided of course that officials were both aware of the attack and willing to investigate, though this cat and mouse engagement would tend to favor the hacker, due to multipath interference and the total area that would need to be covered. (Triangulation and ready discovery is better left to Hollywood.)

In comparison to a hacking attempt via wireline, the sophistication of the tools required to penetrate the wireless network is greater than the wireline access method, but it is far safer for the attacker when it comes to detection, because the attack can occur from virtually any server or PC in the wireless coverage area.

Extensive RF Hardware and Software

Hacking directly via a BBFW radio would require radios that are identical to either the target site CPE or the HE radio. This is far more expensive for a hacker than simply using a desktop or laptop computer for a fraction of the cost. Some BBFW radios cost upward of $50,000, which is easily ten to twenty times the cost of a contemporary desktop computer at the time of this writing.

Further, the hacker would have to load the radio and server or PC into a mobile platform, which would most likely be a van that would require a parabolic or other reasonably noticeable antenna on the roof of the van. This antenna would have to be relatively mobile in terms of being able to both point and elevate the antenna. This method is in contrast to wireline hacking, which enables the hacker to operate from a discrete location from within virtually any building. As indicated earlier, in Figure 11-2, most hackers are located within the building or campus of the network that they are hacking.

Extensive RF Expertise

Extensive expertise is required to hack a BBFW link, because the perpetrator would have to adjust modulation schemes, gain, and other settings in order to synchronize their radio to the target BBFW link.

Again by comparison, and referring to Figure 11-1, virtually no expertise is required to hack into a network via wireline due to the sophistication of the software readily available.

State-of-the-Art Schemes Such As OFDDM, 64 QAM, and so on Are Very Difficult to Crack

Assuming the perpetrator has sufficient expertise to synchronize their radios to the target BBFW link, decrypting the modulation schemes would be difficult and time-consuming, though possible, especially if the hacker has a radio identical to one of the CPE radios in use on the network.

Spreading techniques such as OFDM and DSSS provide what is termed "native encryption," which is to say that a measure of security can be gained by spreading the information over a large number of separate channels. The hacker would have to know the channels, the order in which the channels are transmitted (in the case of DSSS), and the type of information contained in each channel.

State-of-the-Art BBFW Link Security

In state-of-the-art BBFW links, part of the security issue resides in encrypting the traffic flows. This can be accomplished as part of the transmission protocol for point-to-multipoint security, and is intended to provide address data privacy across a BBFW link. In short, this provides user data privacy by encrypting information flows between the CPE and HE radios.

IP Wireless Security Features

Wireless security features provide service providers with basic protection against *theft of service* by establishing a unique and identifiable address to each modem. These modems then provide and request a MAC address authentication access. In some systems, modems can still be *cloned* if a hacker can gain access to an authorized MAC address, which then enables the modem to masquerade as an authorized modem.

There is still an issue of getting a new user subscribed onto a HE LAN, as they often have MAC addresses that are sent in an unencrypted format in order to have them brought into the system and then granted a security association. The units then use a public key management protocol, which enables the unsecured modem to be brought into a secure communication status. Modems that are supposed to have access to a BBFW link must have a factory-installed pair of keys that the HE and CPE link are predetermined such that the both the modems and head ends have common tables of keys. This key pair generation must be a one-time operation.

Major providers of BBFW links that have strong network interests are also more likely to provide software updates for security. This is, once again, an example of where BBFW links that are provided from network companies such as Cisco should be considered preferred for BBFW equipment.

Wireless Equivalent Protocol (WEP)

WEP is a security protocol primarily used in 802.11 radios, which are used in Wireless Local Area Networks (WLANs), to secure wireless communications from eavesdropping and to prevent unauthorized access to a wireless network. 802.11 networks are primarily comprised of *access points* and *client adapters*. Access points interface the network and the client adapters. Client adapters reside in laptops as PCMCIA cards, or as internally installed cards in desktop computers and printers. While Chapter 13 will detail this type of BBFW radio, it is appropriate to review some of the aspects of this security measure in this chapter.

Researchers at the University of California, Berkeley, recently published a document entitled "Intercepting Mobile Communications: The Insecurity of 802.11" (by Borisov, Goldberg, and Wagner). This paper claims that "...security flaws in the 802.11 security protocol (WEP)...seriously undermine the security claims of the system." The authors also state, "More sophisticated key management techniques can be used to help defend from the attacks we describe; however, no commercial system we are aware of has mechanisms to support such techniques."

This paper has caused a ripple effect throughout the market and industry that uses and/or supplies 802.11-type radios. Most providers of 802.11 radios were not prepared for this research paper and its effect on the industry, and have voiced considerable concern in response to the paper. However, not all 802.11 providers were caught off guard, and Cisco Systems was one of those companies well prepared in advance for this devel opment. Cisco provides the Aironet product line, which is sold worldwide, and had already released software in advance of the Berkeley paper that afforded centrally managed, dynamic per-user, per-session WEP.

Many industry experts agree with the authors of the paper that WEP does indeed have certain weaknesses, regardless of the length of the encryption key used. One of the weakest areas of a WLAN system is that the WEP key that does not change from user to user, session to session, or location to location. The Berkeley paper states, "In practice, most installations use a single key that is shared between all mobile stations and access points." What the researchers are stating, in part, is that static keys represent a significant potential for security problems.

In fact, Cisco's Aironet wireless security solution affords the most sophisticated key management techniques outlined by the researchers in their paper, including innovations such as dynamic, per-user, per-session WEP and integrated network logon, which address several of the limitations of WEP while promoting hassle-free enterprise deployment. There is little question that if a single WEP key, regardless of its length, is acquired, it makes a security breach quite simple, especially if possession of that key enables the user to have the same access as all the authorized users.

A dynamic key is far better than a static key simply because static keys become ineffective after a short period of time. A dynamic key significantly reduces the window of time during which a hacker can acquire the key. By tying it to the user session and, optionally, network logon, the key becomes even more difficult to acquire.

The downside to a dynamic key is that while it will change frequently, the users must be accommodated in an efficient manner so as to not hinder their network access or productivity. One of the most elegant ways to accommodate users, which is the method adopted by Cisco, is to create a per-user, per-session, dynamic WEP key that is tied to network logon. In other words, the key changes over time during the session to which the user is logged on. This enables a highly dynamic key that simultaneously provides a security method that is easy for network administers to use.

By employing dynamic WEP keys, the Cisco Aironet security solution enhances WEP to decrease its predictability (to the hacker), significantly minimize any attack windows, and tie it to the user session and, optionally, network logon. All of these properties working together have been architected with large-scale enterprise deployments in mind without compromising on overall network security. Cisco has partnered with several vendors to achieve innovations from an end-to-end standpoint and to develop an open, extensible framework for the future.

There are a number of WEP key enhancements that greatly improve the security of an 802.11 WLAN radio access:

- Mutual authentication
- Secure key derivation
- Dynamic WEP keys
- Re-authentication policies
- Initialization vector changes
- Other network security measures

Mutual Authentication Mutual authentication schemes for keys are more secure than one-way authentication schemes, which are more common. Because standards-based mutual authentication implementations are easily deployable, they are still evolving and will likely continue to evolve on a ongoing basis. An authentication scheme called *Light Extensible Authentication Protocol* (LEAP), which is a version of the *Extensible Authentication Protocol* (EAP), ensures mutual authentication between a wireless client and a server that resides at the network operations center. Communication between the access point and the server is via a secure channel. This eliminates "man in the middle attacks" by access points and servers used by unauthorized individuals, which are termed *rogue access points* or *rogue servers*.

Secure Key Derivation The original shared secret that is used to construct responses to the mutual challenges undergoes irreversible one-way encryptions that make password replay attacks impossible. Password replay attacks are when a hacker attempts to reuse a password. The encryption sent over the wire is useful for one time only at the start of the authentication process, and never after.

Dynamic WEP Keys While dynamic keys are clearly more secure than static keys, one of the biggest concerns is the management of the keys. The preferred methodology is to have the session keys unique to the users and to not share the keys across a group of users. Also, with LEAP authentication, the broadcast WEP key is encrypted using the session key before being delivered to the end client. By having a session key unique to the user, and by tying it to the network logon, vulnerabilities from stolen or lost client cards or devices are eliminated.

Re-authentication Policies Customers can also set policies for re-authentication at the network control server. This will force the users to re-authenticate more often and get new session keys. The smaller the window during which a key exists, the more difficult it is for a hacker to attack where traffic is injected during the session.

Initialization Vector Changes An initialization vector (IV) is an external value needed to start off cipher operations; in other words, a mathematical value upon which the ciphertext depends for encrypting. An IV often can be seen as a form of message key. Generally, an IV must accompany the ciphertext, and so always expands the ciphertext by the size of the IV. Therefore, IV deployed on a per-packet basis results in eliminating a predetermined sequence that hackers can exploit. In particular, this makes it difficult for hackers to write or use attacks that use mathematical tables that simply cycle the number of key combinations until one or more are discovered that work.

IVs can be sniffed off the wireless network, so eliminating the possibility of being able to determine future key codes is a good security tactic.

Other Network Security Measures Recalling once again that a BBFW link is an integral part of a network, not simply an access method that can be deployed independently of the network, other security measures such as VPN, firewall, and IOS services can also enhance the end-to-end security of networks.

Customers can enable VPN clients on their laptops or access devices from public areas and establish secure tunnels to their enterprise networks. Customers can also use the network logon, access control lists in switches and routers, and policies on their firewalls to achieve robust end-to-end network security. End-to-end VPN security can also be deployed in intranets where very high security is essential.

HE Security

Access to HE equipment is generally controlled by the service provider's username plus password login control mechanism. This specifically refers to all gear mounted on a roof, mast, or tower in order to prevent physical damage to the equipment.

Element Manager

In point-to-multipoint systems as well as point-to-point systems, there is software called an *element manager* that requires a username and login identification. More advanced systems also require special login information for network partitioning, task-based access, and configuration of specific user groups.

Command Line Interface (CLI)

The CLI is where the network technicians control the actual radio settings, and resides on the routers or switches at each end of the BBFW. This, too, is username- and password-protected. The CLI must be protected by at least a screensaver that is password-enabled. If an individual had open access to the CLI settings, they would have a much easier time hacking a BBFW link or, at the least, possess information that could enable a DOS attack on a BBFW link or its associated network.

CALEA Cooperation

CALEA (Communications Assistance for Law Enforcement Act) is federal law enacted within the U.S. that requires that all BBFW providers that can carry voice over their networks be able to intercept and deliver to law enforcement agencies detailed information on voice calls that originate or terminate within their area of geographic coverage. This law is controversial in terms of how it is implemented into networks, and no standard has arisen to guide this implementation.

This law includes monitoring and providing information to law enforcement officials, which is generated by the *Call Agent*. The Call Agent is a software package that routes the voice calls as well as being able to provide information carried by or access to other communications devices that may be attached at either end of a BBFW link. CALEA cooperation also requires that the BBFW link provider be able to provide an element for wiretap services through a device called a wiretap server.

In the scenario where law enforcement agencies require information and support in targeting a CPE or HE voice device, the Call Agent will recognize when a target is involved in a call and will pass on detailed information about the call to the wiretap server. The wiretap server will then deliver the information to the law enforcement agency. If call content is also to be delivered, the wiretap server will instruct the appropriate voice communications transport device (for example, an IP router) to replicate and deliver to the WS the voice packets associated with the target's call.

The WS will then deliver the replicated call content, in real time, to the appropriate LEA monitoring center. It is expected that the LEA will be responsible for the setup of collection facilities to receive the call content. The format of the call content delivered from the WS will be dependent on the specific needs of the LEA (for example, call content may be delivered in packet format or over a PSTN connection).

WIRELINE ACCESS AS A HACKING MEDIUM

The primary focus of this book is clearly BBFW as opposed to wireline mediums. However, a network has numerous points of crossover, and not all BBFW issues are mutually exclusive of wireline issues.

As stated earlier, state-of-the-art BBFW links appear, for the most part, as routed links to the network; in other words, they appear to be wireline links. Figure 11-3 illustrates the most common types of security breaches.

According to one study, one of the most striking aspects of this figure is that *approximately 85 percent of hacking is performed not by hackers or outside users, but by authorized users*—employees, employee leaks, and access abuse. These are forms of hacking or unauthorized use that would virtually never be seen as a perpetrator spoofing radios,

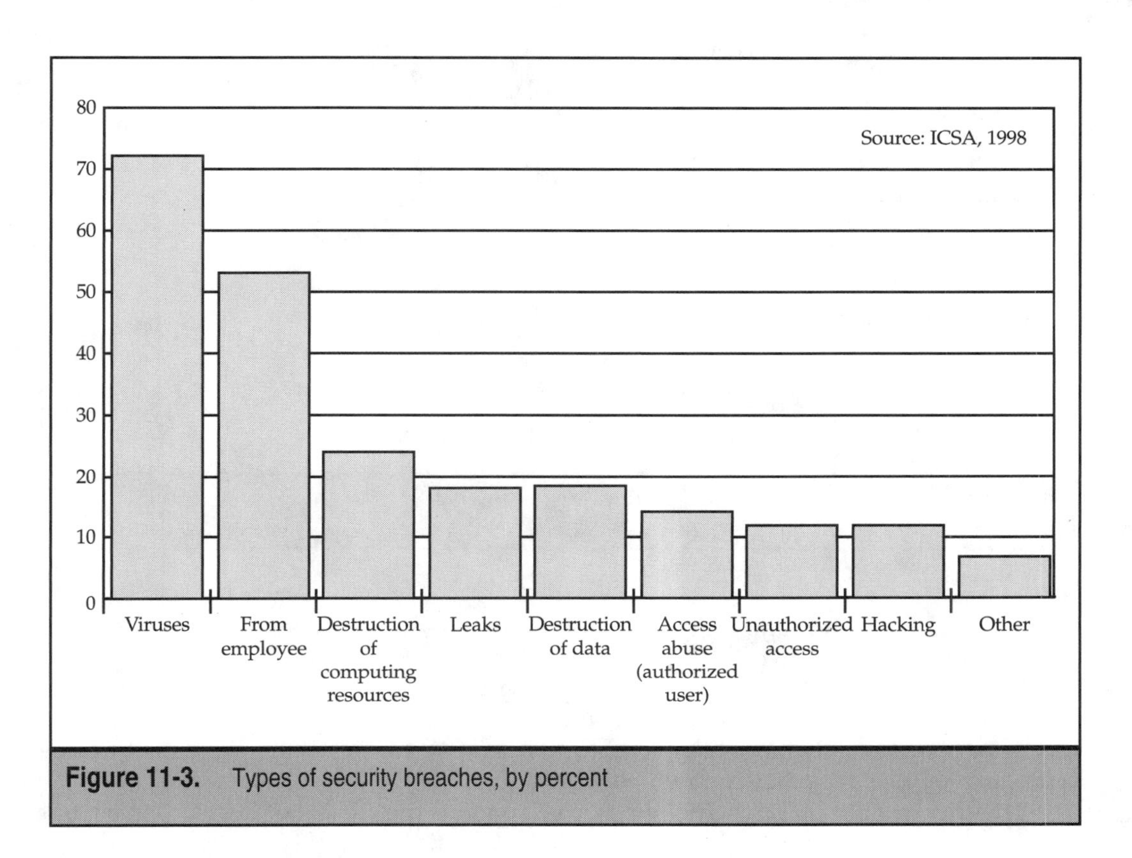

Figure 11-3. Types of security breaches, by percent

because it would be far easier to simply access the network with standard passes or passes taken from fellow employees.

A wide array of secure communications protocols has been widely adopted by the BBFW and network industry. Table 11-1 identifies some of the more common protocols.

There are also cryptographic protocols available on the market, but this author agrees with the advice of industry experts, which is that the safety of a communication is in the *length* of the key, not the complexity of the encryption used.

Unix-Based Hacker Software

Table 11-2 identifies a few of the more common Unix-based hacker software programs.

Of the hacker software tools listed in Table 11-2, perhaps the one that generated the most publicity and fervor was the writing and release to the general Internet community of SATAN (Security Administrator Tool for Analyzing Networks). This software scans a target LAN for a wide range of vulnerabilities. This tool is not used much anymore, because better ones have subsequently arrived on the market, but it was the first. It was written by Dan Farmer and Wietse Venema to be used as an assessment tool for administrators

Name	Use
Secure MIME (S/MIME)	E-mail
Secure Sockets Layer (SSL)	Transactions
Secure HTTP (S-HTTP)	Transactions
Transaction Internet Protocol (TIP)	Transactions
Pretty Good Privacy (PGP)	E-mail
Kerberos	E-mail
Server Gated Cryptography (SGC)	Transactions
Transport Layer Security (TLS)	Transactions

Table 11-1. Common Security Protocols

to help plug holes, recognizing that computer systems were becoming more and more dependent upon the network.

The intent of SATAN's authors, by releasing this software to the open community, was to force network owners to increase the levels of security in their networks. The theory in part was that the network owners and operators would secure copies of this free software and use it to test their own networks.

Whether or not that has actually occurred remains the subject of heated debate, but many consider this software nefarious and problematic because it has clearly landed in the hands of those with ill intent.

Name	Function
SATAN	Network vulnerability scanner
Tcpdump, Ipgrab, sniffit	Displays network traffic
queso	Displays host operating system
nmap	Port scanner (can specify a range of hosts) that identifies operating system detector
tcp_scan	Displays version of services
Rdns	Pings a range of IP addresses

Table 11-2. Common Unix Hacker Tools

Viruses

A virus is potentially destructive software that spreads itself from program to program, from computer to computer, and from LAN to LAN. In today's high-speed world, viruses spread like wildfire simply because of the amount of e-mail, files, and discs that are routinely exchanged between individuals, groups, companies, and countries.

All viruses need both a host to reside on, which is typically a well-used type of software such as e-mail, and a way to spread, meaning they must be transported from computer to computer by disc or e-mail. Viruses either modify programs or destroy information, even hardware in extreme versions. Recall that a security breach is not necessarily intended as unauthorized access to the network, but also includes the intent to either destroy other people's equipment or deny access to others.

The truth about viruses is that your laptop, server, or router is more likely to be damaged from a spilled cup of coffee than from a virus, but viruses commonly occur. More people lose information to bad discs and user error than to viruses. Of the many viruses that are on the Internet, most experts claim that the same 100 or so viruses are responsible for approximately 90 percent of all computers infected. Once the patches for these viruses have been deployed, the viruses are no longer dangerous.

There is a very high degree of monitoring for new viruses, and there is a formal network of professionals who monitor, repair, and deploy patches along with information about viruses. In fact, a whole industry has emerged to address this issue. You cannot acquire a virus from simply being online or visiting the eBay™ site or other Web sites—unless you download information from them or have information sent to you. It bears emphasizing that simply having a virus sent to you will not necessarily cause contamination to your personal computer or network. In most cases, the e-mail or attachment must be opened to activate the virus.

One of the most important things you can do to prevent damage from viruses, as well as user error, is daily or weekly data backups of both information and configuration files.

There are basically only three types of viruses:

- Logic bombs
- Trojan horses
- Worms

Logic Bombs

Similar to a real bomb planted by an individual, logic bombs lie in wait until triggered by an event like a specific date, number of times a program is executed, or even the deletion of a file. These viruses can be very destructive and difficult to locate until they have been executed.

Trojan Horses

Named after the wooden Trojan horse that was delivered as a gift to the city of Troy but that secretly contained Greek soldiers, a Trojan horse is a seemingly harmless computer program that delivers destructive code, such as a logic bomb, and therefore is a carrier, not a virus. This type of attack appears as a useful piece of software until executed. The summary here is that you shouldn't use software on your LAN or PC unless you know it comes from a trusted source such as your administrator.

Worms

A worm is technically not a virus, and does not typically destroy data or hardware; rather, a worm makes copies of itself, and those copies then create even more copies, and so on. What happens, then, is that there can suddenly be such an enormous amount of e-mail generated that a system will shut down.

The Four Types of Viruses

There are four types of viruses, which are classified by where they attack. The first is the *system sector virus*, which is also commonly called a boot sector virus.

Sectors are not files but instead are areas on a PC disc, server, or router that are read in chunks. For example, DOS sectors are 512 bytes in length. Sectors are invisible to your applications but are vital to the operation of a PC, server, or router because they contain the basic information for applications and data. When the sectors are disrupted, the results are terminal for the performance of the PC, server, or router.

In addition to the system sector virus, there is the *sector virus.* Sector viruses modify the data that resides within sectors. These viruses are usually far larger than the 512 bytes available in a sector and therefore usually reside within the RAM portion of a PC, server, or router and then go on to affect the data in sectors. Because this type of virus can take up residence within RAM, it means that even if the disc is repaired, they can recontaminate the disc immediately after being repaired.

File Viruses

File viruses are the most common kind of virus and they typically go after a file with a certain extension, such as .doc or .exe. They attack by overwriting parts of the files such that the file becomes unusable or highly unstable, which then crashes the PC, server, or router. File viruses, as with sector viruses, can often reside in the RAM portion of the PC, server, or router, and thus care needs to be taken in rooting them out during the repair phase.

Macro Viruses

Macro viruses are not as well known as other virus types and most commonly reside in Microsoft Office applications such as Word, Excel, PowerPoint, Access, and so on. This type of virus actually resides within the application, and is executed when loaded onto a hard drive.

Cookies

There are many rumors about cookies, such as what they're about, what damage, if any, they can do to a network, and what confidential information, such as credit card information, they can provide to third parties.

Most of the rumors are unfounded and not even close to being correct. Cookies store information that users provide to a Web server at some point. Thus, each time you access a Web site, a trail about you is left behind. This could include your computer name and IP address, your operating system, and the URL of the last page you visited. While cookies themselves are not gathering data, they can be used as a tracking device. A cookie cannot read data to find out your identity or your home address. But, if you were to provide that information to a site, it could be saved to a cookie. As more information is gathered, it is associated with the value kept in your cookie.

For example, if the user were looking for an automobile, they might go to www.cars.com. The "cars.com" server would then send a cookie to the user's Web browser, which would reside on the user's hard drive. The cookie is not an executable program and cannot do anything to an Internet user's machine. Cookies will not divulge your name or e-mail address, or anything about your computer.

Whenever a browser requests an image or information from the Web server, the server sends a copy of that cookie back to the server along with the request for information or images. Thus, a server sends a cookie to the user's computer, and the Web browser owned by the user sends the cookie back to the server when a request is made for a file or image. The intent of the server owners is to then be able to provide custom information to you about other Web sites that are similar in nature.

When a browser sends a request to a server, it includes its IP address, the type of browser the user has, and the operating system of the user's computer. A cookie sent to the server from the person contains only that information; there is no additional personal information sent to the server.

Cookies serve other useful purposes, such as when a user shops on the Internet. When a person first visits an Internet shopping site, they are sent a cookie with the ID number of a shopping cart. While the user shops, the items are added to the shopping cart. When the user has completed their shopping, there is a checkout function that lists all of the items the person wishes to purchase, including their prices and other relevant information. Without cookies, the shopper would have to keep track of each item, its cost, its part number, its color, and so forth and then type that information into the checkout page, one item at a time.

Another use of cookies is when you set up a custom Web page from Yahoo!™ or another source. A cookie is sent to your browser for each of the custom items you place on your personalized Web page. Whenever the person revisits their custom Web page, the cookies are sent to the server in order to determine which items to display. Without cookies, a server would require the person to identify himself each time so that it would know which items to display on the custom page. The server would also have to preserve copies of the actual custom pages for each visitor, which is a premise that would not scale well.

One of the less desirable uses of cookies, and perhaps one of the most controversial, is to track the habits of the Internet shopper. This does not make it easier to track the user; it actually makes it easier for the server to provide links to other areas within its own Web site or to links the user might have interest in. On multiple server sites that are serviced by a common marketing entity, cookies can be used to track your browsing habits on all the servers. With this in place, when the user visits one of the server sites being managed by the single marketing entity, the advertisement shown is from the marketing firm, not the Web site the person may be visiting.

Web Browser Vulnerabilities

While the ubiquity of today's Web browsers may be one of the key reasons for the popularity of the Internet among residential and other users, browsers are also one of the weakest areas in terms of security. Table 11-3 indicates some of these challenges.

Browser	Definition	Problem
Internet Explorer	Type of browser	Cookies
Netscape	Type of browser	Cookies, file transfers that can circumvent firewall/virus scanners
Push technology	A common browser element that sends data to the browser user as opposed to having the browser user "pull" the information from a site	File transfers that can circumvent firewall/virus scanners
CGI	Computer Generated Image, which is part of a browser application	File transfers that can circumvent firewall/virus scanners
Java/ActiveX	Two examples of common browser languages	File transfers that can circumvent firewall/virus scanners (also known generically as *active content*)

Table 11-3. Common Browser Vulnerabilities

Data Encryption Standard (DES)

DES was introduced in 1977 and was proposed by IBM with either 56- or 128-bit keys. It is probably the most widely used cryptography scheme used on the Internet. To understand DES, you must be familiar with the following basic terminology:

- RSA
- Bit
- Block
- Key
- Cipher
- Plaintext
- Ciphertext
- Round
- Cryptanalysis

RSA RSA was the first major public encryption keying system and its name is a combination of the first letter of the last names of Ron Rivest, Adi Shamir, and Len Adleman, the engineers who created RSA.

An RSA key is at least ten times as long as the information it is securing, and is also one of the most common encryption methods in use today. The number of keys used to convert an RSA encryption from ciphertext to plaintext is so large that the entire range of keys is open to the public. The actual keys used by the transmitting and receiving parties are kept confidential. The RSA system hinges on the fact that the number of possible keys is so large that it would be impractical for a prospective intruder to go through all of the different keys until the correct one was found.

This type of encrypting is called *one-way* or *trap-door* because it is easy to encrypt, hence the term "one-way". It is also called "trap-door" because the conversion back to plaintext by an authorized party is easy to perform, hence the term "trap-door" for the receiving party's access.

Bit Interestingly, although the term *bit* is one of the most elemental terms associated with the Internet and the computer, it is not widely understood. The term is a contraction of *binary digit*, which is the smallest possible unit of information a computer can handle. An alphabetic character or number is generally made up of 8 bits, which comprise 1 *byte* of information. Therefore, a single character, such as the letter *b*, requires a combination of eight 1's and 0's.

Block A *block* of information is that of a certain size, which is treated as a single unit. For example, 64 DES is a common encryption used on the Internet and is termed *64* DES because it encrypts 64 bits at a time. While even a single bit can be encrypted, it would be rather easy to decode; that is, the solution would likely be the opposite of that which is shown. In other words, if the encrypted bit is a 0, the decoded value would be a 1.

For this reason, in part, encryption takes place over a relatively large number of bits, i.e., 64 bits, and there is also a 128 bit version of the DES. In some encryption techniques, a single bit can be represented by two or more bits after being encrypted. This is called *data expansion enciphering,* which means that the number of bits in the *ciphertext* is larger than the number of bits in the *plaintext.*

Key A *key* is used to "unlock" ciphertext, and one may think of it in the same relative terms as a lock and key. However, in the DES standard, and in the world of encryption at large, a single key can generate a large number of different versions of ciphertext from the same plaintext. There are also different kinds of keys, such as the Running Key, which encrypts the sequence of a number of bits, a Message Key, which is different for each and every message. In the use of keys like Message Keys, obviously both the transmission source and receiving source must know the order and specific key that is used on each transmission.

There are also keys such as the User Key, which is the key the user actually uses and remembers, and Alias Keys, for which alias files associate various users with various keys. There are numerous other forms of keys, such as symmetric keys, in which the same key is used to cipher and decipher information, and asymmetric keys, which are different keys that are used to convert the same ciphertext into plaintext.

Cipher A *cipher* is a key that converts plaintext to ciphertext. This is not to be confused with some forms of secret codes in which certain words or phrases are replaced with secret code words or phrases.

Plaintext *Plaintext* is the original, readable information. It is usually a set of alphanumeric characters, but can also be other forms of data, such as values or mathematical symbols.

Ciphertext *Ciphertext* is text that has been *ciphered,* or encrypted. While the ciphertext will contain the same information as the plaintext, it may or may not be the same number of bits. Certain lower-end systems may have difficulty accommodating encryption, the technical term being *data expansion ciphering*. Ciphertext always requires a key to determine the plaintext.

Recalling that a bit is most commonly either a 1 or a 0, a 64-bit block of information can be encrypted in 1.8×10^{19} different ways. This is how many different combinations of 1's and 0's that can be arrayed in a 64-bit block of information. A complete set of these combinations may be considered as a unique *code*. The code is decrypted by a *key*.

Round A *round* is a set of encryption operations performed on a block of information. For example, 64 DES uses 16 rounds of operations to produce the final version of the ciphertext, which can then be transmitted over an open BBFW link or other unsecured method of transmission.

Cryptanalysis Also referred to as code breaking, *cryptanalysis* is an analysis of the strength of the cryptography used to secure information. Cryptanalysts continually evaluate the manner in which codes are broken in order to produce even more sophisticated ciphertext, which is often eventually broken, thereby fueling rounds of "cat and mouse." In the end, however, true data security is established not by the sole use of a highly sophisticated encryption technique, but rather by maintaining a minimum level of security for the entire operation.

In other words, it would be of little value to have the most sophisticated encryption available when copies of plaintext are routinely discarded in ordinary waste bins that are then discarded into publicly accessed dumpsters. The ironic aspect of this statement is that this happens far more often than one would imagine.

And while there is now a DES cracker that can be purchased for approximately $200,000 that can break DES keys in an average of 4.5 days, and a high-performance version that can crack DES keys in less than 24 hours, most data can be protected with a modicum of common sense and uniform adherence to basic security principles.

RC4

RC4 was developed in 1987 by Ron Rivest, for RSA Data Security, and was a propriety algorithm until 1994, when the code was posted to the Internet, and thus to the rest of the world.

RC4 is a *streaming cipher,* which is a cipher that encrypts messages of arbitrary size by ciphering either individual bits or bytes, as opposed to *block ciphering,* where blocks of 40, 64, or 128 bits are ciphered at a time. The classic streaming cipher is very simple and uses a random number generator on a bit-by-bit basis, which produces a random-like confusion sequence or running key.

Certificates

Certificates are a common concept in modern society. We use them for drivers licenses, for club memberships, and as identification. These items bind a public key to an individual, position, or organization. A certificate is a digitally signed statement from one entity saying that the public key of some other entity has some particular value.

Certificates also reside between browsers and servers. With regard to Internet security, certificates provide:

- Identification
- Date of expiration
- Issuing authority
- Serial or identification number
- Policies about how the user was identified
- Limitations on how the certificate may be used

Certificates are very useful on the Internet. As an example, when a shopper goes on-line to a commercial site and makes a purchase with a credit card, this begins an SSL/TLS encryption communication between the server and the user's browser.

The server responds to the certificate sent by the browser, which contains a public key and information including the private user's key. This transaction occurs at a network operations level generally only available to network administrators; in other words, this operation is transparent to the end users.

SECURITY CONCLUSIONS

Security is clearly an issue for today's Internet. It is important to distinguish between the BBFW link and wireline security similarities and differences.

In summary, there are many common elements to both BBFW and wireline security. One of the most common is that the higher the level of security, the greater the impact on productivity and information throughput. This has been the situation for decades and there does not appear to be a workaround at present. Security, by definition, should not be seamless; it must have presence, require extra steps, and all users, both authorized and unauthorized, must pass through certain filters in order to best maintain the integrity and safety of the LAN.

The corollary to this is that employees, along with private users, are reluctant to use security tools that inhibit their productivity or are too difficult to use. Yet, one of the most important things individual users can do is simply to protect their passwords and not exchange them. Another common and easy-to-perform security function is to not allow individuals into buildings and areas who do not carry the appropriate badges or entry devices such as keys.

A third concept for security that is easy to adhere to is to not discuss confidential company plans and information in public places. It is fairly common to overhear very detailed conversations about companies and individuals in very public places like airports, train stations, parking lots, outside buildings on campus grounds, and even in elevators. The author once overheard the entire business plan for a startup company while attending a professional sporting event, due to how loud the individuals were speaking.

For network administrators and technicians, the use of screensavers that require passwords to undo is another recommended asset in ensuring a secure CLI and deterring other unauthorized server access.

Security is a system. It is a system of processes, rules, and enforcement. Security requires layers and it requires updating systems and network elements to remain one step ahead of unauthorized users or the ability to use networks in an unauthorized manner. A balance of security and usability must be obtained so that people within the organization do not circumvent the processes put in place due to frustration with the system. But security must also be tight enough to prevent outsiders from intruding.

The discarding of company information should be reviewed carefully as well. U.S. law stipulates that information found in garbage once it has left the building becomes fair game, because it then resides in the public domain. The author once worked on a commercial project, approximately 20 years prior to writing this book, in which the president of the company visited a key competitor's site, rifled through their dumpsters, and returned with an enormous amount of information in the form of confidential memos, detailed engineering drawings, budget documents, and even employee lists. The author does not now, nor did then, condone this type of activity; it is merely an example of how easy it is to acquire detailed information on competitors or to provide means and opportunities for disruption.

Another concept for BBFW and network owners and operators to consider is that static security measures are easily and eventually overcome. Security measures must be dynamic and upgraded on a continual basis. Security requires a nominal budget, native or contracted expertise, and a permanent commitment to continually upgrade this important corporate or personal asset.

Ultimately, however, a large portion of the responsibility for security rests upon the individual users. With the concept that most hacking and unauthorized usage comes from within the walls of a LAN, there can be no greater issue than to somehow encourage and reward employees for supporting security measures, as well as to take appropriate action against those who willfully violate security measures. The most common security measures being deployed at the time of this writing are shown in Figure 11-4.

ADDITIONAL READING

There are many documents, forums, and companies with security expertise available in addition to what is being planned and ratified by various standards bodies. One document of the many worth studying is the *IETF Site Security Handbook* (RFC 2196), which illuminates topics such as various security policies that are appropriate for network users and service providers, and security architectures, a topic that includes services, firewalls, and topologies. The handbook is widely used, and includes information on authentication, authorization, access, modem usage, cryptography, auditing, and appropriate backup procedures.

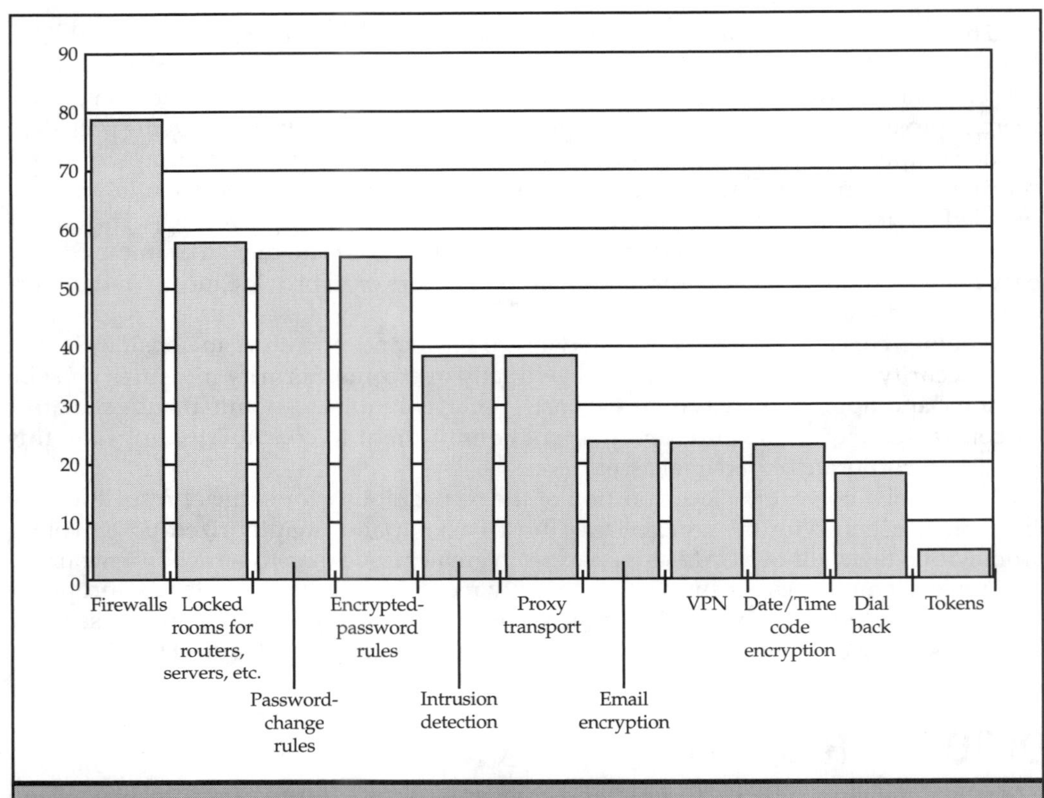

Figure 11-4. Most common security measures being employed

CHAPTER 12

Comparison of Unlicensed vs. Licensed BBFW

Service providers are scrambling to provide final mile access to bandwidth- and speed-hungry customers. Currently, approximately two-thirds of all the potential DSL customers are out of the reach of DSL-based service. Further, the DSL meltdown in areas within reach of the telco plants have generated considerable amounts of bad publicity on the miscues by the telcos attempting to deploy DSL. Among the concerns are the very long waits for installation.

The author in fact waited for nearly six months for a DSL line only to have the installation technicians arrive on the day he was relocating out of state. It took no less than three phone calls to the phone company to get them to discontinue further attempts to install DSL in what was by then a vacant residence. Other miscues involve shoddy DSL installations that commonly fail and require two and sometimes three visits by installation technicians to correctly install the modems. The DSL market misfortunes have placed even greater pressure on broadband providers to get service to both residential and commercial building sites.

ADVANTAGES OF UNLICENSED SPECTRUM

Given that over 80 percent of all of the MMDS spectrum in the United States is owned by Sprint and MCI WorldCom, that LMDS and other millimeter wave equipment is too expensive and suffers from poor range capabilities, and that license holders carry an enormous financial burden simply for the licenses themselves, which must be purchased before any equipment or other assets, unlicensed bands become an attractive frequency.

The U-NII band has an additional feature that is highly attractive to service providers: it does not require the purchase of a broadcast license. Spectrum licenses typically cost millions or even tens of millions of dollars. Further, licensed BBFW links must reside within tightly defined BTAs. Figure 12-1 highlights some of the expenditures made on spectrum licenses.

Additionally, companies like Sprint and MCI that have plans to deploy licensed links over enormous regions of the United States will eventually approach maximum market penetration in their BTAs. At that point, their options to maximize market penetration include purchasing additional spectrum from the government, or purchasing operating companies that own spectrum. In the latter case, U-NII would appear a compelling choice, because it does not require additional cost for the spectrum and, importantly, can be deployed in any strategic BTA selected by the company, thereby making it possible to have as many contiguous BTAs as possible. The strategy of using an unlicensed band also has significant advantages in terms of speedy deployment as there are no delays incurred for negotiating or bidding.

Having BTAs that are contiguous offers significant advantages in terms of backhauling, because the links can be connected to fiber nodes that reside along a common fiber route. Further, having contiguous BTAs reduces the manpower required to physically maintain a network. Finally, this strategy enables companies with licensed spectrum to capture the majority of a common metropolitan area within which reside numerous BTAs.

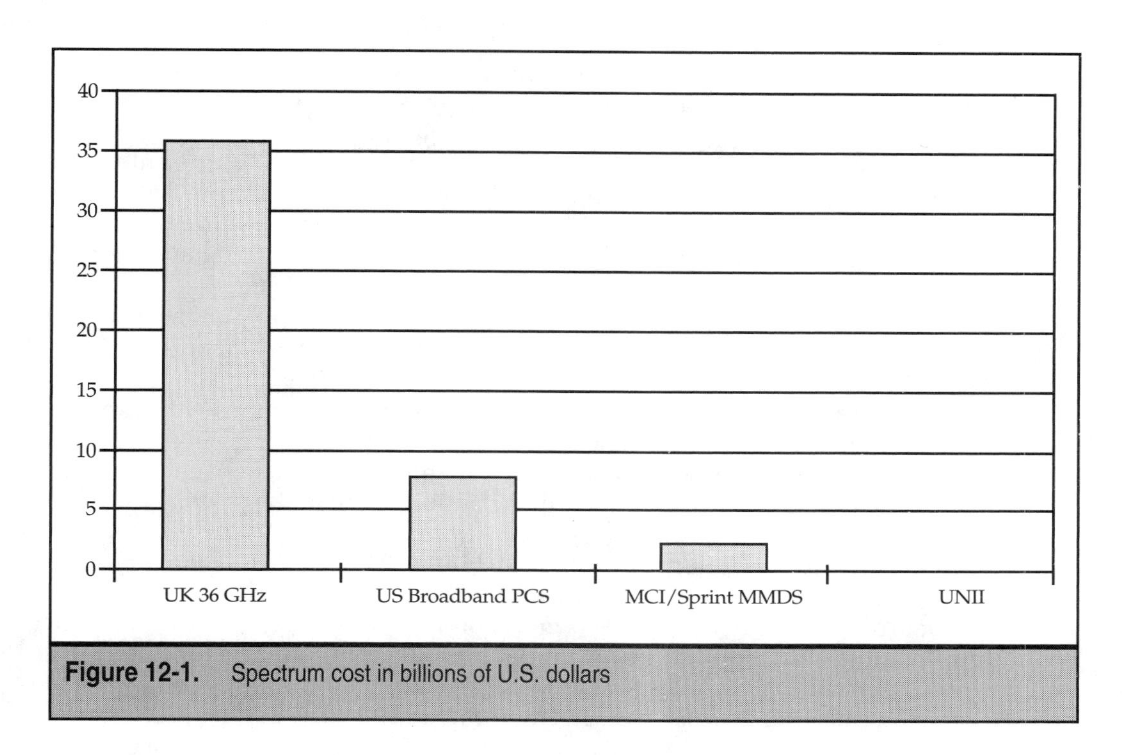

Figure 12-1. Spectrum cost in billions of U.S. dollars

There is no penalty for using the U-NII band as opposed to a licensed frequency in terms of how it can be used by the service provider. U-NII band service providers utilize policy-related features identical in every manner to those that can be offered on a licensed link. This includes QoS, voice, throttled bandwidth, and identical element and network management resources.

Further, as the U-NII band is well below 10 GHz, it is not subject to restrictions based on weather or air pollution. It also has fewer multipath characteristics than the MMDS frequencies, especially given that it operates at a lower power. It can also operate at the same reliability and availability as licensed frequencies; in summary, the U-NII band has all of the benefits of a licensed link without the overhead cost of a license.

Finally, the purchase of an MMDS or other license can be delayed through negotiations, litigation, and variables such as purchasing the license from an entity in bankruptcy. State-of-the-art BBFW systems have indoor gear that is independent of the carrier frequency. This means that the user can install U-NII until the purchase of the licensed spectrum is made available. When the licensed spectrum becomes available, the service provider can either redeploy the U-NII equipment to another site or leave it in place while bringing the licensed equipment online.

While it is true that the 2.4 GHz frequency is being used as an ISP medium, it isn't truly a broadband option in this author's opinion due to the fact that the 2.4 GHz spectrum is very crowded and has less frequency available than U-NII. Further, there is a lack of BBFW equipment in the 2.4 GHz band that makes it competitive to offerings in the U-NII band. While the 2.4 GHz offerings will be covered in detail later in this book, the 802.11a frequency-based equipment should not be confused with the 802.11b equipment, as the former is a 5.7 GHz-based product line and the latter is a 2.4 GHz-based product line.

There are also other bands that are *quick licensed,* which is to say that licenses can be purchased for very little money (on the order of approximately $5000). They generally can be acquired within approximately 30 days of application. The majority of these licenses, however, are limited to point-to-point links. The interested party initiates this engagement by contracting with a site survey company, which then performs a site survey between the two end points.

The GPS addresses, along with the site survey results, are then sent to the FCC, and if there are no other licenses granted in the area, the party is granted a license. A grandfather clause takes effect then, which means that no other entity will be granted a license in the same frequency if they are within radiating range of the party already up and running.

These frequencies in the United States are at 6 GHz and 23 GHz. These frequencies are not included as traditional BBFW links as used by service providers, because they are not to be used for profit and are not for point-to-multipoint links. In other words, they are for connecting campuses and buildings over ranges from 20 or more miles for the 6 GHz equipment to approximately 2 miles for the 23 GHz gear.

There are four major unlicensed bands in the United States. The ISM band covers the spectrum between 2.4 GHz and 2.485 GHz and will be covered in an upcoming chapter. The U-NII spectrum covers three separate allocations, which are indicated in Figure 12-2.

Unlicensed band users in the United States primarily have Apple Computer and the Wireless Information Networks Forum (WINForum) to thank for the existence of U-NII. WINForum requested 250 MHz of spectrum at 5.10 to 5.35 GHz in May 1995. At approximately the same time, Apple Computer requested the allocation of 300 MHz in the 5.10 to 5.35 GHz and 5.725 to 5.875 GHz ranges to establish a new unlicensed radio service to promote full deployment of the U-NII.

The 5.725 to 5.825 band was already in use by amateur HAM radio users, but there hasn't been much publicity about the decision, and with rare exception, U-NII service providers and HAM radio operators seem to coexist peacefully and without interfering with one another. Part of this may be due to the fact that HAM radio operators tend to use their radios in the evenings and on weekends, while most U-NII users at present are in the MxU market space and therefore use the frequency primarily during normal business hours.

This effort to lobby the FCC in the United States to make the two lower U-NII frequencies available took approximately two years of effort and was formally introduced to the public in January 1997. The upper U-NII band (5.725 to 5.825 GHz) was already authorized by the FCC in the United States under its Part 15 Rules.

One of the interesting aspects of the FCC allocation is that it did not specify channeling plans (how the radios break the entire spectrum into subsections), spectrum modulation

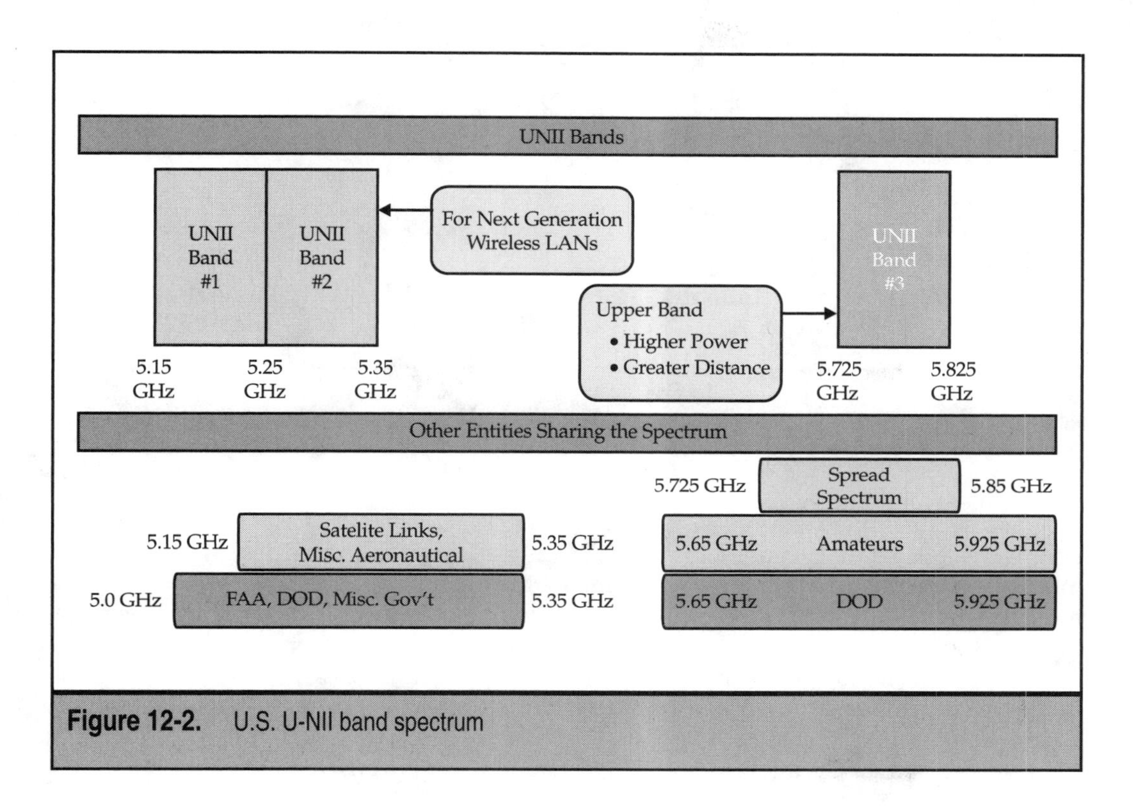

Figure 12-2. U.S. U-NII band spectrum

efficiency requirements, or even spectrum etiquette. Instead, the FCC focused only on limiting the power output and power spectral density (a definition of how many watts of power per hertz of spectrum can be used, or, in other words, how much energy per single burst of frequency can be used). Rather, it left this up to the industry to resolve through technological and community-based diplomatic means.

The initiative by Apple was originally called "Unlicensed Personal Communications Services" (Data-PCS), and was intended not only to provide classroom wireless connectivity to the Internet, but also to have what Apple called *community networks*. It is an essential part of what is now called the National Information Infrastructure, which is a subset of the Global Information Infrastructure (GII).

One portion of the band (5.150 GHz to 5.350 GHz) was already earmarked in Europe for a proposed standard called *HyperLAN*. The U-NII band allocation incorporates the European standard and will allow its use in the United States. It also means that U.S. designers and manufacturers can offer competitive products for the European market. The upper U-NII band is already in use in the United States, Canada, and Australia.

Table 12-1 outlines the summary characteristics of the U-NII band.

	Band 1	Band 2	Band 3
Frequency (GHz)	5.125 – 5.25	5.25 – 5.35	5.725 – 5.825
Radiated Power	50mW	250mW	1W
With 6dB Antenna Gain	200mW	1W	4W
Applications	Indoor	Campus and small neighborhood	20 miles P2P; ~6 miles P2MP

Table 12-1. U-NII Bands Summary Specifications

Band 1, sometimes referred to as the *low U-NII band,* was designated primarily for indoor use, while Band 2, sometimes referred to as the *middle band,* was designated for campus use (in other words, building to building within a campus). Band 3, sometimes referred to as the *upper U-NII band,* was designed for community networks, which through development and substantial processing gain, in addition to modulation schemes such as VOFDM, now has the ability to cover distances up to or exceeding 20 miles in a line-of-site environment or 3 to 5 miles in a non-line-of-site environment, depending on the nature of the obstructions.

The reason the FCC allocated the U-NII bands was to promote the following:

- Sharing of spectrum
- Mobility of wireless applications
- Efficient use of spectrum through intermittent use
- Innovation and research for new fixed wireless devices and applications

SHARING OF SPECTRUM

One of the key differences between the key BBFW resource (spectrum) and most other mediums, such as those transmitted by copper or fiber, is that spectrum is very limited. By offering a small portion of the spectrum that many people can use in relatively common geographies, the government preserves other spectrum blocks.

The sale of radio spectrum is highly valuable to a government, and billions of dollars are generated for the government from the sale of spectrum to commercial interests. Providing even a limited amount of spectrum to the commercial industry that does not generate revenue for the government has the effect of placing a prospective premium on the balance of the spectrum simply because there is less total spectrum available.

The way the FCC in the United States authorized the use of the U-NII band, and by intentionally providing it with as few guidelines as possible, also reduces the governmental costs of monitoring users and hearing and resolving complaints. It also reduces the cost of government participation because the government is less likely to have to pass future laws that feature greater specificity and penalties for use outside the intended boundaries. The continual addition of rules requires a somewhat proportional increase in applying penalties; both of which increase the cost to the government of maintaining this asset.

MOBILITY OF WIRELESS APPLICATIONS

The U-NII band is ideally suited for both broadband mobility and pervasive computing. Broadband mobility is a feature that enables users in an office environment to move their laptops or even office phones from location to location. This is possible because U-NII has relatively good propagation characteristics; that is, it has a decent range for a given power and can penetrate some walls at a close distance to the transmitter, and yet it does not generate as much multipath flow as transmissions at lower frequencies.

EFFICIENT USE OF SPECTRUM THROUGH INTERMITTENT USE

One of the intents of the FCC was to enable the use of the spectrum through intermittent use. This can be overcome by using spreading or other modulation techniques, and it forces the equipment providers to advance research and product development that enables the users to fill the airspace with transmissions at the minimal possible amount of time.

INNOVATION AND RESEARCH FOR NEW FIXED WIRELESS DEVICES AND APPLICATIONS

When this spectrum became available, industry gambled on it being valuable to the service provider and other industries. This set off a wave of development and innovation that not only is applied to this spectrum but also often crosses over to products in other frequencies and products that can be associated with U-NII radios such as PCMCIA cards, wireless bridges, and antennas.

The drawbacks for U-NII include the following:

- High potential for interference
- Coexistence with other U-NII networks

HIGH POTENTIAL FOR INTERFERENCE, AND COEXISTENCE WITH OTHER U-NII NETWORKS

State-of-the-art U-NII systems have an array of "tools" that can be implemented as required for co-channel resolution. These methods are numbered as follows:

- The transmission and reception frequencies can be adjusted at the command line interface (CLI) of the router a total of 12 MHz for the 12 MHz channelized unit or 9 MHz for the 6 MHz channelized unit to avoid interference from other U-NII broadcast sources.
- Interference is more of an issue in the area local to the receiver than in the area local to the transmitter. To accommodate this phenomenon, the duplexor can be inverted at ODU (on both ends of the point-to-point data link) so that the transmit/receive frequencies are inverted to provide frequency spacing from the competing source.
- The duplexor can be replaced on P2P links. A common duplexor configuration splits the U-NII band into two 48 MHz sections allowing some flexibility to the network deployment. Cost of replacement for a duplexor is about $300 USD, and the replacement is accomplished in about ten minutes per ODU.
- The packets include a header address that enables the system to discriminate native transmissions from non-native transmissions. This process is accomplished by the digital signal processor portion of the wireless line card.
- Obstructions (buildings and foliage) can provide substantial protection from competing U-NII transmissions. Transmission power drops by the square of the distance in a line-of-sight link, and drops even faster when obstructions are present (by the fourth or fifth power of distance). Therefore, even if non-native but distant transmitters happen to be in the beam path of the native receiving antenna, obstructions will reduce their signals to insignificant levels in a great many cases. OFDM enables Cisco RF network users to deploy antennas at relatively low elevations, thereby using obstructions to even greater advantage.
- The statistical probability is very slight that two competing U-NII links will operate within a few hundred feet of each other. If they are located that close, refer to point 2 above.
- Error correction is a big part of any commercial radio system. Cisco's system uses both convolutional encoding and Reed-Solomon block encoding to recover bits lost due to interference. These codes are concatenated for even greater interference immunity than afforded by either type of encoding alone.
- OFDM and VOFDM resolve for co-channel interference. For example, Cisco's VOFDM system combines robust forward error-correction coding with time and frequency interleaving for great immunity to narrowband interference and bursty interference.

- The difference between OFDM and VOFDM lies in the use of a second antenna at the receiving end of the data link. The *V* in this case stands for "vector," which is another way of stating "spatial diversity," meaning a second antenna placed at the receiver (diversity is not deployed at the transmitter, although the concept exists and is being developed). Use of the second antenna is how you interleave the time element of a reception (signals arriving via multipath arrive on the order of four to six microseconds later than the intended signal).
- VOFDM systems offer multiple modes of operation with different amounts of error correction and interference immunity. High-robustness modes may be chosen that use extra error-correction coding and require lower signal-to-interference margins. High-robustness modes allow line-of-sight interference sources to be 60 percent closer to each other. However, this trade-off is not without compromise. As robustness is increased, other compromises such as link speeds or distance must be balanced as part of arriving at the best overall design.

As indicated in Figure 12-1, it is the upper U-NII band that companies like Cisco and others have designed into their BBFW gear. The middle and lower bands are at present being used for indoor WLAN use.

U-NII-based BBFW links perform as well as or better than those in the MMDS spectrum and those in the "B" LMDS-licensed spectrum (which has 300 MHz of spectrum). U-NII customers can receive speeds ranging from 128 Kbps up to 45 Mbps depending on equipment type. A report entitled "U.S. Fixed Wireless: Unlicensed Spectrum" by The Strategis Group, dated March 8, 2001, states: "Despite billions of dollars poured into U.S. broadband wireless licenses, unlicensed wireless technologies maintain a clear lead in terms of deployments to date."

This comes as no great surprise to those close to the industry, especially when one considers that, at least in the United States, the majority of the optimal frequency for a nationwide deployment is owned by only two companies as of this writing, Sprint and MCI WorldCom. The combination of these two entities owns more than 80 percent of the total amount of MMDS license in the United States.

At the time of this writing, both of these companies have been bogged down with numerous deployment issues, including the selection of a radio that meets all of their requirements, sufficient deployment resources, and sagging economic market conditions that place a strain on deployment budgets and customer acceptance rates. At the same time, both Sprint and MCI WorldCom face considerable competition from a very wide array of smaller competitors that are more nimble and do not have to amortize the enormous overhead associated with purchasing fixed wireless licenses.

The total amount spent by both of these companies is in excess of $2 billion (refer to Figure 12-1). The bulk of the licenses owned by these companies were acquired through the acquisition of a large number of smaller licenses; in other words, both companies adopted the strategy of purchasing the rights to MMDS in as many BTAs as they could possibly muster. The competitors to Sprint and MCI WorldCom vary in size from a very wide array of smaller startups to multibillion dollar companies such as AT&T and SBC.

Although the amount of spectrum available in unlicensed channels is less than in their licensed alternatives, depending on the equipment purchased and the area covered, U-NII links can outperform licensed links. Figure 12-3 provides a snapshot of the amount of spectrum available in the U-NII band compared to other common allocations.

Many myths exist regarding how unlicensed BBFW links compare to licensed BBFW links. To clear up some of those myths, Table 12-2 provides a snapshot overview of the comparison of the two approaches. Each of the issues is discussed in more detail in the following sections.

100 Percent Availability?

It is virtually impossible to guarantee 100 percent availability over an extended period of approximately 12 months because there are so many elements both in the radiating environment and with the indoor and outdoor equipment. A simple lightning strike would certainly damage or destroy outdoor equipment, and could perhaps destroy some indoor equipment, such as cabling and lightning arrestors.

There can also be failures in the indoor equipment, such as with the modem cards that are used in some systems, and even outages on networks with single points of failure, such as those that do not have auxiliary power supplies that provide uninterruptible power in the event of a power outage to the network.

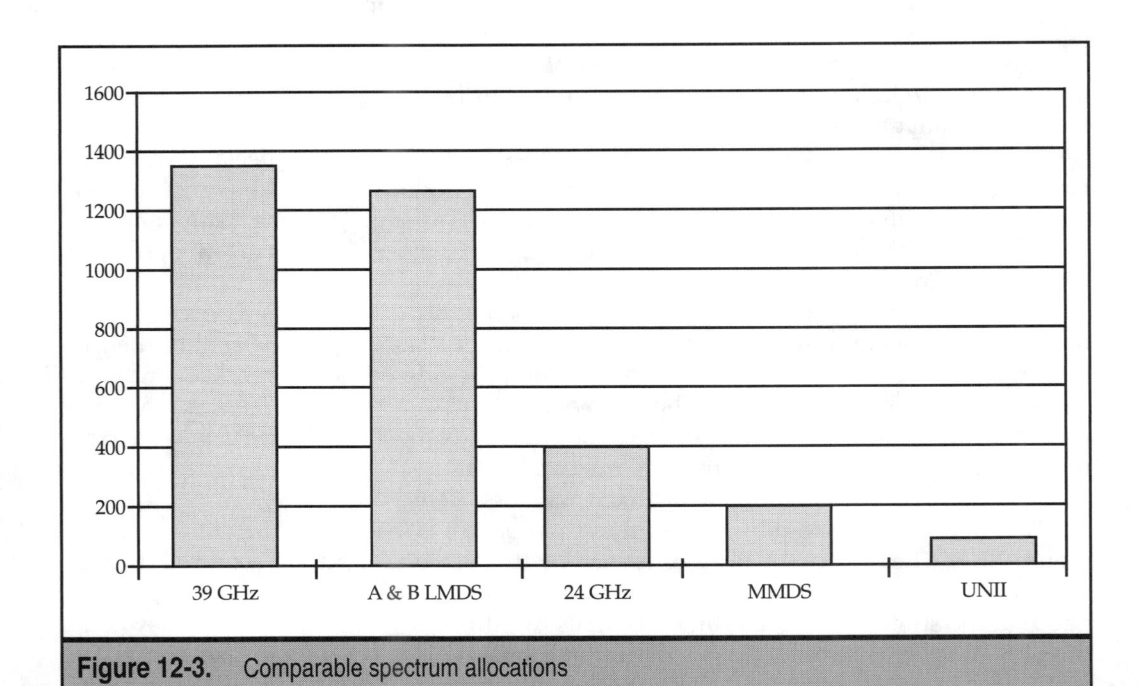

Figure 12-3. Comparable spectrum allocations

Issue	Licensed	Unlicensed
100% availability?	No	No
High pre-equipment investment?	Yes	No
Guarantee from interference?	No	No
Guarantee from link degradation?	No	No
Restricted footprint?	Yes	No
All Layer 3 options?	Yes	Yes
Recourse for co-channel interference?	Primarily legal/diplomatic	Technological; nine different ways to mitigate
In-band market competition	Limited	Yes

Table 12-2. Snapshot Comparison of Licensed vs. Unlicensed BBFW

Events other than network failure can have a significant impact on the overall availability of the network. While most multipath challenges can be accommodated through careful link planning (incorporating sufficient performance margin into the link plan to compensate for random drops in link performance that can occur without notice) and monitoring, there can also be such catastrophic link events as a building being placed directly in the path between the two antennas. This type of event could block the signal entirely or dramatically change the multipath properties, which would require either a recalibration of the equipment or relocation of the outdoor equipment, specifically the antennas. BBFW links are also subject to acts of nature such as fire, earthquakes, tornadoes, and so forth. Neither licensed nor unlicensed links are immune to the effects of one or all of these phenomena.

The catastrophic effects of these items can be mitigated through the use of strategies such as *path diversity*, which is the concept of having two independent paths between two sites. This of course requires independent sets of BBFW gear, as the links are physically separated from each other and can take very different paths. Layer 3 radios can harness path diversity by load balancing at all times when both the primary and backup BBFW links are operating normally.

Increased antenna height, extra gain (sensitivity), and additional link margin (additional link performance margin) can also be utilized, but care must be used not to overcompensate for events with a statistically low probability of occurrence. Excessive insurance measures lead to unnecessarily lower levels of performance. There are also element redundancies that can be built into the links, but ultimately, there is no guarantee that can be issued that stipulates a link will never go offline or incur substantial bandwidth or speed degradation.

High Pre-Equipment Investment?

One of the advantages to having an unlicensed system is that there are no upfront costs associated with purchasing a license. However, upfront costs for evaluating the environment to ensure that the spectrum is clear and available for use should be anticipated in virtually all link deployments. MMDS, LMDS, and other licenses are very expensive and can easily run into the millions or tens of millions of dollars.

Using an unlicensed BBFW link means that fiscal resources can be used directly and immediately on equipment, deployment, and network equipment purposes, as well as for making the necessary investments in personnel and overhead costs such as office sites and so on.

Guarantee from Interference

It is a widely held myth that licensed BBFW links are immune to interference. What they are intended to be immune from is co-channel interference from competitors that are broadcasting in the same frequencies.

BBFW links, both licensed and unlicensed, can be interfered with from the effects of multipath and from high emission sources such as welders used during construction that may be either adjacent to the site or in the site path near one of the antennas. Adjacent channel interference can also be a problem for both types of links where a broadcast does not occur directly over the licensed or unlicensed frequencies, but rather occurs adjacent to those frequencies.

Guarantee from Link Degradation

As stated previously, the probability of link degradation increases over time as the radiating environment changes through which the information is transmitted. Adding to that potential are the effects of water seeping into connectors, damage to cables from activities not associated with the operation of a BBFW network such as construction to the NOC building, resurfacing of roofs, and so on. The multipath environment may also change, which can readily cause link degradation.

The summary of this issue is that a careful establishment of link baseline performance and ongoing monitoring and adjustments thereof will likely eliminate or at least reduce the issues of link degradation. This does not mean that all links in every location suffer degradation as a matter of course; rather, the probability of a link degrading over time, specifically in a metro environment, is high enough that it warrants the constant monitoring of link margins and performance as well as the accumulation of an archived performance record.

Restricted Footprint

One of the most significant differences between a licensed BBFW link and an unlicensed BBFW link is that the licensed link must comply with strict parameters set forth by federal agencies with regard to where it can broadcast. However, in the interest of a balanced presentation, it should be stated that federal regulators place tight restrictions on the output of unlicensed systems.

Unlicensed bands such as U-NII and ISM in the United States generally have no such restrictions, although they must not broadcast in areas or in a manner that causes interference to other users of the unlicensed band, some of which are users of amateur radios. The essence of the unlicensed bands is that they are unregulated, but what that truly means is that they have fewer compliance issues with regard to where, how, and what they broadcast.

Both types of equipment require acceptance testing by laboratories that are licensed by federal agencies, to ensure compliance with the radiating output, both intentional and unintentional. As all electronic devices radiate energy, this energy is measured both in terms of frequency and strength to ensure they do not interfere with other forms of electronic equipment.

Layer 3 Options

The OSI Layer 3 aspect is frequency-independent, which is to say that it occurs at the router as opposed to within the radio. Therefore, both licensed and unlicensed transmitting equipment can operate as a routed link, as opposed to a hub or switch. Cisco, for example, accomplishes this by integrating the radio modem card into a router, both at the HE and the CPE ends of the link in point-to-multipoint links, and at each end of a point-to-point link.

Recourse for Co-Channel Interference

In a licensed link, the primary recourse for broadcasts that occur in the same frequency and in the same BTA is diplomacy, with any luck, but most likely will be resolved by litigation, because one or both of the parties will have paid substantial sums or at least made substantial commitments for the right to broadcast in a reserved frequency. This adversarial approach is expensive and time-consuming, with one party ending up the clear loser, and the only party truly winning being the attorneys, having gained financially through the dispute.

With unlicensed BBFW links, however, the resolution lies within the technology of the units themselves, in the case of state-of-the-art systems, and because the frequencies are essentially unregulated, the opposing parties are more likely to resolve the matter diplomatically. This diplomatic resolution would be based on a common agreement to share the spectrum in such a manner that they would not interfere with each other. Figure 12-4 illustrates how the U-NII band might be broken up to accommodate two parties.

Two parties can readily agree on how to allocate frequency in a common geographic area, and they can also agree to divide a larger common geography among them.

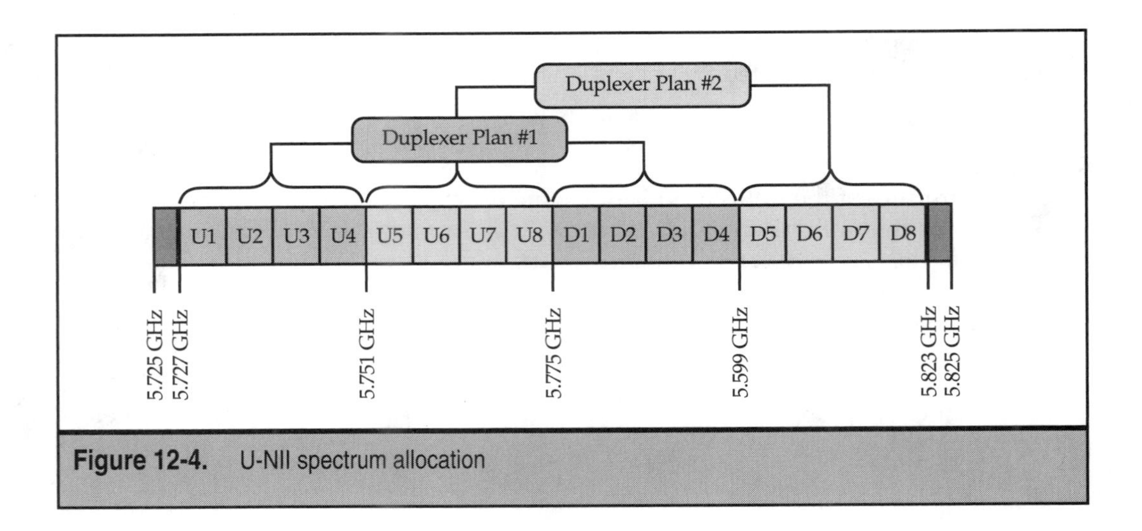

Figure 12-4. U-NII spectrum allocation

In-Band Market Competition

By definition, a licensed band operator has no competition within one or more BTAs *in the same frequencies*. While they are free from competition in the same BTA for a given frequency, such as MMDS, they are not free from competition from the U-NII- and ISM-based service providers, who, with today's equipment, can provide the same services at the same speeds to the same customers. Nor does it release the licensed bandholder from the issue of incumbent transmitters in the licensed band. This is a very real issue in the MMDS band, where ITFS one-way video exists in many communities.

NOTE: Neither ISM nor U-NII is typically competitive with LMDS, as LMDS has much more spectrum with its "A" license. LMDS "B" license spectrum, which is much more on the order of MMDS and U-NII, may find competition within its BTAs from service providers capable of offering services in the unlicensed bands.

The 802.11b ISM band, while commonly used for data by ISPs, generally at the time of this writing does not carry voice traffic and is quickly running out of favor with ISPs and other users of that frequency because it has a steadily rising number of competitors in that frequency. This issue is treated in detail later in this book. The 802.11b band, however, will likely be found to be very competitive with U-NII, MMDS, and the LMDS "B" block of spectrum.

U-NII AS FINAL MILE FOR FIBER

One of the architectures being touted by some of the major equipment providers is connecting fiber nodes to MxUs. While fiber trunks are virtually ubiquitous in every major metropolitan area, trenching a fiber connection between a trunk and an MxU site even over short distances requires no less than 90 days and can take as long as six months or more at a cost that can exceed $1 million USD per mile.

Connecting BBFW to a fiber node to a BBFW HE, which then communicates to either a point-to-point device at an MxU site or an array of point-to-multipoint CPEs, is a compelling architecture. This solves a number of problems rather elegantly:

- The fiber provider can't access large buildings fast enough
- The BBFW service provider often requires a large backhaul pipe
- The tenants in the large building wish to bypass the local telco exchange

The Fiber Provider Can't Access Large Buildings Fast Enough

One of the largest problems for fiber providers is increasing their rate of market penetration. Fiber is virtually always laid in areas where there is existing competition from DSL, and it has significant advantages over DSL in both cost and speed. Trenching is expensive and time-consuming, and it is reasonably straightforward for a BBFW HE to connect to a fiber node. This architecture provides immediate access to large buildings within a three to five mile radius, assuming there are line-of-sight issues. For point-to-point architectures, assuming the service provider has access to suitable elevation, buildings can be accessed at distances of 20 miles or more on a point-to-point basis.

The BBFW Service Provider Often Requires a Large Backhaul Pipe

One of the largest challenges for BBFW service providers is aggregating backhaul for either P2P or P2MP architectures. Having access to fiber nodes virtually eliminates this problem and also eliminates the need to lease copper from the local telco exchange, which has a material effect on the profitability of the service providers; that is, fiber is generally less expensive to lease than copper.

The Tenants in the Large Building Wish to Bypass the Local Telco Exchange

MxU tenants using the local telco exchanges for DSL broadband typically pay a premium for their speed and for speeds that are typically less than those afforded by a BBFW/fiber hybrid architecture. The cost savings are significant and the transfer to BBFW/fiber access can occur in a very short time, often less than a week.

The speed at which a BBFW link can be deployed is occasionally the basis for a business plan by some companies as a very high reliability broadband circuit from the telco's can take 60 days or longer to install as compared to the very short time in deploying a broadband link, which can be anywhere from half a day to several days once the access to the buildings is granted and given that only a minimal amount of construction is required to install the antenna masts.

U-NII FOR EXTENSION OF DSL AND CABLE

One of more common uses for U-NII is and will continue to be the extension of service areas by DSL and cable modem operators. A P2P link can be installed at the edge of a DSL or cable service area and the service can be extended to other areas as far as 20 or more miles away. This eliminates the need for trenching or leasing copper lines from a third party, which would require a monthly ongoing fee.

Often, BBFW point-to-point links can amortize in the 18 to 24 month range. While this is sometimes a slightly longer amortization schedule than some companies desire, once the equipment is paid for, the monthly cost of operating the equipment is significantly reduced.

The DSL and cable modem operators can also use a BBFW point-to-point link as an interim medium until they have other copper or trenched assets in place. The BBFW equipment can either be redeployed or used as a redundant link to ensure highly reliable service.

VENTURE CAPITAL ASPECTS OF UNLICENSED VS. LICENSED SPECTRUM

The issue of venture capital (VC) is well within the scope of comparing unlicensed to licensed spectrum because financing of unlicensed links would not be likely without the advent of unlicensed links. However, there are elements associated with capitalizing unlicensed links that are unique to that sector of the BBFW industry.

There is an increasing level of communication between the industry that provides complex communications equipment and the VC industry. This stemmed originally from the fact that virtually all of the major Internet equipment companies are publicly traded, and the financial industry played an active role in observing the industry to gather intelligence on which they based investments.

However, even with the largest companies such as Cisco and Lucent, the great preponderance of the sales are not multibillion dollar ventures, but rather are much smaller than that and commonly in the millions or tens of millions of dollars.

Currently, even Cisco makes most of its revenue from the enterprise market, and it's a rare enterprise deal that goes beyond the tens of millions of dollars. This will be changing with the newly and rapidly growing utility companies adding Internet equipment to their lists of assets in order for them to become telecommunications players, but most of the revenue for Internet equipment companies currently comes from relatively small deals.

What this means to the VC community is that a considerably large number of deals are made with VC funds from the tier-two and tier-three venture capital companies, as opposed to the major houses like Smith Barney and others. VC firms and, to a lesser degree, formal banking institutions are routinely engaged by emerging service providers and enterprise businesses to finance BBFW operations.

It is the author's opinion that the equipment manufacturers, the largest of which are also common sources for financing of equipment, must play an ever-closer role to the VC and banking communities. This not only will better facilitate the funding of emerging companies, but will also help prime the pump in terms of getting greater amounts of BBFW U-NII links into the marketplace and online. It will also assist the emerging companies in making a transition from VC sourcing to routine and conventional banking sources, which are generally more cost-effective and less onerous to the emerging companies.

There are four major entities required to make a telecom venture operate. While there are arguably additional entities, such as deployment and integration entities, the primary ones are the following:

- Equipment manufacturers
- Service providers or enterprise customers
- Retail, or, end users
- Financiers (the venture capitalists and bankers)

Of these four entities, the equipment manufacturers, in particular the smaller ones, are generally the first to advance the state of the art in equipment. Either these smaller entities are purchased by the larger entities or the larger entities build similar equipment. The people in this industry fully understand not only the technological advantages of their equipment, but also the economic and social ramifications brought about by their technology offerings.

There are typically two types of service providers and enterprise customers: the early adopters of technology, and those who wait a year or so until the technology has been proven. Service providers and enterprise customers—both of the early adopter mentality and the more conservative variety—are in constant contact with the equipment providers and, therefore, also are well informed about the economic and other potentials afforded by state-of-the-art equipment. The service providers and enterprise customers of the equipment providers commonly provide guidance in terms of technology features, price, and other key aspects. The concept of early adopters will be covered in greater detail in the upcoming chapter on Standards.

This leaves the remaining two entities of the four previously listed, both of which generally are not technically literate, or at least are late in coming to the table in terms of gaining literacy on U-NII BBFW.

Quite arguably, end users should find the technology as transparent as possible, and not require a deep or even cursory knowledge of BBFW U-NII equipment. Their focus should primarily be on the services and traffic carried by the system, as opposed to having a deep technical knowledge of the equipment. It is important to remember that it is this entity that ultimately pays for the equipment and services, because no company can be grown on venture capital alone; it must be grown from the profitable engagement of end users.

The final entity of these four is the VC industry and, in general, it arrives last or second to last in terms of being literate on both the U-NII equipment and how the technology translates to profitable business strategies.

On one hand, the mission of VCs is to successfully invest funds from which they grow their capital base and then redeploy it in similar or other ventures. On the other hand, they must possess at least a working knowledge of the technology they are interested in. This is more difficult than it seems, as the breadth and depth of knowledge required to fully understand a technology or a technology strategy are quite deep.

The VCs are also compelled to move very quickly, and therefore often do not have the time to truly understand BBFW technology in general, much less the differences between funding a licensed operation versus an entity that wishes to use U-NII as its primary carrier band. While VC firms are continually increasing the numbers of personnel with backgrounds in engineering and technology management, most of them currently only have personnel with business backgrounds. This requires the manufacturers as well as the service providers or enterprise players to have uniquely talented personnel who understand how to reduce key technological elements such as VOFDM to straightforward business operating terms that are native to the investment community.

The author has spent many hours with venture capitalists from all sizes of VC firms, and has found the bulk of them substantially lacking in understanding the value of a U-NII-based BBFW operation. The VC community lacks expertise in the BBFW unlicensed market for the following reasons:

- BBFW is still an emerging broadband medium
- VCs view a broadcast license as an asset
- Resolution of co-channel interference
- Competition from numerous U-NII regional service providers
- Underestimating the value of management execution over frequency selection

BBFW Is Still an Emerging Broadband Medium

Most VC firms are still catching up to the fundamental concepts of how to run a copper- or fiber-based broadband operation. The recent market crash of the spring of 2001 occurred during the writing of this book and it is clear that at least part of the blame resides with the VC community, as they made investments in BBFW and other businesses with poor business plans, weak management teams, and, in too many cases, virtually no BBFW expertise.

BBFW is, at this writing, still very much an emerging technology. The unlicensed version of this medium is only one of the latest aspects of a medium that, at best, is not well understood by the VC community. In other words, the VC community has had to work hard at becoming conversant in copper- and fiber-based broadband technologies; they're generally even farther behind in their understanding of BBFW.

They View a Broadcast License As an Asset

VCs rightfully view a broadcast license, such as that for MMDS, as an asset. Again, these assets are tangible in that they can be sold and have a market value for many years. A U-NII operator has no such asset, but the advantage there is that capital can be applied directly to the deployment of assets that generate revenue, such as equipment, personnel, and right of way access.

Resolution of Co-Channel Interference

State-of-the-art equipment can mitigate the issue of co-channel interference with a very high degree of resolution, as noted previously in this chapter. Perhaps no issue is less understood by the VC market than this, and it is this issue upon which the author has spent most of his time briefing VC personnel.

Competition from Numerous U-NII Regional Service Providers

This is an appropriate concern for VCs but one that is commonly overestimated because demand for broadband is pent up. The primary question currently asked of telcos at this writing by customers is "When can I get broadband to my home?" This translates to a high market penetration rate for the first U-NII providers. If the service provider can retain a reasonably high degree of customer satisfaction (though easier said than done), there is a minimal risk of churn, primarily because once users experience broadband, they find it difficult to return to dial-up speeds and Internet access that is not always on.

This is one of the key reasons why U-NII providers should target markets in which there is no broadband service, and do so where there is not likely to be competition from copper within 12 to 18 months. The smaller the geographic market, the better, and the farther from DSL and cable modem service, the better. This translates to distances of at least 20 to 30 miles at a minimum.

Finally on this point, there is more market opportunity than can be serviced by even the largest service provider, and having BBFW competition in the same region actually validates the concept as opposed to eroding customer bases.

Underestimating the Value of Management Execution Over Frequency Selection

A U-NII-based BBFW operation is an execution operation, which means that the potential for success is more incumbent on a winning and experienced management team that is well funded than on having either a licensed BBFW asset or an unlicensed BBFW asset. Even the largest and best-funded BBFW players like Sprint and MCI are struggling, while many smaller BBFW service providers are thriving. As stated earlier in this chapter, according to a March 2001 report by The Strategis Group, unlicensed BBFW deployments far outnumber licensed deployments.

For the reasons outlined in this chapter, one of the most important elements in the worldwide BBFW market will be the use of unlicensed spectrum links either as the primary BBFW medium or to augment licensed BBFW architectures. In spite of its current relative obscurity at this writing, especially given the headlines generated when companies pay billions of dollars for a spectrum like MMDS, the U-NII band is poised for considerable press exposure. For virtually all but a very few service providers, U-NII is the only spectrum available.

CHAPTER 13

802.11 Wireless Local Area Networks

A fact not commonly known to many people outside the Internet industry is that most of the worldwide Internet is yet to be built. This premise is based on two principles: China, as the world's most populous country, has not been built out, nor have second-world countries, and most of the Internet will be comprised of final mile technologies and access mediums that cover the final "400 feet," which is an approximation intended to refer to the LAN, or in this case Wireless Local Area Networks (WLANs).

To better understand this concept, consider as an analogy the number of freeways and main highways that exist compared to the number of small side streets, and the number of side streets compared to the number of driveways and entrances to individual condominiums and single-family homes. There are far more driveways than streets, and far more side streets than freeways or main highways.

WIRELESS LOCAL AREA NETWORKS: THE FINAL 400 FEET

WLAN technology and products represent the driveways in the aforementioned analogy. Most of the Internet "driveways" have yet to be built. Indeed, relatively very few have been built, and that is part of the reason why WLAN is so exciting; it will become the part of the Internet most people will experience, because that will be the part of the Internet through which they interface the world at large via an online experience.

Prior to WLAN, the LAN had to be created. This came in two primary protocols: Token Ring LANs and Ethernet. While there are many Token Ring LANs still in existence, it is the Ethernet standard that has become the de facto standard, in the opinion of this writer.

Ethernet was developed by Xerox at its Palo Alto Research Center (PARC), which also happens to be the birthplace of Apple Computer's graphic user interface (GUI), along with many other distinguishing accomplishments, many of which we scarcely give a thought about in today's computing and networking world.

Ethernet is largely the standard in today's LANs, although there is still a sizeable deployment of Token Ring. Ethernet enabled companies and groups of people to communicate to each other via a set of wires that were threaded through buildings. This is still essentially the standard way corporations and large groups of people communicate. What was needed was the ability to have the same connectivity to LANs, but without the requirement of being tethered to an Ethernet port. This was the impetus for the wireless LAN.

WLAN radios provide two mission-critical networking capabilities to the Internet:

- ▼ Broadband mobility
- ▲ Pervasive computing

Broadband Mobility

Broadband mobility is defined as the ability to engage a network at rates in excess of 1.5 Mbps. 802.11 radios accomplish this while moving from one physical location to another as long as the user remains in an area where the RF signal from that network can be received and transmitted. By having a card within a laptop computer, a user can move from their office to a conference room or another office. If at home, a user can roam from their home office location to virtually anywhere in the house, and in most cases, coverage is very adequate within the yard surrounding the home. This makes the possibility of doing work and taking business phone calls while sun tanning by the pool or on a person's deck or lawn a reality.

WLANs are most common in three markets: the home/SOHO, the office, and public places such as airports, libraries, hotels, and convention centers. While the markets for the residential/SOHO and business office are apparent, it's interesting to note the surge in deployments in public locations. Two examples come quickly to mind, which are both 802.11 deployments at major airports. Companies like Mobilestar Network Corporation have been placing 802.11 *access points* (APs) from Cisco Systems in American Airlines lounges since approximately 1999. At the time of this writing, another company, called Wayport, has begun installing APs in airline terminals so that any passenger with a client adapter (definitions are provided later in this chapter) can access their Internet service provider or office LAN.

This movement in the industry poses a genuine threat to the mobile industry, which intends to provide Internet access via its phones. One of the several tremendous challenges is that an infrastructure called 3G (referring to "third generation cellular phone network") has to be deployed, which will enable access speeds of up to 14.4 Kbps. One of the other challenges is the device itself; phones in their present form factor do not lend themselves to the broadband experience because of the small screens and the keypads that are not intended for high-speed keying.

Broadband mobility ensures that the user retains connectivity as they move from location to location within a common building or even campus. U.S. colleges have begun detailed plans to issue 802.11 radios and laptop computers to all students and staff. This enables the users to remain current with schedules, test results, class changes, assignments, course registration, and dropping from courses all possible from virtually any location at the dorms or on campus.

The ability to move from an office to a conference room with one's laptop has an enormous effect on productivity and corporate profitability because the users have all of the necessary information either on their laptops or on the servers to which they are connected wirelessly to one or more local APs and from the AP to the appropriate server by Ethernet. With the advent of instant messaging and paging via laptops, connectivity and communication between employees has never been greater.

Pervasive Computing

Pervasive computing means that an individual can have access to their files in many more locations than prior to the advent of WLAN. While broadband mobility hinges on the concept of maintaining continuous connectivity with the network at all times, pervasive computing implies that one can perform communications and processing functions in many more physical locations. Examples of this include working from anywhere within the office building or campus as well as anywhere within their home or yard. This concept also implies that your work portal goes with you, as opposed to transporting yourself to work portals.

Mobile IP

During the earliest days when the IEEE standards body was considering how to resolve the problem of bringing mobility to Ethernet users, the concept of Mobile IP was considered. Mobile IP operates at Layer 3 of the OSI stack (the networking layer) as opposed to Layer 2, which is where 802.11 primarily operates (more on this later in the chapter). The premise behind Mobile IP is similar to the way 802.11 handles the movement of a user from one AP to another, but it has the drawback of being compatible with IPv4, which is the prevalent release at the time of this writing, but not compatible with the newer IPv6 due out shortly.

WLAN RADIOS

One of the most exciting and practical developments in the BBFW world is the great proliferation of WLAN radios that adhere to the IEEE 802.11 standard. These 2.4 GHz radios are sold in approximately 60 countries worldwide at the time of this writing, and the rise in the number of these radios is just beginning to hit its full stride.

Table 13-1 provides a quick summary as to what a WLAN is, and more importantly, what it isn't.

While wired LANs such as an Ethernet LAN require the users to be tethered to the nodes, a WLAN enables the user to roam freely around a facility or campus and have access of speeds up to 7 Mbps one way or 11 Mbps both ways.

A WLAN Is	A WLAN Is Not
Local, not wide area	Cellular phones
In-building or campus-area for mobile users	Pagers
Up to 25 miles for P2P	Packet data
RF or IR	Ardis, CDPD, RAM mobile data
No FCC license required	PCS
Customer owns the equipment	Wireless PBX or cordless phones
Shared bandwidth	Dedicated bandwidth

Table 13-1. Definition of a WLAN

There is at least one major competitive standard to the 802.11 radio, which is the HomeRF standard radio. However, the 802.11 radio is the most prolific of the WLAN radios on the world market today, and therefore the focus of this chapter will be on this radio and its associated architectures. Radios that meet this standard are more commonly used on indoor LANs, but also see duty as bridges between buildings for campus use.

Outdoor WLAN radios, better known and more commonly deployed as bridges, generally connect two or more buildings and may be the most cost-effective way to connect up to three buildings while providing true broadband connectivity. Outdoor WLAN bridges are typically used for the final 400 feet by indoor WLAN equipment as they work together seamlessly and whether or not they are provided by the same vendor, operate at the same frequencies, and have identical operating systems.

Outdoor WLAN Radios

WLAN bridges are designed to connect two or more LANs together, which in effect creates one continuous LAN. Distances of up to 25 miles can be obtained, provided the link is line of sight, the appropriate antennas are used, and the system is configured correctly for reduced data rates. To the network, a WLAN bridge looks more like a cable between two ether nodes than a bridge, and a WLAN bridge does not add to the hop count.

The advantages for WLAN bridges are similar in most respects to those of conventional BBFW links:

- Rapid installation
- No requirement for trenching
- Can provide high-speed access between two or more sites

There are also additional advantages to a WLAN bridge, which include:

- No FCC license (although the same can be said for BBFW UNII)
- Approximately one-fourth or less the cost of a BBFW link
- Much easier to install and configure
- No recurring charges based on time and distance as with T1/E1 telco lines

A common outdoor WLAN architecture is shown in Figure 13-1. A common WLAN architecture features the following elements on each building:

- A transmitter/receiver (a bridge)
- An antenna
- Cabling to the inside of the building
- Clients
- Management software

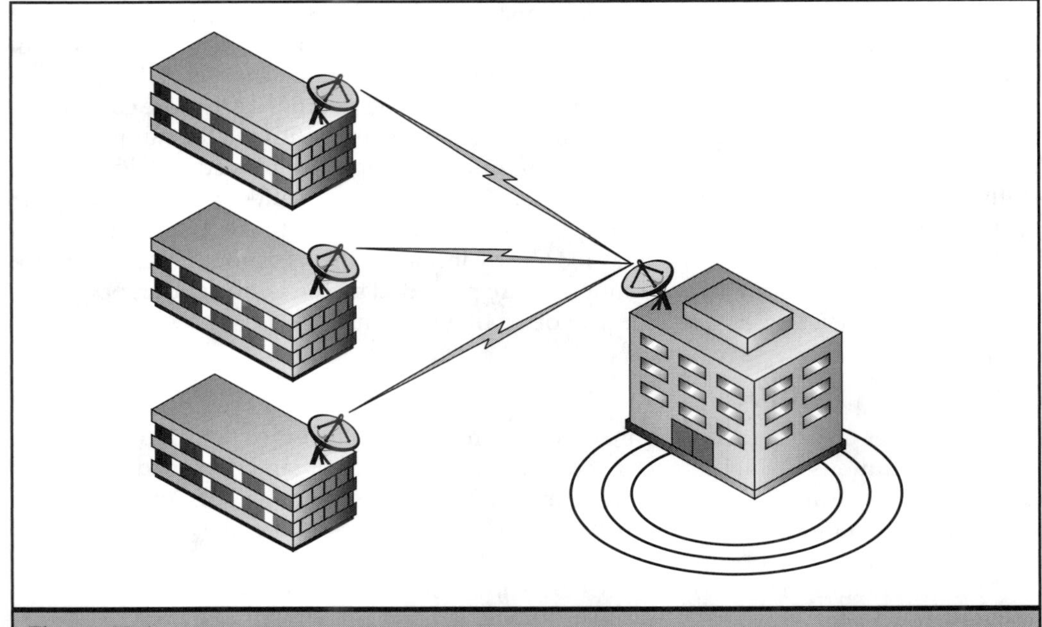

Figure 13-1. Typical outdoor WLAN architecture

The Bridge

The bridge is an indoor-rated device that has an externally mounted antenna that can service up to three buildings at 11 Mbps per building, or multiple buildings that will share the 11 Mbps. While some companies are providing the electronics package that is outdoor-rated, most companies at this writing do not.

The Antenna

When only two buildings are connected with a WLAN bridge, a pair of either parabolic or Yagi antennas can be used. If there is to be more than two buildings connected by the bridge, then an omnidirectional antenna is commonly used; however, patch antennas are gaining popularity for campus or short-range applications because they are more esthetically pleasing for some customers.

Cabling

Coaxial cabling exists between the externally mounted antenna on the roof of the building (or typically near the building) and the bridge itself, which is not rated for outdoor use. The coaxial cabling is therefore located in an electronics closets or another location as appropriate. Having said that, there are deployments where the indoor equipment is being placed in weatherproof and heated/cooled enclosures. This is often done to shorten the distance between the equipment and the antennas.

Clients

One of the features on the more advanced versions of this radio is that the bridges will communicate directly with client adapters, such as PCMCIA cards, from laptops that might be used on the campus outside a building. Another configuration might be for desktop computers that require a PCI or ISA card for WLAN interface. This configuration for PCs will generally require an antenna that is externally mounted to the building.

Management Software

The software required to configure and manage the bridges is the same software required to configure and manage access points. The software in the best of the products includes utilities such as link analysis and site survey software.

Data versus Distances

For deployments where the range between the buildings is less than one mile and also line of sight, a user could deploy access points and be able to get a full 11 Mbps of data rate. For distances that are not very short range, a bridge is strongly recommended for reliability and higher data rates. This recommendation is due to the timing constraints the IEEE 802.11 specifications puts on acknowledging packets that have been successfully sent from one side of the bridge to the other.

In other words, the FCC has set a maximum allowable time limit for packets to be acknowledged as having been successfully received. Because of this maximum time limit, WLAN equipment generally operates within the FCC specifications at distances of less than 1000 feet between the antennas. As acknowledging transmissions over longer distances require more time for the received packets to be acknowledged, 802.11 equipment in that application typically violates the applicable FCC, which may cause compatibility problems between equipment from different vendors. It's important to emphasize that the FCC in the U.S. does not have rules on maximum transmission distances, rather, the issue is with the maximum allowable time allocated to acknowledge received packets.

Table 13-2 illustrates the various distances with maximum-length cables. Maximum-length cables that run between the bridges that are inside the building and the antennas that are external to the building are important items because the signal strength typically drops by half for every 50 feet of cable. This loss translates directly to usable range. This loss is added to the loss incurred during the transmission, which is a normal event as the energy is transmitted through the atmosphere.

A standard 21 dBi parabolic dish is assumed in the calculations indicated in Table 13-2. From this table, you can see that the performance of the link varies very substantially, and is based on the combination of distance between antennas, data rates, and the amount of cable used at each end of the link between the antenna and indoor mounted

Data Rate	Maximum Distance	Standard Cable
11 Mbps	12 miles	50 ft. per side
11 Mbps	18 miles	20 ft. per side
5.5 Mbps	13 miles	50 ft. per side
4 Mbps	6 miles	50 ft. per side
2 Mbps	25+ miles	50 ft. per side
1 Mbps	25+ miles	50 ft. per side

Table 13-2. WLAN Bridge Configurations

bridge. The customer must therefore be able to make the following assumptions prior to installing the links:

- What data rates are required?
- What is the distance between the buildings?
- What is the amount of cable necessary between the bridges and the antennas

Of the three questions, the first one is perhaps the most difficult and will be addressed in detail later in the book. At this point, however, it is sufficient to state that when calculating data rates, the current data rates should not be used as the basis for designing the system. Rather, the customer should calculate or project what the data rates are anticipated approximately 18 months following the installation of the system. This time frame will generally allow for the system to pay for itself given that T-1 lines cost anywhere from $500 to over $1,000 per month. Second, this time frame is sufficiently short that the equipment will not have fully depreciated. (A bit of accounting magic there, but it is important to the owners of the system).

Indoor WLAN Radios and the Five Elements of WLAN Communication

Indoor 802.11 radios have two primary components as stated in Chapter 11, a *client card* and an *access point* (AP). The client card is a device that resides either as an Ethernet type card in the Ethernet slot in a laptop computer or as an internal card on either a desktop computer or printer. The AP performs the function of the transceiver between the client card and the rest of the network and is attached to either a high point on a wall or the ceiling.

Five separate elements of communication occur between two or more 802.11 radios:

- Association
- Disassociation
- Re-association
- Distribution
- Integration

Association

The association makes what is called a *logical* connection between two WLAN elements, which can be a pair of access points, client adaptors, or a combination of the two, as long as none of the APs is a *root AP*, which is the AP that connects a group of clients or APs to the Ethernet. A logical connection is one that is independent of a physical location; in other words, it has a *logical address* as opposed to a *physical address*. These logical connections, which contain address, identification, and security information, most often occur between the client adapters, which are mobile, and one or more access points or bridges.

An association, which is a one-time event, must be established before information can be transmitted. This association is always initiated by the client adapter, which is the mobile device, such as a PCMCIA card in a laptop computer. This association cannot occur with more than a single access point due to FCC constraints, even though two or more access points may be within the transmission range at the same time; the association can be changed from one AP to another based on load-balancing requirements. The term *load balancing* refers to when a network will take traffic overloads from one network path and redirect this traffic overload to another path, or, more specifically, to a device along a path with a lesser amount of use. It can also occur between two or more PCMCIA cards in what is called *ad hoc* network architecture. An ad hoc architecture is where peer-to-peer communication occurs directly between two or more laptop or desktop PCs where there is no access point to connect the ad hoc network to the network backbone, which is typically an Ethernet line.

Disassociation

A disassociation occurs when a mobile client breaks off its association with an access point. When a client card breaks off its association with either one or more other PCMCIA cards, for example, or an access point, it must re-associate with another WLAN network member before information can once again be transmitted through it to its host (typically a laptop). While it is virtually always the mobile client that disassociates itself with the access point or other mobile client, an access point can be taken offline, which then initiates the disassociation. When an AP supports load balancing, it can initiate the disassociation and force the client to move from one cell and reassociate it with another AP. This can occur for a variety of reasons, such as a new access point being placed closer to the client adapter, the access point being taken offline, or an access point being overloaded.

Re-association

Re-association is initiated by the mobile client and occurs when a mobile client or access point changes its association from one WLAN element to another. This is a highly common occurrence in a WLAN architecture and is indeed one of the most compelling reasons for having such a LAN.

In most WLAN systems, the re-association includes information about the access point or client adapter it was last in contact with. Information that was formerly routed to the last access point (for example) will then be routed to the access point where the mobile client resides. This is an important feature, because the former access point may be holding information in its buffer that needs to be sent to the mobile client.

Distribution

Distribution is the service used by a stationary WLAN element and it depends on the information provided through the associations, disassociations, and re-associations. Distribution is that which guides the information to the optimal access point. The distribution process should not be confused with the transmission of actual information; rather, its purpose is that of traffic director, which means that it directs the information to the appropriate access points.

Integration

Virtually all networks must engage other networks that are logically adjacent. This is, a certain amount of information is relayed within a specific LAN, such as that between two employees in a common building or campus. However, there is also typically a great deal of traffic that arrives on and departs from the LAN.

The integration process takes the information off the WLAN and gets it onto the network core and then to an access medium of some description off the LAN. A key function of the integration process is that it converts the MAC frames to a format that can be used by a non-native LAN, and of course it performs the same function as frames from a foreign LAN at the WLAN.

Infrared 802.11

Not commonly known, the 802.11 specification also covers transmissions made in the infrared (IR) spectrum. Instead of using a DSSS or FHSS modulation scheme, the IR medium relies on pulse modulation techniques to transmit data at either 1 or 2 Mbps. This type of transmission is typically used for very close range communications such as that between a cell phone and a laptop computer or PDA. Very few commercial 802.11 infrared products currently exist.

Common Uses for a WLAN

State of the art access points or bridges such as those supplied by Cisco can support approximately 2000 users, although there are other vendors that can support no more than

256 users, and some can support no more than 32 users. The more important question is how many users can use the system simultaneously, which is typically between 25 and 30 per access point before significant performance degradation occurs. This is primarily due to the fact that data is very "bursty." In other words, users tend to send a file or e-mail and then nothing for a period of time that can last seconds, minutes, or even hours.

By overlapping the transmitting range of access points, the number of users in a given room or area will grow at the rate of approximately 25 to 30 users per access point up to about 90 users for a given area or room. The key thing to remember is that the issue is not truly one of user count, but rather of usage. On a standard Ethernet domain, the 10 Mbps is shared and is relatively independent of the user count. Figure 13-2 indicates how an array of access points can be deployed given the different speeds provided at different ranges.

One of the many advantages of a WLAN network is that it is readily deployed and, even better, can be readily moved from one physical location to another. This is an attractive feature in a fluctuating economy where businesses grow and shrink and where even in stable companies, individuals or entire groups are moved from one location to another.

Even in stable economies with stable companies, occasionally leases expire on buildings, or companies are acquired. There can be little question that having networks that adhere to common standards makes the integration of new locations and personnel that much easier.

A WLAN is also a very appropriate communications medium at conventions and corporate retreats and offsites, because it delivers maximum flexibility in terms of users, is very easy and quick to set up, and does not require large packing crates or special handling of equipment.

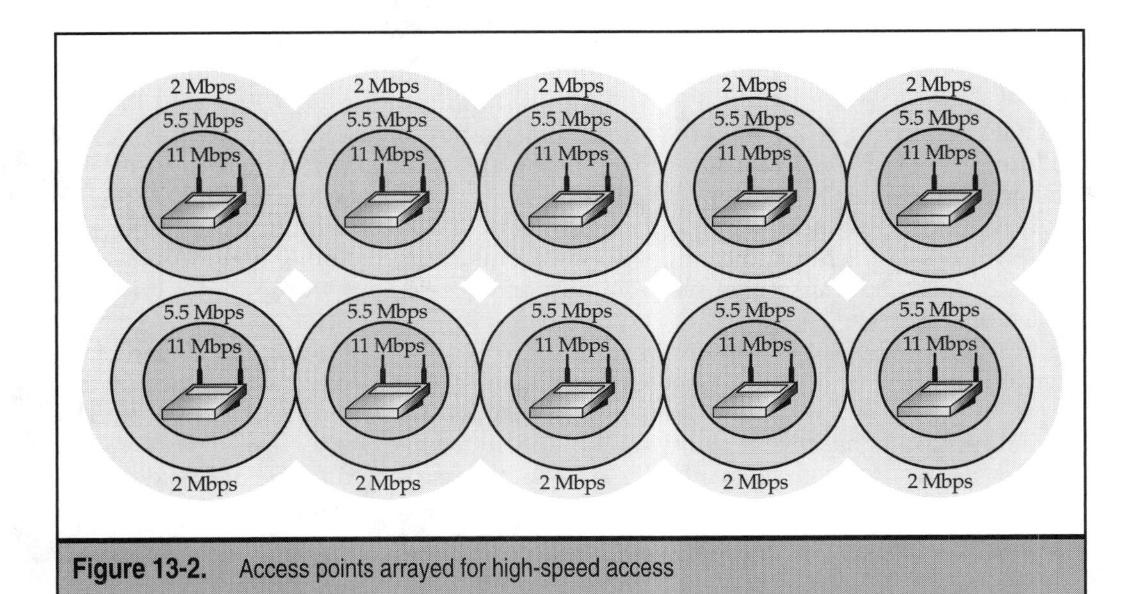

Figure 13-2. Access points arrayed for high-speed access

WLAN gear is well entrenched in these settings:

- SMB offices
- Homes
- Schools
- Hospitals
- Factories

SMB Offices

A WLAN is often much easier to install than running wires through walls and ceilings. While WLANs often require some wiring in the ceilings, they can communicate without Ethernet connections between the access points. The most common application for a WLAN is to provide broadband mobility and pervasive computing to an existing Ethernet LAN, as mentioned at the beginning of this chapter.

Homes

Homeowners are often reluctant to rip open walls and crawl along attics and under floor crawlspaces to install LAN wiring. This is of course less a problem in a home being constructed, but there is still a considerable cost savings in installing a WLAN in a home as opposed to a wired Ethernet.

One of the additional benefits to a WLAN network is that the homeowner can use their laptop outdoors in their yard, should they so desire and given that there is sufficient signal strength. Home offices are the norm in major cities, which has brought along with it an increasing use of WLAN equipment.

Schools

Schools are proving to be one of the hottest markets for WLAN equipment. As school buildings are often older structures that are insulated with asbestos, there is generally a tremendous aversion to opening up walls or ceilings. A WLAN network can be deployed on a school campus such that it can keep the student in constant touch with schedules, homework assignments, and changes in classrooms. Tests and other homework assignments can also be transmitted electronically and wirelessly, with grades returned or posted on a Web site.

The author has had experience in working with major colleges in the United States that are rolling out plans to issue every student a laptop computer loaded with a PCMCIA card and a cell phone that operates on a private school cellular network. This not only greatly enhances the ability for the student to have a better learning experience, but it is also a source of revenue for the schools because they can have the students use their phone and information networks as opposed to those provided by third parties.

Hospitals

Hospitals, more often than not, are operated as profitable businesses. The average revenue generated by a recovery room is between $1000 and $2500 per day. This means that the rooms cannot remain in a decommissioned state for very long. Further, the cable trays in most hospitals are filled to capacity. A WLAN system requires virtually no construction tools for installation, is easy for the hospital network managers to maintain, and can be installed in a floor within a wing in a day or two, depending on the size of the wing.

Some hospitals have a strict "no wireless equipment" prohibition even though some 802.11 equipment is built into some EEG machines that have been certified by the FDA. In most cases, moving the antennas one to two meters from even the most sensitive equipment is usually sufficient to significantly reduce the chances of impairing the sensitive equipment.

Some WLAN gear has been tested with pacemakers and hearing aid devices. Some tests involved placing a 250 mW antenna right next to a pacemaker, and no ill effect was detected by the doctors monitoring the pacemaker. This statement should be taken as anecdotal, as testing that results in FDA certification is rigorous, time consuming, and thorough, to say the least. Where there have been problems, they seemed to stem from the fact that the medical devices had deficiencies with regard to being resilient to modest amounts of RF energy.

However, a "no wireless" policy occasionally runs counter to the hospital IT and IS staffs and varies from hospital to hospital, as opposed to originating from a legal or authoritative body. The probability of interference with sensitive hospital equipment such as EKGs, ECGs, and other monitoring devices is low given the power output and range of WLAN devices.

One workaround for this is to provide tests in approved areas that do not have life-sustaining equipment such as the administrative offices. Further, all WLAN equipment has to pass stringent FCC tests for emissions. These levels of emissions should be acceptable to hospitals. Laboratory tests can be performed with sensitive hospital equipment to ensure compatibility.

Factories

There can be little question as to the effect a WLAN system has on improving efficiency in very large workspaces such as factories and warehouses. This not only extends from the administrative areas, but to all parts of the floor, assembly, storage, and development areas. Time and money are saved where communications can occur seamlessly from any location within a plant.

Indoor versus Outdoor WLAN

802.11 radios perform two types of functions: indoor wireless LAN and outdoor wireless bridges. Of those two, indoor WLAN is far more common.

Indoor WLAN Architectures

WLANs have three architectures that enable broadband mobility and pervasive computing:

- Infrastructure
- Ad hoc
- Repeater

Infrastructure Architecture A WLAN infrastructure architecture is one that has connectivity with the network via one or more APs while a number of client adapters interface with one or more APs. Again, state of the art systems can support up to 2084 clients, which can connect to the same AP, though of course not at the same time, because one of the features of 802.11 radios is that the bandwidth is shared among the users. Figure 13-3 illustrates an infrastructure architecture.

Ad Hoc Architecture An ad hoc architecture is one in which two or more clients, typically laptop computers, communicate with each other on a *peer-to-peer basis*, which in this context refers to the fact that the laptop computers can communicate to each other directly, without going through one or more APs. Figure 13-4 represents an ad hoc architecture.

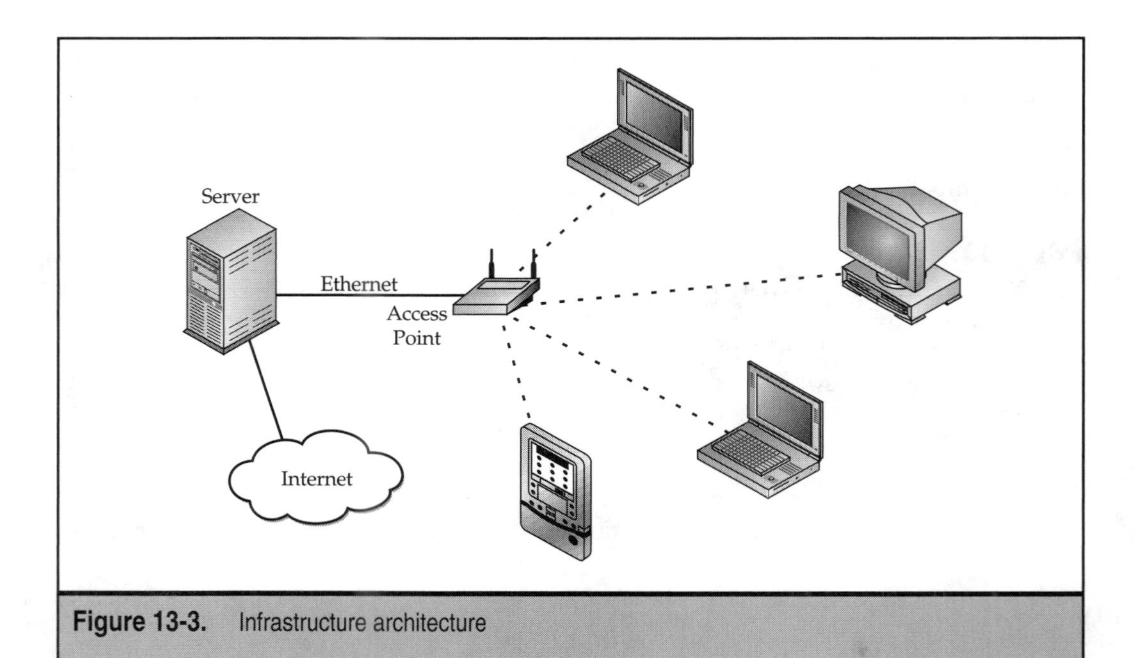

Figure 13-3. Infrastructure architecture

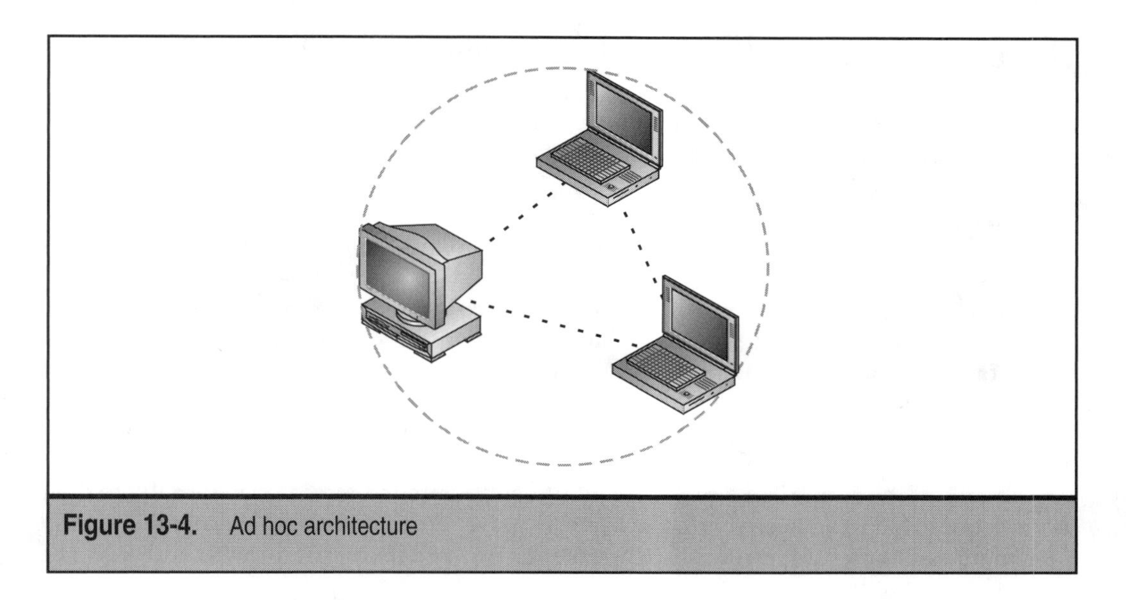

Figure 13-4. Ad hoc architecture

Repeater Architecture Some APs support repeater functions that can repeat data from one cell to another. This is an appropriate architecture where the network managers wish to extend the coverage but do not have access to a backbone such as Ethernet to connect the APs. Figure 13-5 represents a repeater architecture.

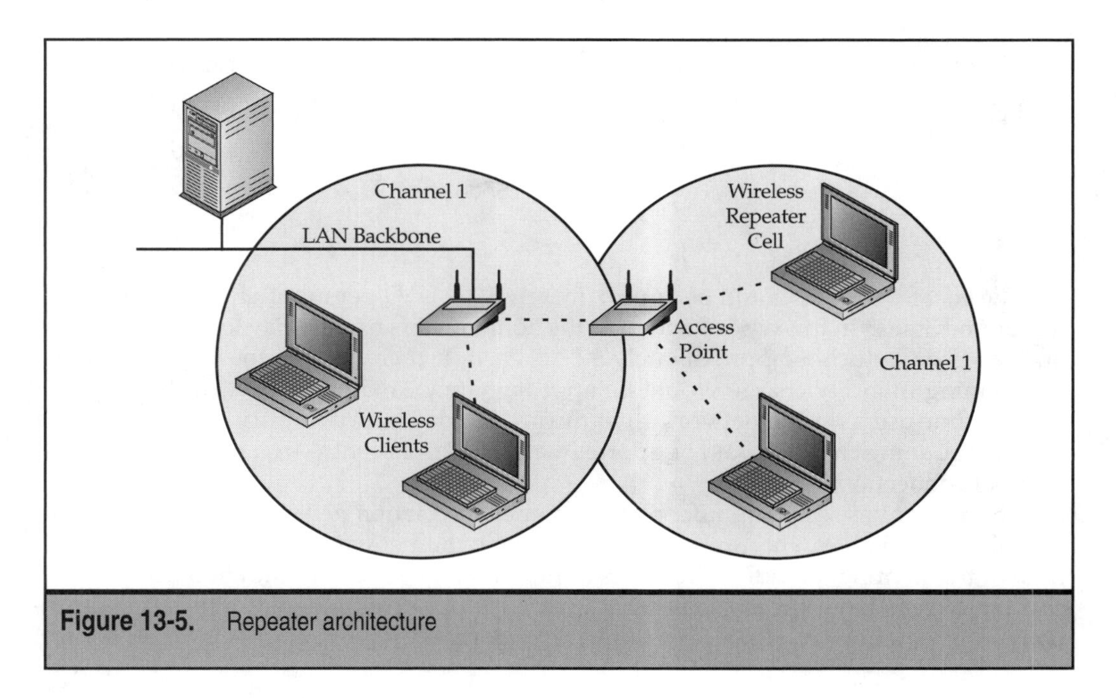

Figure 13-5. Repeater architecture

Homeowners Association 802.11 Architecture

There is an array of service providers that use these radios to provide broadband service to an area, which, under certain circumstances, is an appropriate use of this type of radio. In other words, an appropriate use of an 802.11 radio for a homeowners network is where one owner has a T-1 line brought into their house and acts as the head end for several other users. Figure 13-6 outlines this architecture.

As shown in Figure 13-6, a group of homeowners could easily set up their own broadband network. The residence with the incoming T-3 would connect the line to a hub, which would then split the line to the T-3 residence owner's personal computer, with a second line going to an 802.11 bridge. The bridge would then have a conductor to an antenna on their roof, which would have a line of sight to two other homes nearby in the neighborhood. The network could also be set up with four homes if two of the homes were using the BBFW links during the day as SOHO users while reserving the bandwidth in the evening for nonprofessional use.

At each of the subscriber sites, there would be an antenna mounted externally to their house connected to an access point inside the house. From there, an Ethernet line would be run into either a hub or router within the house for data distribution to and from the LAN within that house.

The preceding illustration is only one example of the speeds and size of the incoming line to the first home, which can be anywhere from a T-1 at 1.5 Mbps to a DS-3 at 45 Mbps, all distributed between the users in the homeowners association network. Keep in mind that the 802.11 links only have an 11 Mbps line rate. The cost for operating this type of network would be approximately as follows:

Equipment	Quantity	List Cost Each	Line Item Cost
Bridges	3	$2000	$4500
Routers	3	$350	$1050
Misc.	1	$500	$500
TOTAL	--	--	$6050

The cost for a DS-3 would be approximately $500 USD per month in most regions of the United States. In this type of network, the homeowners usually pay for the equipment on their own, which is approximately $2,000 each. If there were a total of three users, the running monthly charge would be approximately $165 per month. If there were a total of four users on the network (two during the day and two during the night), the monthly running charge would be approximately $83 per month—for an 11 Mbps shared Internet connection.

The author has seen a number of homeowner association networks like this, particularly in Silicon Valley where a surprising number of homeowners cannot get DSL. At the time of this writing, broadband cable is approximately $45 per month in Silicon Valley, and most users have speeds somewhat under 1 Mbps. This speed diminishes substantially as the number of cable users in the neighborhood increases. This is what, in part,

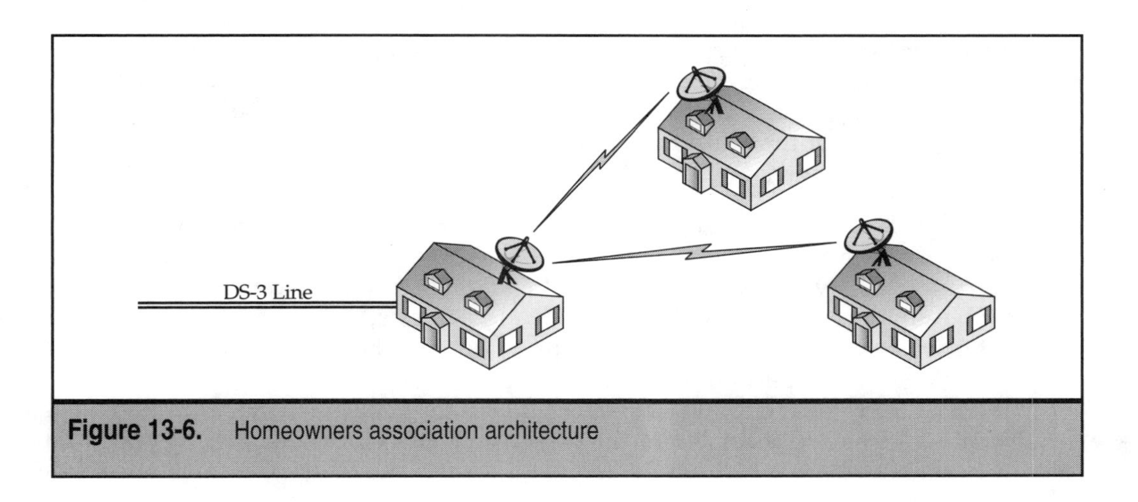

Figure 13-6. Homeowners association architecture

makes the homeowners association network attractive; for 2 to 4 times the rate of a broadband cable subscription, the user can have up to 11 times the speed!

Admittedly, this type of architecture is not widespread, but with an upfront cost approximately that of a new computer, and with the assumption that those interested in a broadband connection would already have one to three (or more) personal computers in their homes, it is not surprising to see this type of architecture.

802.11 Radios As a Service Provider Medium

More often than not, however, this radio is deployed for service provider networks. The service provider charges a flat monthly fee for broadband access to each of the users, because it has no other means of charging for a set amount of bandwidth. While some equipment providers encourage the use of an 802.11 radio for use as a medium for providing broadband service to a geographical area, the radio was not intended for this purpose and in fact has the following shortcomings:

- Does not scale
- Bandwidth cannot generally be throttled
- Does not allow for differentiated service
- Elementary software management system for ISP use
- Requires line of sight

Does Not Scale With Direct Sequence Spread Spectrum (DSSS) modulation (to be covered in greater detail later in this chapter; also see Chapter 8), which is the most common modulation scheme used on this type of radio, there are only three nonoverlapping RF channels. This means that a deployment can only have a maximum of three receivers local to an access point unless the users begin sharing bandwidth on a shared (per sector) basis.

This means that a service provider must deploy a head end for every three radios. The author has seen deployments where there were more than three radios interfacing a single head end, but if one user was online and using a band, the other user would have to wait until the first user relinquished use, which is one of the basic premises of Collision Sensing Multiple Avoidance (CSMA) modulation schemes.

Bandwidth Cannot Be Throttled An 802.11 radio operates in such a manner that the user on a channel gets a certain amount of bandwidth, which cannot generally be adjusted, although interestingly, this is typically not offered on state of the art systems for which reasons the author can only speculate. On BBFW systems designed for service providers, the amount of bandwidth is controlled by the amount of channels granted to a specific user, or for a specific application such as voice.

If the service provider deploys a network without the ability to manage the amount of bandwidth available to specific users, the person who gets online first gets the spectrum until they relinquish their turn on the radio. 802.11 radios are designed with a control mechanism that requires them to "listen" for an open radio channel before they will broadcast a signal.

Does Not Allow for Differentiated Service The 802.11 radio currently works in a bridge mode, which means that it cannot provide differentiated services that can be allocated by router port, user, or application. In other words, a state-of-the-art BBFW system operates at Layer 3 of the OSI stack, which is the routing layer. There is not a strong demand for Layer 3 differentiated services at the time of this writing; however, this demand will increase in the near future in order to provide the services users will demand and operators will be required to offer in order to mitigate user turnover.

Elementary Software Management System for ISP Use The current 802.11 radios do not have highly sophisticated network-operations type software. This is not necessarily a negative thing, as the radios were not designed for service provider use.

In a BBFW system designed for service providers, a considerable amount of effort is put into monitoring the links and other items such as being able to provision a user. Provisioning a user includes events such as granting a user authorization, resolving for billing, generating trouble tickets in the even the end user requires service or technical support, as well as being able to grant the user additional bandwidth, upon the user's request.

When a service provider deploys 802.11 radios to supply broadband access to users, it does not have the ready and sophisticated means to know a link is either down or performing poorly until it receives a call from the end user. When this occurs, the service provider generally always needs to send a technician to the site, because the provider cannot troubleshoot the link from the network operations center. State-of-the-art systems such as the one offered by Cisco, can track and report errors through its Standard Network Management Protocol software (SNMP), and this again is because Cisco views all elements as part of a network, not merely radios attached as an adjunct to a network.

Requires Line of Sight 802.11b radios require line of sight between the head ends and subscribers. This will change as the 802.11a radios come onto the market with non-line-of-sight capabilities (see next section), but that won't occur in a major way until approximately 2003, possibly sooner.

In fairness, it should be stated that only a very few BBFW providers with systems designed for service providers can resolve the problem of having occlusions in the paths between the end user and the head end; however, it can also be said that radios designed for the service provider market are generally more robust in terms of maintaining a link because they typically have the software management tools to optimize the link in real time, such as packet-by-packet power and gain adjustments and other adjustments that can be made at the routers or BBFW radios themselves in addition to far more complicated modulation and spreading schemes.

Most Common Modulation Schemes for 802.11

Initially, most 802.11 radios used Frequency Hopping Direct Spread Spectrum (FHSS), but have since migrated to Direct Sequence Spread Spectrum (DSSS). Table 13-3 outlines the differences between the two most common types of modulation at this writing. There is another 802.11 modulation called Wide Band Frequency Hopping (WBFH), which may become quite prevalent due to its balance of relative simplicity and high performance.

802.11 DSSS and Chipping

The subject of DSSS is covered in Chapter 9, but there is one aspect of DSSS this is not well known, which is the concept of *chipping*. This fundamentally involves making one bit into a series of numbers that are distributed throughout the spread spectrum signal. Chipping techniques can vary from company to company.

	DSSS	FHSS	WBFH
Frequency	2.4 GHz	2.4 GHz	2.4 GHz
Data rates	1, 2, 5, 11 Mbps	1, 2 Mbps	10 Mbps
Advantages	Range, throughput	Resistance to interference	Resistance to interference
Processing gain	Yes	No	No
Maximum number of APs	3	15	3
Maximum throughput per cell with max. number of APs	33 Mbps	16 Mbps	3

Table 13-3. Comparison Between DSSS and FHSS

The minimum chip rate mandated by the FCC is 10 per bit. The better systems on the market use 11 for the lower data rates of 1 and 2 Mbps. An example follows:

1. If the data is: 1001
2. The data is comprised of a 1 followed by two 0s and then a second 1.
3. Aironet for example uses 11 chips per number. For every 1 in the data, it actually sends 00110011011. For every 0 in the data, it actually sends 11001100100.
4. The transmitted data would then be as follows:

   00110011011 11001100100 11001100100 00110011011
   1 0 0 1

By using this chipping code, the system sends very robust data, because in order for the receiver to not be able to interpret the transmission, more than 5 of the 11 numbers would have to be wrong or missing. The downside to chipping is that it takes a lot more spectrum, which gets back to the premise stated earlier in the book: the more information sent for a given modulation scheme, the greater the amount of spectrum required to send the data. If a single bit is transmitted, it requires 1 MHz of bandwidth. With 11 bits, the bandwidth required is 22 MHz.

This leads to the issue of how the DSSS bandwidth is set up in a market-leading system. As the example for Aironet indicates, there are 11 chips transmitted for every 1 or 0. Again, these 11 chips take up 22 MHz of spectrum.

Recall from Chapter 9 that DSSS spreads its energy over a wide area of band. The Aironet products use three 22 MHz-wide channels or eleven 7 MHz channels in 83 MHz of ISM spectrum (2.4 GHz). Of these 11 channels that overlap each other from one end of the ISM spectrum (2.4 GHz) to the other end of the ISM spectrum (2.483 GHz), only 3 channels, which again are 22 MHz wide, do not overlap (and therefore do not interfere with each other).

802.11 FHSS

The FCC requires that 802.11 systems that use Narrow Band FHSS use at least 75 independent channels (or frequencies) before repeating any channel. The maximum time that can be spent on any one frequency is 400 mS, which is just under half of one second in any 30-second period of time. The hopping from one frequency to another is done in a pseudo-random order, which is to say that while the order appears to be random, there are actually a limited number of combinations that can be used. 802.11 rules mandate that there are 26 different hopping patterns in three different sets.

These 26 patterns were calculated to have the minimum amount of time where the channels step on each other. When the channels do not step on each other, they are deemed to be orthogonal, hence the patterns are orthogonal patterns because they have a sequence that does not interfere with itself.

The situation becomes complex when more than one access point or client adapter begins to transmit in a common area. Even with 26 different channels available, there will be some interference, and therefore some data lost. What makes the system work, in part, is that the data is retransmitted.

One of the aspects of FHSS is that you cannot simply increase the total number of FHSS channels to increase data rates, because as the number of channels increases, so does the amount of interference, and therefore the data either has to be retransmitted or is lost altogether.

Wide Band Frequency Hopping

The Home RF Working Group (HRFWG) submitted the Wide Band Frequency Hopping proposal (FCC Proposed Rule 99-231) to the FCC in 1999. The proposal was a request to allow wider bandwidth channels for Frequency Hopping (FH) spread spectrum systems, which would enable these products to go from a maximum data rate of 2Mbps to as much as 10Mbps thereby reaching near parity with 11Mpbs Direct Sequence (DS) products. The rules state that at 5MHz wide channels, you must use at least 15 channels, and they cannot overlap. With only 83MHz of overall spectrum, it limits the user to 15 channels.

WBFH has even worse scaling problems than Narrow Band Frequency Hopping (FBFH). With the fewer channels and wider channels, the maximum number of APs that can be placed in a single area is 3 before channel collisions and interference between the systems occur causing a negative increase in available bandwidth.

802.11 DSSS vs. FHSS

You have already learned that DSSS will enable more throughput to a BBFW link simply because it has more processing gain, which is to say, it can recover more bits for a given frequency (in this case 2.4 GHz) than FHSS. It is a fact that the over-the-air data rate in an identical link will favor DSSS by a factor of 2:1. It would also stand to reason, therefore, that if both modulation approaches were operating on identical links, the DSSS would have a 2:1 advantage in range. We can also, therefore, deduce that 2 Mbps per second in a DSSS system will have the same range as 1 Mbps transmitted over an FHSS link. For these reasons, DSSS prevails in every nominal configuration for a given link.

Transmission Range

A radio wave is a radio wave, regardless of the modulation scheme implemented, given that the antennas, path, and power are the same. What makes the difference, however, is the modulation scheme, because its complexity directly affects the receiver's ability to demodulate the energy into usable bits of information.

Transmission range in general is dependent on four items:

- Transmit power
- Antenna selection
- Receiver sensitivity
- Processing gain

Transmit power being relative to range is an intuitive concept to grasp; the greater the power transmitted, the greater the distance covered. However, non-line-of-sight path correlation between range and power can be dramatically different than when compared to power versus range in a line of sight link, as drops in power over range tend to be logarithmically greater in non-line-of-sight links.

The difference between receiver sensitivity and processing gain is important to understand. Advances in processing gain is the fundamental premise as to why the best 802.11 radios use DSSS as opposed to FHSS. Receiver sensitivity means essentially that while the radio can pick up a weak transmitted signal, it does not distinguish between information that is corrupt, or in fact that any information is missing. High-sensitivity receivers are not difficult to manufacture, and having a radio system that can demodulate received energy to usable bits is a far more complex matter and is one of the issues that separates the best radios from the others.

Processing gain resides in the DSP portion of the radio, and this is where the DSSS advantage resides. The fundamental concept is that a well-designed radio should be able to recover bits quite well from the noise that is also received. The best radios can even interpolate what the missing or corrupt bits are and enter them into their appropriate locations—this is the fundamental premise behind OFDM that will be made available in the 802.11a radios, which are beginning to enter the market at the time of this writing.

The amount of processing gain is added to receiver sensitivity. The FCC has mandated that DSSS radios be able to provide at minimum of 10 dB gain through processing. This is relevant because the range for a given data rate doubles with only a 6 dB gain. Figure 13-7 illustrates the relative ranges of both DSSS and FHSS systems.

It is noteworthy that at this writing, the FCC and IEEE are looking at OFDM for 2.4GHz and have designated a task group entitled "802.11g." If this effort is successful, the requirement for processing gain will likely disappear as processing gain will be become a "digital modulation" as opposed to spread spectrum. A number of interesting technical issues have arisen out of this effort, although the author sees this development as positive.

State of the art DSSS systems that use the Intersil chipset set can receive signals with delay spread of 140nS in comparison to the older radios, which were rated at 70nS at best. If the newer DSSS radios also use antenna diversity, the DSSS will provide the same or better performance of any FH system available at this writing. As further evidence, both General Motors and Ford have standardized on Cisco's 802.11b DSSS for their manufacturing plants, which is further evidence of the progress DSSS has made in high levels of performance in very challenging environments.

The newer DSSS systems outperform the FHSS systems in every deployment where range, path, and data rates are common. The content of this chapter should convince any reader who is considering a WLAN system that a very compelling case is made in favor of the DSSS version over FHSS, in all but a very few environments. These environments are typically those that possess a very high degree of multipath and interference, such as those found at auto-manufacturing plants where the robotics systems radiate an enormous amount of RF energy through unshielded, or poorly shielded, motors and cables.

An FHSS system may also prevail over a DSSS system where there is a very high degree of multipath. An auto-manufacturing plant again is a prime example because of

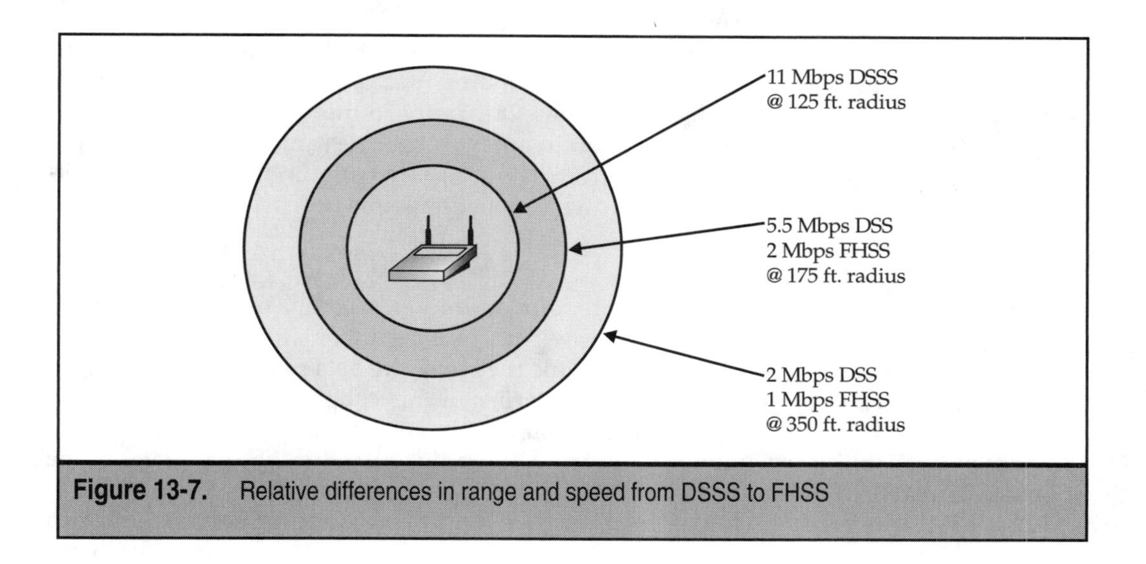

Figure 13-7. Relative differences in range and speed from DSSS to FHSS

the large collection of RF reflective panels in proximity to the radiating area. A simple site survey will inform the user as to whether or not they should consider an FHSS system, even though its data rates will be lower.

Throughput versus Data Rate

It is common to see confusion between link *speed* and link *bandwidth*. Speed is the rate at which data is transferred between two points, while bandwidth is the amount of spectrum required to achieve those speeds. Modulation, link environment, and other issues affect the ratio between bandwidth and speed, but it is sufficient at this point to simply point out the basic differences between the two. The trick one can use to differentiate the two is that speed is referred to in units per second, which at this writing is megabits per second (Mbps), although eventually it will surely be in gigabits per second (Gbps) in the not-too-distant future. Bandwidth is measured in terms of frequency, meaning MHz or GHz.

There is considerable difference between *throughput* (which is sometimes also referred to as *goodput)* and *data rate*. The data rate is the rate at which the BBFW link can send data. However, it's critical to understand that data rates also include *overhead,* which is the portion of the communication between two or more links that provides control of the links, monitoring information, and association packets, which are packets that identify and link two ends of an 802.11 link along with other data that traverses the pipe such as *retries,* which is when information is re-sent.

Throughput can be anywhere from 50 to 70 percent of what is quoted as data rate. In 802.11 links, throughput is generally on the order of 65 percent of the quoted data rate. Hence, even in the best systems, an 11 Mbps link will actually perform at 6.5 Mbps aggregate in terms of what the user sees. The summary here is that while indeed the link will accommodate 11 Mbps, a lot of the speed is used to operate the link.

There is also a noteworthy difference between the speed of a link in a point-to-point architecture versus a link in a point-to-multipoint architecture. In point-to-point links, each user accesses the pipe on a dedicated basis. On a point-to-multipoint basis, a number of subscribers access the same head end or, in other words, the same pipe. Envision the difference between one person drinking from a hose and a large number of people drinking from the same hose, each with their own connection to the hose.

Autorate

In an 802.11 radio, whether the architecture is for one mobile user or a number of users, or even for a number of bridges, the data rates will not be constant or similar for all users because the farther away the user is from the link, the lower the data rate because the signal strength decreases with distance. As stated earlier in this chapter, the signal typically decreases in power by the square of the distance—a very significant drop in power. Because link reliability is very important, an 802.11 radio will reduce the data rates as the distance increases, as this improves link reliability.

As all 802.11b systems require line-of-sight for longer-range applications, this drop in signal rate is a reasonably accurate rule of thumb. When the 802.11a equipment becomes the de facto standard on the market, non-line-of-site links will become the norm at which users will see the power drop at rates of the 5th power to the distance, or greater, depending on how much obstruction there is in the link.

Another scenario that currently is fairly common is where one person has a client card that operates at a slower rate than the newer ones on the market. In this case, the person with the older card will not be able to exceed their maximum data rates even though they are close enough to the AP or bridge to receive data rates that are higher. They may witness that colleagues in the same area will be receiving data at higher rates if they have the higher-data rate capacity equipment.

802.11 Site Surveys

BBFW site surveys must take a relatively large geography into consideration. Changes in the environment within two miles or more of a receiver can have an impact on link performance and reliability. With an 802.11 survey, the site survey generally takes into consideration areas that are generally, but not always, limited to a single room, although the "room" can be as large as a warehouse or large manufacturing facility. Having stated that, in general, physical elements at ranges farther than 400 feet don't affect the performance of either access points or client adapters.

State-of-the-art 802.11 systems include site survey software, and it's far easier to use than that required in outdoor service provider links. A sample of the Aironet Site Survey Client Utility is shown in Figure 13-8.

Figure 13-8 shows that there are two important criteria with regard to a link: signal strength and signal quality. This again proves that a signal can be strong, but include a considerable amount of noise that must be resolved by the receiving processing capabilities.

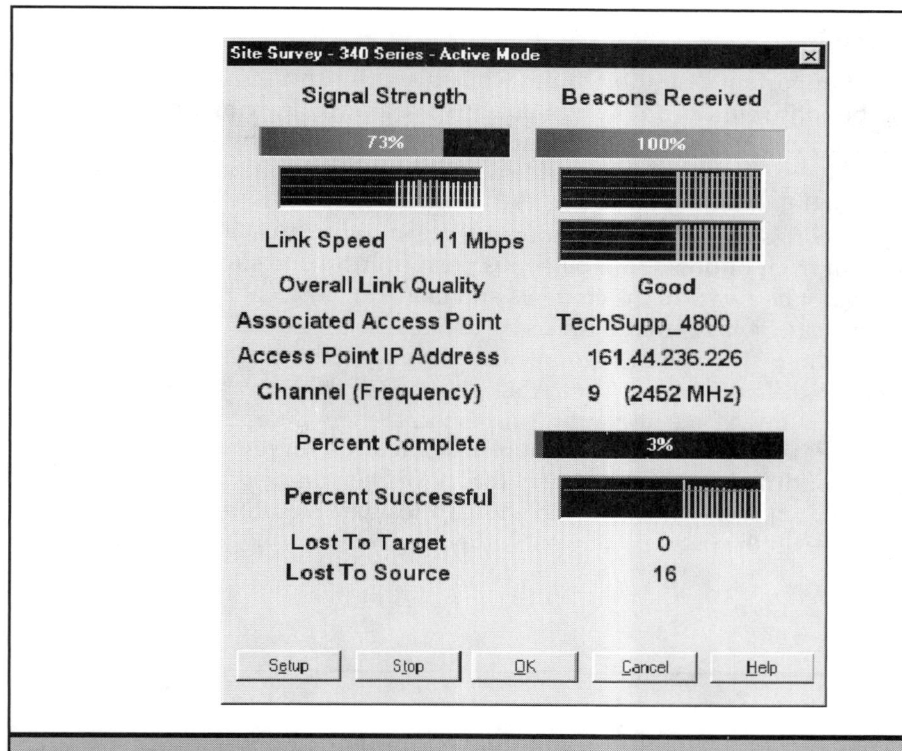

Figure 13-8. Aironet Site Survey Client Utility

With the continuing reduction in prices, equipment and site survey vendors, along with end users, are putting slightly less emphasis on highly detailed site surveys, because it's cost-effective to simply install a greater number of APs. This statement should not be construed to imply that site surveys are not important. A great number of elements affect how RF works both indoors and outdoors, and many of these elements are not visible to the human eye.

The following items are essential for a successful 802.11 network installation:

- Building drawings
- Site tour
- Test equipment and antenna selection
- Planning sessions with network managers and technicians

Building Drawings

One of the most important assets to acquire while planning an 802.11 network is a set of drawings for the entire building. It's preferable to have a set of drawings that can be dedicated to the project and therefore can be marked up. Even if the network will only be deployed on one floor, it is essential to understand what exists on floors both above and below the intended floor.

Building drawings can also show rooms that have large cabinets or other large devices placed in front of doors. If a person is performing a site survey on a particular room, they may not be aware of the contents and use of an adjacent room. Both adjacent rooms and floors can have dramatic effects on antenna selection and link performance.

Building drawings will also indicate relative distances between areas and the size of rooms to be covered. This will assist both the experienced and inexperienced installer in determining how many APs to use and where to locate them for optimal WLAN performance. Figure 13-9 is typical of what might be used in a site survey.

Having a set of drawings for the project that have been marked up during the planning of the WLAN will also serve as an aid should a team be required to return to the site later to either scale up the network or troubleshoot a problem.

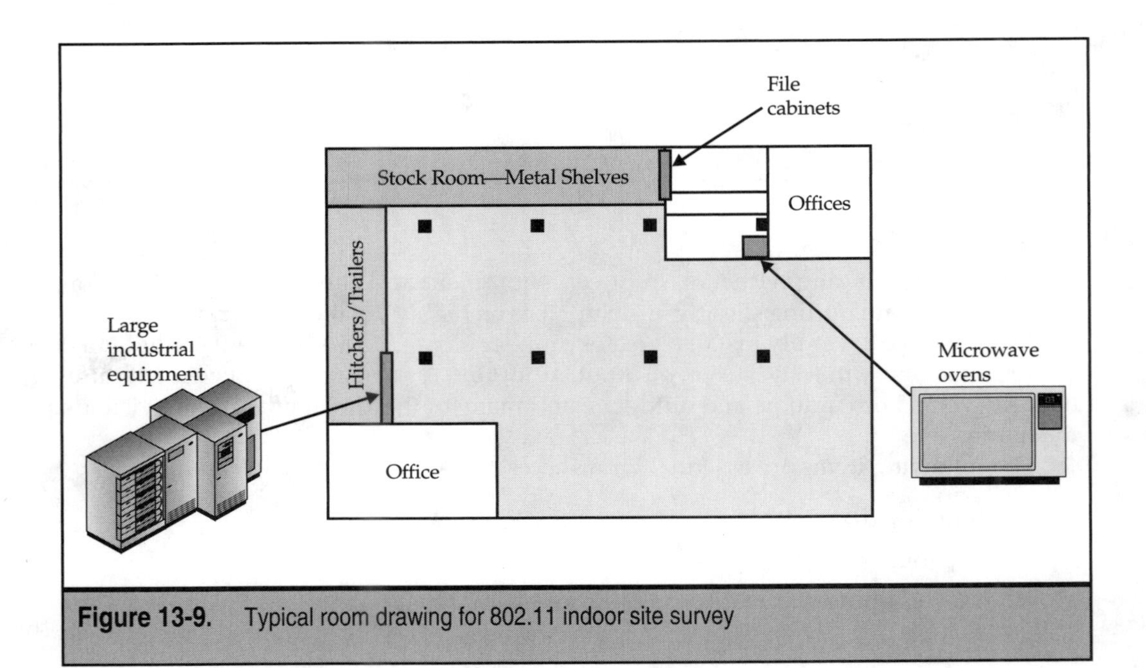

Figure 13-9. Typical room drawing for 802.11 indoor site survey

Site Tour

A site tour with a knowledgeable person will enable the survey and installation team to be aware of what operations occur in adjacent rooms and floors, as well as the contents of adjacent rooms and floors, because building drawings typically don't indicate installed equipment and are limited to floor plans. Also, not all sets of building drawings reflect *as built* configurations in the larger factories and facilities that are located in the more rural locations. As the access points will require AC power in most cases, outlets may have been installed without updating the drawings. In some cases, no building drawings are available. This issue is of less concern with the newest releases of 802.11 equipment as one of their features includes inline power, which means they source their power directly from the Ethernet.

As the WLAN industry continues to mature, so do the products. Higher-end APs, such as Cisco's 350 product line, do not require a separate AC source and acquire their power from the ether line. It is this author's prediction that eventually most deployed APs will use ether line power as opposed to requiring a separate AC source.

Test Equipment

802.11 site survey equipment is much more manageable, much easier to use, and far less expensive than that required for service provider–type BBFW links. While this type of gear can be used, it is not generally necessary. 802.11 site surveys are typically performed with the actual equipment that will be placed into service. What is required is an access point, an array of antennas, and a laptop or palmtop computer with a PCMCIA card installed, along with the client utilities.

It's also a good idea to procure some duct tape and tie wraps to hold antennas in place during the site survey; doing this will mean that the site survey team will be able to use one less person. Based on the author's experience, two other useful items are a common flashlight to see above the ceiling area, which typically isn't lit (and, in some cases, to see the space between floors), and an auxiliary battery pack for access points that will be placed in locations for site surveys where there is no local AC outlet.

Planning Sessions with Network Managers and Technicians

Having at least an initial meeting with the network managers and technicians is crucial, because the installed 802.11 equipment will become an extension of the existing network. These individuals will also know where the ether lines are and the best places to connect the access points to the existing Ethernet or Token Ring.

The network managers will also be able to inform the site survey team where the areas of coverage will be. While the users will be able to roam significant distances from the access points, the system should be designed with a focus on the areas where the users will spend most of their time. High-use areas may call for the installation of extra access points in order to accommodate a greater number of users or to cover the area appropriately if it is an odd shape, such as a long rectangle, or an area divided by floor-to-ceiling metal shelves. Even the location of microwaves, such as in employee lunch areas, should be considered for high-use areas.

Typical Site Survey Steps

A typical site survey occurs in the following order:

1. Commit to purchase equipment from the vendor.
2. Study the building drawings with building and network managers.
3. Identify possible obstructions not shown on drawings, and mark up drawings.
4. Build layout of site on drawings.
5. Perform site survey.
6. Select antennas.
7. Write final report.

Commit to Purchase Equipment from the Vendor

It doesn't happen very often, but occasionally a site survey occurs that is not followed up by the purchase and installation of equipment. It is this author's opinion that because site surveys require a lot of time and effort, if no guarantee has been made by an account to purchase the equipment, the survey team may want to prioritize that account below accounts that have purchased the equipment or at least placed orders for the equipment.

Site survey teams charge approximately $5000 to perform a survey of a facility that requires 20 or more access points. This figure can vary depending on whether the site survey team is the same one that will be selling the equipment. Other factors can also affect the survey cost, such as geographic location, difficulty of installation, market demand, and expediting fees. In no event should the site survey team perform the work without being paid for the survey or as part of a turnkey package (in other words, as part of the equipment purchase).

While the occasional customer will demand the site survey be done at no charge, to determine how an 802.11 system might perform in their facility and to then base a purchase decision on that, it is this author's opinion that a customer that balks at paying a site survey fee of $5000 probably isn't a solid customer in any event, because the customer will have very substantial expenditures in other areas such as networking equipment, laptops for employees, and modest construction costs to accommodate the new network devices.

Build Layout of Site on Drawings

Once the preliminary meetings have occurred with the appropriate building and network managers, and possible obstructions have been identified and marked up on the drawings, it is time to build a layout of the system on the drawings.

This is generally done by indicating the location of the access points and their sources of AC power and network connection points. As stated earlier, the newest access points currently offered do not require separate power supplies, most of the access points will require them, and there will be a considerable number of access points remaining on the market for some time that require AC power.

The build layout should initially focus on the high-use areas, which are typically the administrative areas. The site survey and installation team will want to be especially aware of certain physical elements such as elevator shafts, large metal shelving structures, walk-in coolers, x-ray rooms, and large metal cabinets. These items can effectively seal off a room from local access points and may well require dedicated access points. Once the layout has been indicated on the drawings, it will be time to perform the actual site measurements.

Perform Site Survey

One of the most important aspects to remember when doing an indoor 802.11 site survey is that the team will want to perform the measurements by placing the access points high in the corners of the room and directing their energy to the center of the room—in most cases. They will want do this because those are the spots where the access points will most likely be placed. By taking measurements with a laptop or palmtop computer and by walking around with them while taking measurements, a site survey team is most likely to assemble readings that are most like those that will occur during the actual use of the system.

After walking around the room with the computers, mark the areas of coverage on the building drawings, which the user will want to bring with them. If no formal drawings are available, sketch simple drawings with appropriate amounts of detail and orientation. This information will become part of the final report and thus is important.

The site survey team will want to locate access points such that each access point overlaps at least one other access point. This will ensure there are no *nulls,* or areas with reception that is too weak for use. Figure 13-10 illustrates what a drawing might look like after mapping the coverage from various access points. This figure shows four access points mounted, one in each corner. The drawing also indicates which channels would be used in which areas.

Select Antennas

During the site survey, an experienced team will generally know what antennas will work best for a given area or use. However, as each site survey is unique, the antennas are generally not selected until after the site survey has been completed for each area. The selected antenna will have been installed and tested, with the areas of coverage indicated accordingly on the building drawings. Figure 13-11 indicates the most common antennas used in an 802.11 indoor application.

Write Final Report

Following the site survey, a final report is written. Copies are distributed among key members of the customer's business and a master copy is kept by the site survey team. If the installation team is not the same as the site survey team, then a copy can also be given to the installation team, or the installation team can acquire a copy from the customer.

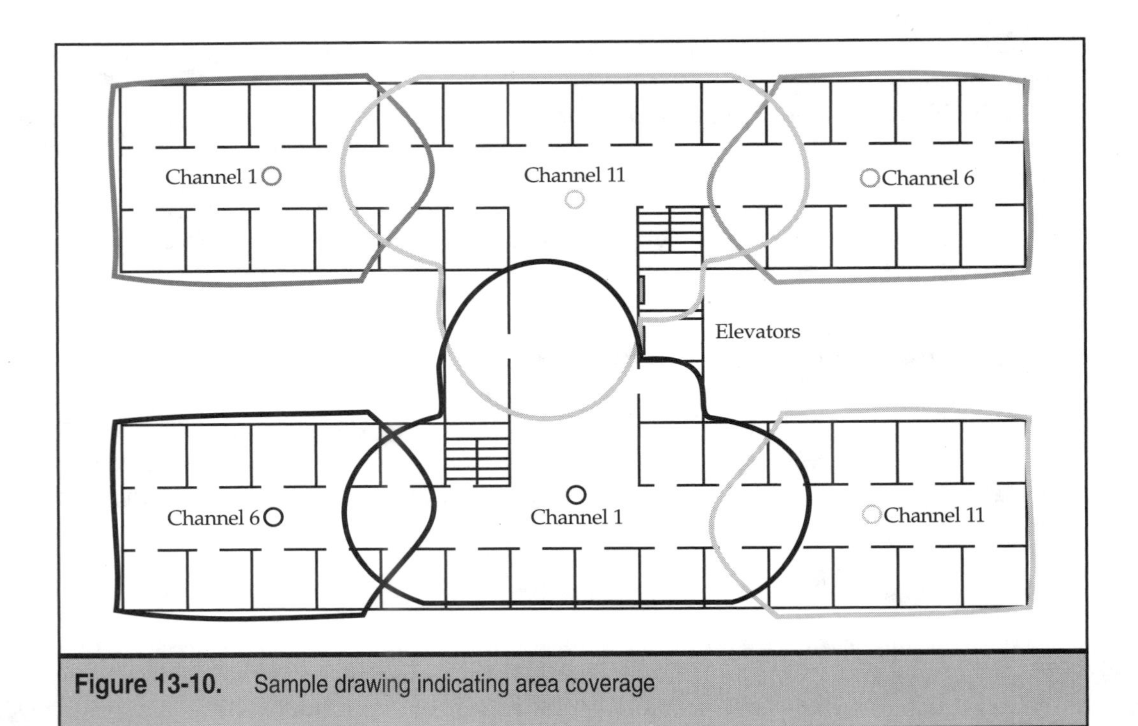

Figure 13-10. Sample drawing indicating area coverage

	Rubber DiPole	Pillar Mount	Ground Plane	Patch Wall	Ceiling Mount	Ceiling Mount High Gain
Type	Omni	Directional	Omni	Directional	Omni	Omni
Gain	2.15 dBi	5.2 dBi	5.2 dBi	8.5 dBi	2.2dBi	5.2dBi
Beam Width	360° H 75° V	360° H 75° V	360° H 75° V	60° H 55° V	360° H 75° V	360° H 75° V
~Outdoor Range at 2 Mbps	2000′	5000′	5000′	2 Miles	2500′	5000′
~Outdoor Range at 11 Mbps	500′	1580′	500′	3390′	790′	1580′
Cable Length	N/A	3′	3′	3′	9′	3′

Figure 13-11. Common 802.11 indoor antennas

It is often a good idea to include in the report writing a senior member or project manager from the installation team, whether or not it's the same team as the site survey team, because they will have information to contribute along with the customer network manager or technician.

A well-written final report will include at least the following information:

- Location of access points, as shown on the coverage drawings. Distances from at least two walls or other permanent elements are necessary to pinpoint location. Some site survey teams place small markers with felt-tipped pen, or use other permanent or semi-permanent marking, such as pins with tags.
- Antennas designated for each access point- critical to ensure performance and coverage as indicated in drawings. It is important to remember that access points can only be ordered with specified antennas.
- Suggested configuration of each access point, including speed, channel, and WEP setting
- Drawings of coverage areas
- Photographs of the coverage areas
- MAC addresses of each access point
- Number of packets sent to each access point during the test
- Packet size
- Data retries for each access point
- Delay between packets for each access point
- Packet success threshold

The configuration, packet size, and other information, such as delay between packets, will be gathered from the site survey utility. The report will also indicate AC power sources and cable routings to each access point, and site survey team contact information. The objective of a good report is to minimize the number of calls from either the installation team or the customer.

Another good reason to assemble a final report with as much detail and clarity as possible is that the installation may not begin until months following the site survey. This is also justification for the use of many photographs, as the physical layout of the facility can easily change over a short period of time, and the movement of large objects over even a few feet can have a substantial impact on the area of coverage. The movement of these objects need not take place in the area of *intended* coverage; rather, movement of the large objects in an adjacent room or floor can also have an impact on the area of coverage.

The good news about having a lag period of weeks or even a few months from the time of site survey to the time of installation is that the system will be tested during installation and checked against the site survey information. Adjustments to optimize areas of coverage are typical, but should not occur as a matter of routine during the installation as long as there have been few changes to the physical area that has undergone the site survey.

Sample Deployments

Currently, there are well over one million 802.11 devices worldwide, which makes for tens of thousands of deployments, if not hundreds of thousands of deployments. The following examples illustrate some of the different types of deployments. These are real-world examples, although the names and locations of these deployments are not disclosed.

- Major retailer design
- Hardware warehouse design
- Educational facility design
- Multifloor medical facility designs 1 and 2

Major Retailer Design

This major retailer, part of a U.S. national chain, requires coverage over an immense floor with over 17,000 square feet of coverage. High-gain omnidirectional antennas were spaced equally in the ceiling to ensure that the retail racks, which were made of metal and as tall as seven feet, did not provide blockage between the client adapters and the access points. The store had cabling available only at the center of the ceiling to which the access points were connected for both AC power as well as to the Ethernet backbone.

In this type of retail operation, the cash register stations are not along one wall but instead are relatively equally distributed along the entire floor space (see Figure 13-12).

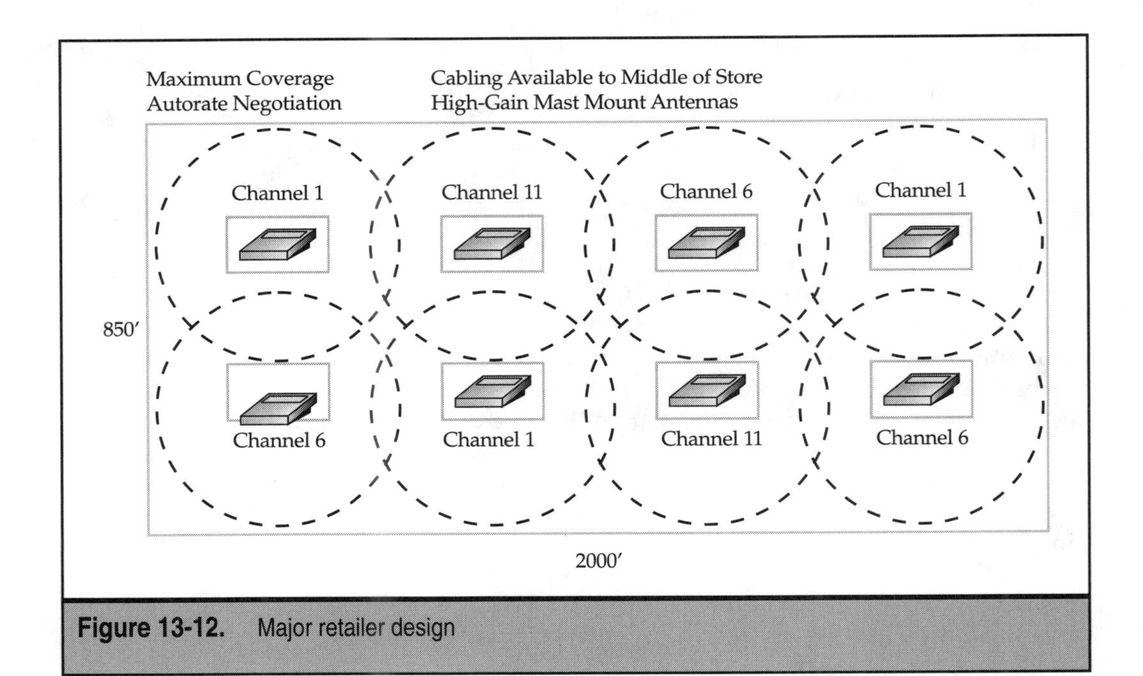

Figure 13-12. Major retailer design

In this deployment, the only power and Ethernet cables available were located at the middle of the building. Dipole antennas were used to ensure ubiquitous coverage around the access points.

Warehouse Design

This design is for a U.S. national warehouse chain that sells hardware. As in the last example, approximately 17,000 square feet of floor space had to be covered, and the only available cabling was at the front of the building. This meant that the access points had to be wired to the front of the building. This necessitated that the access points be mounted not in the ceiling, which would have been ideal, but along the front wall, close to the cabling for both AC and Ethernet.

This is a common design for retail deployments when a good portion of the cabling is placed against one wall, which is generally brought on because of the long rows of cashier stations. Note that the access points had to be mounted very high along the wall for the same reason as in the first example: to clear tall metal retail shelves.

For this reason also, there is more overlapping of channels in a given physical area so that each client adapter has a greater probability of having a line of sight to at least one, if not two, access points. In this application also, there were two types of antennas used on each access point, a dipole and a Yagi (Figure 13-13). A Yagi antenna has a lower amount of angle as its directional, but in trade it has a higher amount of gain, which proved optimal

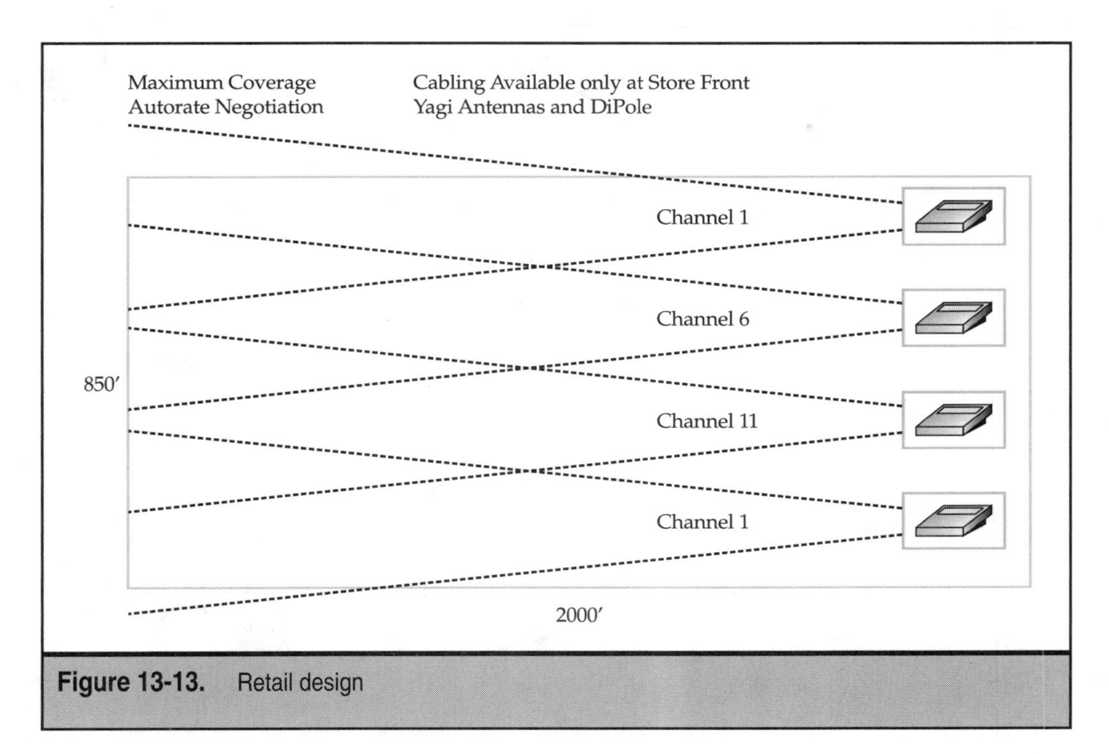

Figure 13-13. Retail design

for this deployment because it increased the probability of providing and receiving sufficient signal strength to and from the client adapters on the floor.

This deployment is for a retail design store where both the power and Ethernet cables were available only at the front of the store. The access points and clients auto negotiated the data rates depending on the range and path occlusions. Both Yagi and dipole antennas were used to ensure a balance between range and ubiquitous coverage.

Educational Facility Design

This example features access points that are utilized by client adapters in adjacent rooms. Also, note that two types of antennas were used in this deployment: dipoles for the indoors, and patch antennas for the outdoors. This antenna selection was based on the need for coverage that would be sufficiently wide enough to ensure the client adapters from two adjacent rooms would have sufficient signal strength to engage the access points. In this example, as in the previous examples, the speeds were auto-rated based on signal degradation from the walls and distance from the access points.

The patch antennas were selected for the outdoor site because this choice proved to be the most esthetically pleasing (in other words, caused the least negative visual impact), while still being able to provide excellent coverage over a wide area in the courtyard (see Figure 13-14).

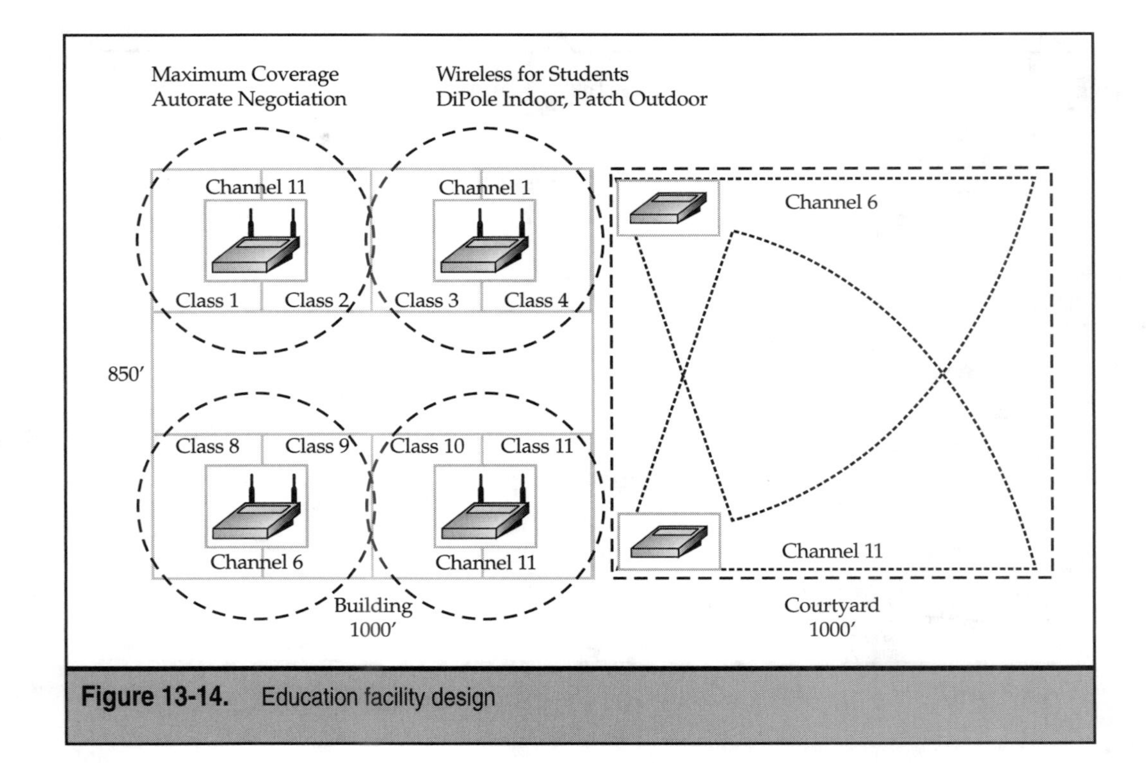

Figure 13-14. Education facility design

Multifloor Medical Facility Designs 1 & 2

This example is the most unique of the examples as well as the most complex, because this deployment had to take several key issues into account:

- Prospective impact on sensitive medical devices
- Complex coverage area
- Identical floors both above and below

The issue of sensitivity to medical devices by 802.11 radios was covered earlier in this chapter. After the hospital staff confirmed that the 802.11 radios did not interfere with sensitive medical devices, the advantages during the deployment phase, in addition to ease of use and exceptional area coverage, made this technology compelling.

The complex coverage area was made even more challenging because of the fact that the access points were close enough to windows that the floors both above and below could pick up signals, thus making channel assignment mission-critical. This meant that the channels had to be selected based on not only the complexity of the floor design, but also had to be different from the floor directly above and below the area of coverage. This deployment had to take into consideration that the total area of coverage was cubic feet, not square feet, because the access points would provide signal to adjacent floors.

The diagram on the left in Figure 13-15 indicates the area of coverage on a single floor. The diagram on the right reveals a look at the deployment from the side—that is, over multiple floors.

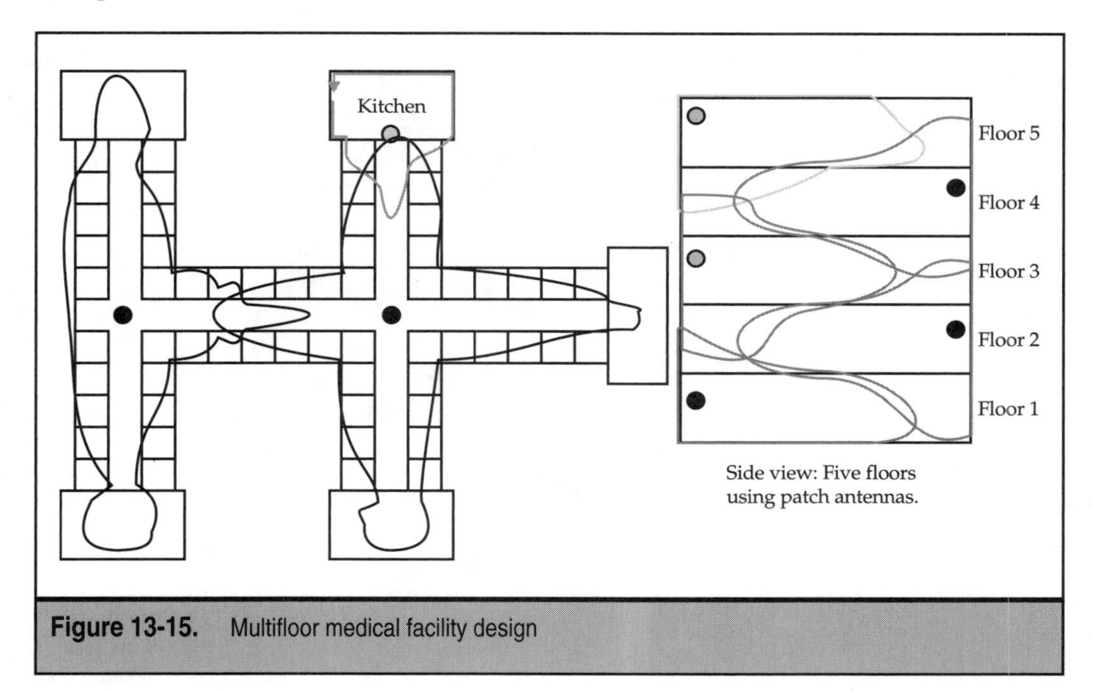

Figure 13-15. Multifloor medical facility design

802.11 WLAN SUMMARY: A COMPELLING TECHNOLOGY AND STANDARD

The successful deployments of the challenging opportunities outlined in this chapter verify the flexible and highly capable nature of the 802.11 radio. In indoor as well as outdoor designs where data rates on the order of 11 Mbps are sufficient, this radio architecture is quite compelling. When other key elements are factored in, such as very low cost per megabit, ease and speed of installation, lack of license requirements, and the fact that the state-of-the-art versions are shipped turnkey with antennas and all other components included, this is truly one of the most compelling and exciting radio architectures available today.

On a final note, this radio is rapidly becoming the de facto standard, not only in the United States, but also worldwide where data rates do not need to exceed 11 Mbps. Even this factor will be mitigated close to the time this book is published with the advent of the 802.11a standard, which will initially ship radios with speeds of 48 to 54 Mbps. While non-line-of-sight capabilities may not be delivered initially with the 802.11a specification, there can be little doubt it will eventually become a standard feature.

Projections by companies like Cisco Systems, with its Aironet product line, are looking at speeds of 54 Mbps near the first half of 2002, and later of 100 Mbps. These speeds will require development teams to overcome some challenging technical hurdles, and it is fully expected that they will rise to meet the challenge.

In the meantime, the market can enjoy a radio that is arguably, dollar for dollar, a tremendously attractive investment and asset.

PART V

Comparison of Technologies

CHAPTER 14

Frequency Selection Issues

Today's service provider has a very wide range of choices in terms of what BBFW technology is available. These options include bands from 2.46GHz, 3.5GHz (for non-U.S. use) and up to the millimeter wave spectrum and also includes the quick-license 6GHz and 23GHz options in the U.S. The real question is not whether a suitable BBFW technology can be selected from a broad array, but rather which technology offers the most favorable set of compromises.

LAYING THE FOUNDATION FOR THE COMPARISON

To make a selection as important as which BBFW technology to pursue and deploy, and the equally important decision of which frequency band to choose for the actual transmission, the user must consider a wide range of issues. Once these issues are resolved, the selection of which technology and transmission band to deploy becomes reasonably apparent.

It is tempting in today's market to simply go with the vendor who appears to have the best "story." Many company heads have had the experience of listening to sales pitches from various vendors only to become increasingly disappointed in the level of BBFW knowledge possessed by the sales team. Small radio vendors often have sales personnel that provide the most accurate information with a commonly a solid understanding of radio requirements, but if the opportunity is larger than several million dollars total, most companies won't risk partnering with a small vendor.

The larger vendors, on the other hand, seem to have most if not all of the entire package, which includes a site survey, deployment, financing, network expertise, and the ability to illuminate the numerous business case issues. Because these issues vary greatly from ILEC to CLEC to Greenfield to Tier 2 and 3 ISP customers, there is a considerable amount of expertise that must be mastered by not only the vendor but also the service-providing customers themselves.

Technology Is Easy—Business Is Hard

Technology in general is easy because it is primarily conformance to the requirements. Given the difficulties of developing BBFW hardware or developing a business, developing the hardware is far less complicated than developing a business or providing a proven value proposition. In general, technical objectives are easier to meet than business objectives.

Technology developers have the distinct advantage over business managers in providing an appropriate deliverable because of the applied science, mathematics and engineering available, in addition to an enormous amount of experience, which is either held within the development team or reasonably accessible outside the team. Further, there is no shortage of expertise that can be hired for a period as short as one hour to a program that extends over several years. This expertise is available both locally and internationally.

Technology is also easy compared to business for other fundamental reasons—it is much more predictable, easy to test, and generally does not have catastrophic results if it fails. In fact, an entire industry is built on failure analysis, the study of the ways in which technology systems fail.

Business is much more difficult than technology, and unfortunately (or perhaps fortunately), it is the ultimate test of a technology, because technology is little more than a catalyst for a successful business. Every person who has experience selling technology knows that offering a technology with the potential of making their customer considerable amounts of money does little more than get them in the door. At that point, the hard work begins. Rarely does someone buy technology strictly for technology's sake. To quote Mitch Taylor from Cisco Systems, "What good is a better mouse trap if your customers don't have mice?" The key issue is that technology providers have a responsibility to provide value propositions to their customers; in other words, they have to be able to make accurate projections in terms of how the BBFW equipment and its associated network elements will generate profit. Technology providers need to be able to accurately model, and therefore predict, the financial outcome of deploying a technology. The bottom line is that customers are now far less willing to verify business plans by large scale deployments as opposed to having a firm grasp of how, how much and when the equipment will provide a profitable return on their investment *prior* to purchasing the equipment.

Business, then, is more difficult simply because it deals with the issues of psychology and politics. Winning at this aspect of the engagement is far more difficult, far more risky, and far less predictable. Business issues are also more difficult than technology issues because they lack accurate models to predict the possible outcomes of an engagement between a vendor and a customer.

Having established that business issues are more difficult to determine than technology issues, business issues nevertheless have to be dealt with to the best of the vendor's and customer's ability. Fortunately, measurable criteria are available by which a customer can select the best of the three common U.S. spectrum-based products.

However, there is no "best" of the three. It doesn't even truly matter which one has more deployments than the others. It doesn't necessarily matter that one of them has far more spectrum than the others, that one of them requires far less capital up front, and that one of them is the best performing, dollar for dollar.

The key metric is what is right for the only business that truly counts—yours. The balance of this chapter provides you with a set of issues that must be resolved to help make this decision.

The Primary Business Issues

The industry recently has suffered a tremendous downturn in purchasing. Companies across the world that deal in technology are downsizing. At the time of this writing, approximately 100,000 individuals recently have lost their jobs in the Internet industry due to layoffs. This is a considerable figure by any measure, and most of the reasons for these layoffs are essentially outside the scope of this book. What has been learned, and what falls within the scope of this book, are the business considerations of each of the aforementioned BBFW technologies.

The Internet industry has learned a very harsh lesson in the winter of 2000 and the spring of 2001, and it is one that will not soon be forgotten. The lesson is how businesses can tightly align conventional fiscal metrics to network designs. The ruthless lesson is rooted in the fact that the Internet industry at large is now a major contributor to the overall Gross National Product (GNP) of the United States and any other highly developed country.

The Internet industry has been a "prime time" contributor at the GNP level for a longer period of time than many people realize. While these metrics continue to evolve, particularly in terms of market cycles, and the Internet industry is contributing to this evolution, the fundamentals must and will remain in place.

This means that the metrics used in the "boom-boom" days of the Internet, which occurred in the late 1990s, such as "sticky Web pages" and "eyeballs," have gone the way of the dodo bird. These metrics were poorly adapted to ensuring the health and growth of the industry, and the industry suffered, in part from using these tools, which were poorly designed. Since the Internet industry "crash" of 2001, the entire industry—from the venture capitalists and bankers to development teams and all the way out to the pundits and consultants—has agreed that the only metrics that are truly reliable and useful are the time-honored and well-honed metrics of revenue and profit, and the supporting forms of measuring tools. However, even companies that retained excellent balance sheets, like Cisco Systems, saw reductions in stock price and company valuations, as the general market had expectations that were not sustainable.

There are essentially two types of companies that use the Internet: those that use the Internet to generate revenues, such as service providers, storage area network providers, and content providers, and those that use the Internet to improve corporate operations. While there are tremendous differences in speed in how Cisco Systems operates compared to a major bank, both companies will use the Internet in approximately the same way.

Fundamental to the growth and technology development of the Internet industry is how well the equipment does the following:

- Reduces the cost of doing business
- Increases revenue and profit
- Increases the profit contribution of employees
- Reduces the time to market for products and services
- Improves the way individuals live, work, and play

Reducing the Cost of Doing Business

During times of economic downturn, businesses tend to focus more on improving profitability than implementing rapid growth. Properly deployed and managed network equipment can provide enormous gains in efficiency, can reduce the time required for overhead tasks such as human resources, billing, and accounts payable, can facilitate handling customer orders, and can even help to troubleshoot customer problems.

While a large number of retailers and service providers will look to BBFW and other final mile mediums to generate revenue, virtually all companies will eventually migrate to higher-capability LANs for their internal use with applications that enable a business to reduce its operating costs. It's important to note that networking equipment does not provide reduction in costs in and of itself; rather, it is the improved efficiency gained through the applications that travel over a higher speed or generally improved LAN that reduces costs. LANs and the Internet at large merely (though importantly) provide the mediums for these applications and their associated data and storage requirements.

BBFW can reduce the costs of doing business by providing a more efficient means of transportation, such as a telco bypass between offices or storage area network sites, or a toll bypass for telecommuters. It can also reduce costs for companies that must lease fiber over long distances by installing a system that they own as opposed to leasing from another entity. The decision matrix for which BBFW technology would be optimal is covered later in this chapter.

Increasing Revenue and Profit

The concept of increasing revenue and profit is mostly germane to service providers, but also extends to virtually any retail operation. Some retail services can also be provided over the Internet, such as income tax filing, credit repair, movies on demand, and automobile insurance.

Each of these entities relies on two items: advertising and the exchange of products or services for revenue. This can be accommodated as nicely over a BBFW link as over any of the more traditional wired formats. A BBFW link should be used to augment other mediums between the retailers and the customers, and should be used where it can enhance the revenue stream. The key point here is that a BBFW link can be used as an additional medium to engage customers.

Increasing the Profit Contribution of Employees

Increasing profit contribution is primarily based on improving two items relative to employees: speed and efficiency. If an employee can exchange a 90-minute commute by working out of a home office, they gain nearly three hours *per day,* which they could invest back into profit contribution to the business by making extra sales and completing the tasks associated with sales.

Even if the employee gains only one hour per day as a profit contributor because of a BBFW link that enables them to telecommute, they would add an addition *240 hours per year*—the equivalent of more than an extra month of profitable time. This factor multiplied across even a small number of employees can quickly add up to an appreciable effect on corporate earnings.

In terms of gaining efficiency, there is little in the business world that can compare in value to being able to access relevant information in real time or close to real time. This information can be on anything from resolving a personal computer problem to being able to research a customer prior to a sales call, or in having efficient communications between employees and their managers. Another example many can relate to is real-time stock ticker information.

Reducing the Time to Market for Products and Services

Having a highly efficient medium for offering products and services is one of the keys to the success of a business. In countries like the former Soviet Union, one of their greatest challenges is that they possess a deficient infrastructure to get raw goods to market. Study after study indicates that countries with highly efficient infrastructures offer their citizens a higher standard of living.

This concept extends to BBFW links. If a service provider, or movie-on-demand provider, were able to blanket a city with BBFW links, they could quickly move from a buildout mode to a revenue generation and profitability mode. While installing BBFW links is no assurance of success, John Chambers, CEO of Cisco Systems, has stated repeatedly, "The competition will not be between the big and the small, it will be between the slow and the fast." Having a high-performance, broadband medium to carry services provides one of the most effective and fastest strategies to reducing time to market for products and services.

Improving the Way Individuals Live, Work, and Play

The Internet will continue to change and improve the quality of our lives. The Internet unquestionably has had an impact on today's society, on a global basis. The Web browser has been deemed by some as the killer application that everyone has been looking for. High-speed fixed wireless links provide an effective medium in addition to DSL, cable, satellite, and optical connections used at and between places of work.

So how do you decide whether BBFW is the right medium for your business? And if it is, what technology should be used and what frequency band will it operate on? You must also address questions such as whether BBFW will be profitable, whether it will provide suitable geographic coverage, and, even more basic, whether it will meet the functional and operational requirements of your business and network. The next few sections will attempt to answer some of these questions to help you determine whether BBFW can be deployed profitably by service providers in the market of choice. As stated earlier, the final decision may be more art than science, but there are some basic decisions, based on "concrete" areas such as physics, that can help the decision along.

Decision Criteria for Selecting the Most Appropriate Frequency

The decision as to which BBFW technology and spectrum to incorporate can best be made following a comprehensive review of a wide array of issues, ranging from careful financial analysis to the various nontechnical issues such as licensing. Prior to deciding which frequency should be used, the following issues should be analyzed:

- License
- Distance
- Data rates
- Local geography/path obstructions

- Interference
- Capitalization
- Expertise
- Vendor offerings
- Vendor analysis
- Equipment analysis
- Hybrid system or not
- Scaling
- Maximizing profit
- Migration to next generation

License

The selection between the offerings truly comes down to whether you select a licensed or an unlicensed BBFW product. Some networks use hybrids of both licensed and unlicensed BBFW links, but the preeminent factor is whether or not the buyer will purchase a system requiring a broadcast license. Recall that Chapter 12 goes into detail on this subject, but the general summary is that if the service provider does not have a license, it is generally not likely be able to afford a license or acquire one from an entity that purchased it during the original auction.

Further, some licenses like the MMDS license are not without their own peculiar conditions. For example, the ITFS band, which is owned by schools and television stations, is right in the middle of the MMDS band. This means that companies that acquired the MMDS spectrum through auction will have to negotiate with every commercial broadcast station and school in every BTA that has an ITFS license, and do so to the mutual advantage of both parties. Because the ITFS spectrum is a 60 MHz contiguous band, this would represent a loss of over 25 percent of the MMDS spectrum that the companies purchased. A delicate and light diplomatic touch, alloyed with financial ramifications favorable to the ITFS owner, is likely in order. This will add to the already staggering cost of the licenses themselves.

What this author has seen are joint ventures between schools that own ITFS licenses and local service providers. Some schools and broadcast entities would rather deal with locally owned service providers than lease or lend the spectrum to a large, national carrier. What typically happens in these joint ventures is that the school lends the spectrum to the service provider in exchange for connecting the school, including all equipment and maintenance, at no charge to the school for the duration of the agreement.

This can be a favorable arrangement for both entities and is popular in some regions, such as Virginia and parts of California. Because only 60 MHz of ITFS spectrum is available, the BBFW network is typically augmented with U-NII for backhaul where the pipes need to be at their largest, while the ITFS is usually deployed at the P2MP portions of the network.

The quick-license BBFW products, such as the 6 and 23 GHz products offered in the United States, should be considered as having the benefits of a license and the low cost and ready availability of unlicensed spectrum. However, while they are indeed licensed and therefore provide the protection, assurances, and increased asset value to a business, they cannot be used for revenue-generation purposes. Other limitations include a grandfathering clause that stipulates that the first business to be granted one of these licenses in between two points (not for a specific geography) has the sole rights to that transmission band.

These bands are almost always limited to short-haul, point-to-point links, seldom ever point-to-multipoint, and not for deployment over the last mile for revenue situation, and should be considered as possible solutions for backhaul situations.

This can be a very positive development if it is your business acquiring the license, and not such a positive development if you want one of these licenses but one has already been allocated near your business such that another license will not be granted. It should be stated that the likelihood of being prohibited from purchasing either a 6 GHz or 23 GHz quick-license is possible, though unlikely, because all an applicant has to do is verify noninterference with any existing license holders.

On the other hand, there are many advantages to deploying a network with unlicensed spectrum, with the prime caveat being that the technology deployed to carry and manage the unlicensed spectrum has the ability to mitigate interference.

Having stated that caveat, the most significant element of deploying a BBFW network without a license is that the service provider or business owner can immediately move to a revenue generation mode upon getting the system up and running, as opposed to making substantial investments in a license and then having to utilize additional significant amounts of money to deploy the hardware, software, facilities, and personnel to enable a BBFW network.

Simply being able deploy a network without the enormous fiscal allocation to purchase a right to broadcast makes U-NII the prime contender for outdoor BBFW use, in the opinion of this author. Having stated that, it should be noted that more ISM systems are deployed for outdoor use than U-NII, but it is the opinion of this author that the number of deployments in U-NII will be greater than that in the ISM band by 2004.

Distance

One of the more important decision criteria is that of distance. The higher-range frequencies such as LMDS do not lend themselves well to distances over two to three miles. The same is true for NLOS features in the U-NII band, which can have ranges as short as three-tenths of a mile in highly obstructed paths in a point-to-multipoint deployment. This is due to the relatively low maximum power allowed by the FCC.

On the other hand, a U-NII point-to-point link can provide high data rate links at ranges over 20 miles in a line-of-sight deployment. The lower the frequency, the longer the distance for a given power, as discussed earlier in this book.

Data Rate

Data rates are an important issue to determine and, as stated earlier in this book, the data rates should be calculated based on future needs, not on what the network is generating at the time of BBFW selection. In general, a good rule of thumb is to calculate what the required data rate will be in 18 to 24 months from the time the decision has been made to deploy BBFW. You can also refer to the Table 4-4 in Chapter 4 on nominal data rates for various frequencies, but in general, the higher the frequency, the greater the data rates.

Local Geography/Path Obstruction

The lower the frequency, the better the inherent signal propagation, as stated earlier in this book. This is why mobile networks don't have frequencies higher than 1.9 GHz, and most of it is in the 800 MHz range. If the deployment is to take place in an area with many valleys and hills greater than several hundred feet in elevation, the cost associated with placing towers and accessing the towers will have to be factored into the decision matrix.

Interference

One of the issues well understood in fixed wireless is that there is more interference at 2.4 GHz than in the higher frequencies. This should be an issue taken into consideration, and in the North American market, this works to the favor of deploying a network in the U-NII or MMDS licensed frequencies.

Interference is of course a greater issue in the unlicensed frequencies than in the licensed frequencies, and you should refer to Chapter 12 on mitigating interference in unlicensed spectrum.

Capitalization

The cost of a license is extraordinary in most cases, typically in the tens of millions of dollars depending on the spectrum, because some licenses like the quick license BBFW links are on the order of $5,000. At the other extreme, a combo A and B LMDS license would provide 1,300 MHz of spectrum. In a highly populated metroplex like Dallas, New York City, or Los Angeles, this license would easily go for tens of millions of dollars.

Sprint and MCI WorldCom collectively spent over $1.3 billion to acquire approximately 83 percent of all the MMDS licenses in the U.S. They have precious few relative deployments of links at this writing, and the FCC has a mandate to reacquire licenses that are not used within three years of purchase. At this writing, both Sprint and MCI are approximately halfway through that grace period, although it is highly doubtful the FCC will retract a licensing of that magnitude, and if an attempt were made, the legal wrangling associated with that attempt would continue for years, at which point the deployments should be effectively completed.

The challenge with purchasing a license is that a very large sum of money is spent before any means is deployed to recoup this expense. Consequently, a very large outlay of capital is made without a direct means to repay it. This calls for both lenient and very long-sighted loan and other capitalization sources, such as the sale of publicly traded stock. Therefore, the opportunity to own a licensed BBFW link is essentially out of the reach of small companies.

Expertise

One of the issues to weigh when considering which BBFW spectrum or technology to purchase is what BBFW expertise resides within the company, because although many individuals are available who have some wireless experience, most of them have only mobile or amateur radio experience. Few have BBFW experience as it relates to a network.

What many companies do to resolve this issue is hire a consultant on a contract basis and then train their current network managers to also handle the BBFW links. Many customers rely heavily on the vendor for expertise and even for the training of the customer employees. In general, some BBFW products are easier to manage than others. Specifically, this refers to BBFW products that are fully integrated at Layer 3, as opposed to operating as a bridge with unique software and management requirements.

Vendor Offerings

In the United States, the majority of the hardware vendors offer solutions in the LMDS, MMDS, and U-NII bands. At this writing, both 3Com and Lucent are also offering the ISM WLAN gear bridges for service provider use, but, as detailed earlier in this book, this author discourages the use of 802.11 radios for service provider use. This will change as the 802.11a products come to the market and are integrated at Layer 3 with more sophisticated network and element management tools.

Vendor Analysis

As stated previously in this book, the customer is well advised to consider whether the vendor understands the end-to-end network requirements, as opposed to having very deep knowledge that is restricted to the radio link itself. If the vendor doesn't understand the end-to-end network requirements, the customer would be well advised to engage a vendor who does understand, or plan to hire a consultant or integration partner who understands, the full end-to-end network and the implications of end to end issues such as security, quality of service, homogenous network and element software management systems and ease of integration in addition to scaling and migration.

In the best possible scenario, the vendor will provide a product that integrates at Layer 3, because this radio will most likely be the easiest to manage on a continuing basis. The reason for this is that it will appear to the network and the network managers as nothing more than a link between a pair of routers.

Equipment Analysis

Once the preliminary analyses have been made, the service provider or other user should have at least two or three radios to select from. If they have no license or prospect of acquiring a license from a joint venture with the local ITVS holder, or leasing it from an auction winner, the customer will likely be evaluating several versions of the unlicensed U-NII product and possibly several versions that function in the ISM band.

It is the author's recommendation that before purchasing gear, the customer should visit the company sites to witness the BBFW links in operation. There is the theory that a customer can simply visit other customers from their prospective vendors, but this rarely occurs, because companies with BBFW links in operation tend to very carefully guard the hard-earned information they acquire about operating and optimizing a BBFW link. Further, training prospective customers from a vendor is not a source of profit for the customer with the equipment up and running.

One possible exception to this is Cisco Systems' Cisco Powered Network (CPN) customer base. A Cisco CPN customer is one that purchases a certain amount of equipment per year and also has Cisco equipment from end to end in its network. One of the requirements of a Cisco CPN customer at the time of this writing is to assist other CPN customers in evaluating both equipment and best practices. This is one more advantage to acquiring equipment from a large networking company.

Once the customer has evaluated the equipment at the vendor site, and possibly at other customer sites, they can begin to trial the equipment at sites of their own selection. At this juncture of the program, the customer will likely have selected the BBFW technology and spectrum. This is the optimal time to have a "bake off" between the vendors to determine not only which gear is the most appropriate for the customer's operation, but also which vendor provides the best understanding of the equipment as well as the best support.

Hybrid System or Not

The odds are fairly high that the installed elements will at first, if not eventually, be a hybrid of at least two BBFW technologies and spectrums. In some cases, the system will be comprised of an outdoor BBFW link and an indoor WLAN system. While these systems have very different requirements and uses, they are, in essence, just two parts of a network (WAN and LAN) that have a commonality, which is the Ethernet port on the CPE side of the BBFW link and the infrastructure end of the WLAN. The BBFW link can also interface directly into the LAN core and not be directly associated with the 802.11 or HomeRF WLAN gear.

The system may also be a hybrid of U-NII and ITFS or either LMDS/MMDS and U-NII, for a number of reasons. This rationale includes market penetration into BTAs for which the service provider does not hold a license, or in the case of where a business has one end of the link that is outside the BTA for which they have a license.

The U-NII link can also be used for backhaul purposes to preserve licensed spectrum. This enables more flexibility on the matter of frequency reuse, as backhaul links tend to require large pipes with high speeds, and therefore large amounts of spectrum.

Scaling

Scaling is a key issue that has close associations with frequency reuse, available spectrum, and the ability to deploy a large number of radios that are relatively cost effective. This would effectively rule out LMDS, because their radios are too expensive to scale for large operations. At this writing, there are no large-scale LMDS networks online in the U.S., primarily for this reason. LMDS license holders were some of the first companies to file for bankruptcy in the "dot-com crash" of 2001.

Of the three major U.S. spectrums—LMDS, MMDS, and U-NII—MMDS scales best because it has 200 MHz of frequency and can provide channels as narrow as 1.5 MHz. U-NII products would finish second on this list, with 100 MHz of spectrum and channels as narrow as 6 MHz, although there are commonly other options such as polarization of antennas, inversion of transmission versus reception frequencies, and the use of different areas of the spectrum, which generally are made available in contiguous 25 MHz blocks.

Maximizing Profit

Addressing the issue of how to maximize the profit generated by BBFW equipment varies tremendously in terms of time as much as quantity. The amortization of a license can take as long as ten years. Again, witness the Sprint/MCI situation, where they are more than two years into their projects with very little revenue generated from their licenses. It often requires at least one to two years before a meaningful revenue generation phase is achieved. Increased profitability will occur on the order of 6 to 12 months as policies and procedures are implemented to streamline the installation phase, which should be dwindling at that point of the program.

Once the target geography has reached most of its potential, which will be on the order of 10 to 15 percent of the market opportunity, profitability becomes easier as the program shifts from a heavy deployment phase to troubleshooting and scaling the core and edge portions of the network.

The advantage in this case would go to the unlicensed BBFW products for the simple reason that they do not carry the overhead of a license.

Migration to Next Generation

Migration to next-generation devices generally occurs at the core and then the edge of a network before ultimately extends to the CPE devices. While migration to newer equipment that is often better, , changing out CPE equipment in a P2MP configuration is a task that occurs with the least degree of regularity.

Where CPE and headend upgrades are required, the optimal network will not be frequency-dependent, but rather architecturally based. In other words, the BBFW links that are designed with outdoor headends and externally mounted radios at the CPE side that provide an intermediate frequency to indoor gear will be able to make the physical upgrades more easily and with fewer unexpected challenges. Using these architectures with intermediate frequencies between the outdoor units and the indoor units means that there will be much less disruption to the indoor equipment.

This is especially critical at the CPE side, because reconfiguring indoor CPE gear could easily add one to three hours per site, which would have negative effects on the profitability of the operation.

Decision Table: Which BBFW Technology Has the Advantage?

Table 14-1 provides a snapshot as to which technology or spectrum may be optimal for a user. The spectrums are given a 1, 2, or 3 rating. The lowest score at the end of the comparison indicates which spectrum may be the most preferable for the greatest number of users.

	U-NII	MMDS	LMDS
License	1	2	3(1)
Capitalization	1	2	2
Expertise	2	1	3
Availability	2	1	1
Testability	1	1	1
Hybrid	1	1	1

Table 14-1. Relative Scoring of Frequencies (lower is better)

The following are some notes regarding Table 14-1:

- LMDS licenses are auctioned in two blocks of contiguous spectrum in the U.S. The "A" block is much more expensive than the "B" block because it accesses approximately 1000 MHz of spectrum versus 300 MHz of spectrum. Price also depends greatly on whether or not the BTA is in a major metropolitan area.
- If the licensee possessed the full 1300 MHz of spectrum, it would be able to scale quite well. Important to remember is that this spectrum and technology are primarily intended for SMB and large businesses as opposed to residential use.
- LMDS has arguably the least amount of resources applied in the industry in terms of providing next-generation systems. This is primarily due to the fact that there are not a lot of companies with extensive millimeter wave radio development experience, which tends to help keep the prices high and the available product low.
- This table is primarily aimed at the bulk of the BBFW buyer market. There are only a handful of very large buyers, such as Sprint, MCI, SBC Communications, AT&T, and the like. The bulk of BBFW purchasers are small- to medium-sized businesses.

THE FINAL ANALYSIS

In the final analysis, if your company has either an MMDS or LMDS license or can get access through a joint venture with an ITFS license holder, then the advantages of having a license are clear because your company will have a long-term right to broadcast without competition in a defined geography. Further, if the company has the license, it likely has already been purchased. The major drawback to a license is its cost. The secondary drawback is that the owner has a relatively limited amount of time to begin deploying the licensed equipment. Given that this period is approximately three years, there is ample opportunity to move well into the deployment phase.

If, on the other hand, the entity is considering augmenting its network with BBFW links for storage area networks, toll bypass, or even to provide services to SMBs, the only real choice is U-NII. While this equipment is generally not designed for residential use because of its high cost, especially at the CPE end of the link, this cost will most certainly continue to come down, especially as the demand for the equipment increases.

Because the bulk of the BBFW users in the U.S. and around the world will likely be small- to medium-sized businesses in terms of bandwidth used although eventually residential deployments may ultimately have more deployments using more bandwidth. Having stated that, the most important selection criteria of a U-NII product will be whether it has the best resolution for co-channel interference. Equally critical, the issue will be how well the unlicensed band will operate in a NLOS environment, as the allowable power is almost always considerably less than that allowed for licensed deployments.

The user is well advised to compare all vendors that offer U-NII gear to determine how these mission-critical elements are managed, along with the many other issues, such as vendor solvency with a keen eye to the concept that today's BBFW links are network elements, not merely radios tacked onto the edge of a network.

PART VI

Standards

CHAPTER 15

WLAN Standards Comparison

Wireless local area networks (WLANs) are big business. Intel has published these statistics on its HomeRF Web site from studies performed in 2000 by both Parks Associates and In-Stat Group:

- WLAN sales currently exceed $10 billion.
- More than 25 percent of all U.S. households will be networked by 2004.
- $4 billion of home networking gear will be sold in 2004 alone.
- More than $10 billion in home networking appliances will be sold in 2005.

COMPETING WLAN STANDARDS

Currently, three competing WAN standards exist, and there is certainly some confusion and inaccurate information in the marketplace about the issue of an indoor "standard."

Intel, through its constellation of cooperating companies in the HomeRF standard, has selected the residential market as its primary target, while the 802.11 standard is more commonly deployed in the SMB, Enterprise, and SOHO markets. Having stated this, 802.11 is rapidly becoming the WLAN standard of choice for residential users, and there are many low-cost retail products that utilize this standard. IBM, with its Bluetooth market, has essentially pursued the commercial/retail market. HomeRF has placed products on the market with prices as low as $80 at Buy.com, while 802.11 equipment has been sold and currently is operating in more than 60 countries worldwide at prices ranging from $300 (list) for a PCMCIA card up to $1500 (list) for an access point. Bluetooth has had perhaps the largest press buildup by far of the three standards but has been utilized in the least number of devices at this writing. Bluetooth devices are not generally sold to the retail customer, but they're embedded in devices such as a PDA, laptop, or cellular telephone. The long-standing target price for Bluetooth chipsets is $20 to the manufacturers who then sell their products to consumers.

These three competing standards are represented summarily in Table 15-1.

The following are some notes regarding Table 15-1:

- 40- and 128-bit RC4 are very robust data security algorithms.
- The 802.11 range of 1000 feet is at outdoor conditions. Indoor conditions are more difficult for these types of RF systems. Using an 802.11 bridge will enable much greater distances than using an access point.
- 802.11 power output of 1W is substantial.
- The maximum number of devices supported depends on the data rate per device.
- Bluetooth is currently heavily restricted by the FCC for power, due to excessive spectral mask leaks. Optimal range for the revised version is estimated at 300 feet LOS, but is currently approximately 1 meter.

Both HomeRF and Bluetooth are standards that are called *clustered*. This means that the companies that support the standard are clustered around the lead development company. In the case of HomeRF, the company central to the effort is Intel, while IBM is the primary sponsor of Bluetooth. The constellations around each of these companies

	HomeRF	Bluetooth	802.11b
Physical layer	FHSS	FHSS	FHSS, DSSS, IR
Hop frequency	50 hops per sec	1600 hops per sec	2.5 hops per sec
Max. transmitting power	100 mW	100 mW	1W
Data rates	1 or 2 Mbps	1 Mbps	11 Mbps
Max. number of devices	127	26	2084
Security	Blowfish format	0-, 40-, and 64-bit	40- and 128-bit RC4
Range	150 ft.	30 ft.	400 ft. indoors 1000 ft. LOS
Current version	V1.0	V1.0	V1.0
Cost	Neither least nor most expensive	Least expensive	Most expensive
Physical size	Neither largest nor smallest	Smallest	Largest
Roaming outside of home	No	No	Yes

Table 15-1. Competing Indoor RF Standards

include channel resellers, which are groups that resell the products that use these standards, as well as silicon and other discrete electronics suppliers.

The 802.11 standard, on the other hand, was written and ratified by the IEEE. This standards body has no close association with any one company over any others. In terms of published papers, it excels the other two by order of magnitudes, and it is well known at every level of the engineering community, from college graduate to the most senior engineer. Rare is the engineer, if any, who hasn't studied at least one IEEE paper during their career.

One of the key reasons for having a standard is to ensure that equipment provided by supplier A will work with the equipment of supplier B. The company that fosters the standard often stands to make considerable financial gain, because it often offers a key element or feature that is required as part of the standard. Large companies like standards-based products because they can often sell more products to a single customer, especially if the customer is a large one. Large customers like standards because they provide stability to the basic product designs and ensure interoperability as they scale and migrate their networks.

It is important to note that only one of the three standards supports what can truly be called broadband, and that is the 802.11 standard. The author makes this claim because although technically HomeRF meets the commonly accepted broadband definition of 1.5 Mbps of data rate and while the data rate of 1 Mbps for Bluetooth (not goodput/throughput) is close to the broadband definition, the 802.11 standard far exceeds this by providing up to 11 Mbps of data rate, though it is shared among the users in the area of coverage. 802.11 incoming standards will quickly move the data rates to 54 Mbps by the year 2002.

There are two other major differences between the standards:

- Range
- Adoption rate by market

The 802.11 equipment clearly has a major advantage in range, the Bluetooth has the least range, and HomeRF is somewhere between the two. It should be noted in fairness that the Bluetooth's primary design was for very short-range use; approximately three feet or less, as it was designed to act as a wireless interface between devices like cellular phones and laptop or desktop computers, PDAs.

The 802.11 Standard

The IEEE adopted IEEE 802.11 in 1997, and it became the first WLAN standard. 802.11 as defined by the IEEE primarily controls Layers 1 and 2 of the OSI reference stack, which are the physical layer and the data link layer (sometimes referred to as the link layer), respectively.

The 802.11 Media Access Control Layer

As stated previously, the MAC layer is a subset of the link layer, which is adjacent to the physical layer in an 802.11 network or any other network. Layer 1 in an 802.11 network performs at least three essential functions:

- Serves as the interface between the MAC layer at two or more geographic locations. These locations are generally only a few hundred feet or less apart.
- Performs the actual sensing for the CSMA/CDCDevents, which occur within the MAC layer.
- Performs the modulation and demodulation of the signal between two geographic points where 802.11 equipment resides. This modulation scheme can be either DSSS or FHSS.

Importantly, the 802.11 standard defines a *rate-shifting technique* that enables networks to reduce data rates as changes in distances, signal quality, and strength occur. 802.11b data rates may be as high as 11 Mbps or as low as 1 Mbps with DSSS modulation, while FHSS-modulated data rates are either 1 or 2 Mbps. The standard also enables compatibility between 802.11a and 802.11b radios. The portion of an 802.11a network that uses 802.11b equipment will result in the lower data rates of the older standard.

The MAC layer is a sublayer of Layer 2 of the OSI stack and it controls the connectivity of two or more points through an address scheme. Each laptop or access point has a MAC address. The IEEE 802.11 standard defines how this addressing works, along with how certain aspects of Layer 1 operate. This standard is similar in many respects to the Ethernet standard established by the same standards body. In effect, it defines the following:

- Functions required for an 802.11-compliant device to operate either in a peer-to-peer fashion or integrated with an existing WLAN
- Operation of the 802.11 device within the range of other 802.11 devices, and how a client card will physically migrate from one access point to another
- MAC-level access control and data delivery services to upper layers of the network protocol stack
- Several physical layer signaling techniques and interfaces
- Privacy and security of user data being transferred over the wireless media

What makes a WLAN different from an Ethernet LAN is obviously the ability for users to roam from one point in the network to another. This is the most important feature of a WLAN as well as the one that distinguishes it the most from an Ethernet LAN. The way that the 802.11 MAC operates under this standard is what allows the higher levels of the OSI stack to function normally. In other words, the MAC layer is the one that handles the mobility issues of an 802.11 network.

It is for this reason that an 802.11 MAC layer is forced to take on certain functionalities normally left to layers higher in the OSI stack such as the session layer (Layer 5), which controls session initiation and termination. In the 802.11 MAC standard, the flow of information is performed on a best-effort basis, which is also called *connectionless*. Connectionless links are those in which the receiving end of the link does not verify the receipt of data by the transmitting link. The technique used by the MAC layer is called Carrier Sense Multiple Access with Collision Detection (CSMA/CD), which is a technique that requires the transmitter to "listen" to the local environment to ensure there are no other transmissions in the frequency to which it is assigned. The actual sensing is performed at Layer 1, but the timing for the transmissions is controlled at the MAC layer.

CSMA/CD is a protocol intended to resolve transmission conflicts. As stated, the transmitter will determine if there is a transmission in the assigned frequency of an access point or client adapter. If a transmission is in progress, the access point or bridge will wait for a specified period of time, following which it will determine whether or not the radio channel is clear. The radios are programmed such that the timing between the attempts to determine whether or not a particular radio channel is clear is random. Some simple statistics are employed that state that the highest probably for a channel to remain in use is just after an attempt to transmit was halted because the radio channel was in use by another transmitter. It is for this reason that time between attempts to transmit is randomly spaced. The amount of time between repeated attempts is often called *back off time*.

In most 802.11 systems, however, the back off time steadily diminishes until the transmitter determines there is an open channel. By having steadily diminishing but unequal time periods, a WLAN gains efficiency. It is easy to understand that network efficiency would suffer if all the radios on a common channel were to wait for an increasingly longer period. Because the radios listen for increasingly longer, though randomly selected, durations, the radios waiting to transmit traffic will wait the least amount of time.

In a best-effort architecture, there can be no guarantee that the data sent will be received successfully. One of the things an 802.11 system does to help ensure the successful receipt of information is to send the information repeatedly, which is called *chipping*.

One other function provided by an 802.11 MAC layer is that of security, which is typically handled at the presentation layer (Layer 6). The security measure compliant with this standard is Wireless Equivalent Privacy (WEP), which is a method for handling keys and encrypting the data. Refer to Chapter 11 for more information about WEP.

802.11b Versus 802.11a

At present, there are a number of 802.11 specification variants as proposed and ratified by the IEEE, which include standards such as 802.11b and 802.11a among others. The 802.11b specification was released to the market several years before the 802.11a specification. Table 15-2 illustrates the primary differences between the two standards.

The HomeRF Standard

The HomeRF standard is rooted in the Digital Enhanced Cordless Telephone (DECT) standard, and perhaps for good reason, because there are more than 200 million handsets that comply to this standard, with more than 100 suppliers worldwide. This also explains why the HomeRF standard is the only one that currently can carry toll-quality voice traffic, and in fact is taking the opposite migration path of 802.11 and Bluetooth, which is to go from voice to data.

It would seem clear that the intent of the HomeRF group is to provide an appliance that keys off a very large customer base of cordless phone users. Its intent seems to be to provide a central appliance that connects the devices within a home to the cordless phone, to enable phone calls to and from the residence, and then to provide broadband to the personal computer by interfacing a broadband connection within the home.

	802.11b	802.11a
Frequency	2.4 GHz	5.1 GHz, 5.2 GHz, 5.7 GHz
OFDM	No	Yes
Non Line of Sight	No	Yes
Data rates	11 Mbps	54 Mbps

Table 15-2. Snapshot Comparison Between 802.11b and 802.11a Specifications

This is accomplished with an approach similar to 802.11, which is to have both MAC and physical layers of the OSI stack that comply with this standard. The HomeRF standard utilizes a combination of CSMA/CD for packet data and TDMA for voice and video traffic to optimize traffic flows on a priority basis.

The physical layer utilizes frequency shift keying (FSK) to supply variable bit rates of .8, 1.6, 5, and 10 Mbps at 2.4 GHz. The throttling is accomplished through the utilization of seventy-five 1 MHz channels for voice and data channels at 1.6 Mbps. The standard also uses fifteen 5 MHz channels for 5 and 10 Mbps data channels. The physical layer also features smart hopping to avoid transmission residency in channels that are heavily congested with interference.

Streaming media takes the top priority for transmission and is identified by application headers, while voice and data traffic take up the balance of the bandwidth, again identified and based on application headers. The streaming media uses time and physical buffers to allow voice and then data traffic.

Of considerable interest, the HomeRF standard includes an impressive array of voice features such as call line ID, call waiting, call forwarding, and residential intercom. This is directly attributable to this standard's genesis in a telco voice standard.

For security, HomeRF uses 128-bit encryption augmented by the native security enhancement of an FHSS modulation at the physical layer. This combination should provide a high level of resistance to denial of service attacks, though the author was not able to determine whether or not the keys are static or dynamic.

The Bluetooth Standard

While Bluetooth is indeed a WLAN standard, substantial confusion exists as to whether or not it competes directly with 802.11 and/or HomeRF. The sum of the matter is that Bluetooth does not compete directly with 802.11, and competes only in a peripheral manner with HomeRF. The reason is that Bluetooth is intended to be a standard with a nominal range of approximately one to three meters. Its intent is to connect laptop computers with cellular phones, PDAs with laptops and cellular phones, and so on.

The first versions of the Bluetooth specification were released in early 1999, with version 2 promised for release in either 2001 or 2002. This duration of time between the original specification's release and the second version's release could spell problems for the standard, especially as there has been a considerable amount of media exposure to Bluetooth 1 which was released in the mid-1990s. In other words, it took approximately four years for version 1 of the specification to be released, and an excessive delay in the release of Bluetooth 2 may erode the support base as the supporting cast of engineers and companies may pursue other standards that are closer to release or are already on the market. At this writing, there are nearly 2,500 Bluetooth members listed in the Special Interest Group.

A company purchases the rights to use the Bluetooth standard at no cost. The design is a single chip, which is elegant, small, and has very low power consumption. This combination of features lends the design and the standard to certain applications, but is hardly a panacea for all broadband concerns.

While it does indeed approach the threshold of being a true broadband device, according to its homepage, the maximum data rate is 1 Mbps and is indicated as a gross data rate. Throughput should be expected on the order of approximately 750 Kbps, depending on how much overhead is used for security and frequency hopping management.

The Bluetooth standard has two strong points:

- **Size** The form factor (size) afforded by Bluetooth enables it to be embedded in wrist watches, PDAs, and other small electronic devices where size is an important design criteria.
- **Power savings** Bluetooth uses 30 micro amps, which is a very small amount of power. It uses a fraction of the power used by a standard wristwatch, and uses orders of magnitude below that used by cellular phones. This feature also plays very well to the industry, which often builds devices such as cordless headsets based on how much battery is required to provide a meaningful duty cycle (period of continuous operation).

In terms of security, Bluetooth has an encryption method, but it is not identified on its Web page. It should be mentioned that an FHSS hopping scheme of 1,600 hops per second and a very limited range of one to three meters would make it difficult to be eavesdropped upon from any distance.

If there is any negative aspect of the Bluetooth standard it is that it has been the recipient of an enormous amount of hype. The Bluetooth Special Interest Group (SIG) predicted in the mid 1990's that there would be over 200 million PCs with Bluetooth devices built into them as an original manufactured device by the year 2000. It turns out that what PC manufacturers are relying on are connections with speeds that are well in excess of that which Bluetooth can provide in addition to transmission and reception ranges well beyond that which the Bluetooth standard provides at the time of this writing. This is why both Apple Computer and Dell Computers can be provided with an 802.11 option.

Just prior to this writing, Microsoft announced it was backing away from the HomeRF standard. While the standard from which HomeRF is derived is very widely distributed throughout the world, it is the opinion of this author that the standard that undoubtedly is making the most rapid growth is 802.11.

In summary, Bluetooth will likely find its way into many small devices such as cellular phones, wristwatches, PDAs, and perhaps some computers, and this is an appropriate technological home for it. As far as networks go, broadband capability, capacity and range is central to whether or not technology providers will support a given standard., With the 802.11a standard promising 22 to 54 Mbps (and 100 Mbps in the not too distant future), the WLAN battle will likely be won by 802.11 for broadband devices and by Bluetooth or a similar standard, for connectivity over ranges of approximately 1 meter or less.

CHAPTER 16

Technology Standards

The issue of standards is one that many of us don't think about very often, yet it has been important to humankind for as long as people have been communicating with each other. Standards exist primarily to ensure a common basis and agreement for the form and function of the devices and services by which we are surrounded. Additionally, in modern times, standards are also used to protect those who come in contact with products and services.

Standards also reduce the number of variables and challenges with interoperability and information exchange. The more broadly adopted a standard is, the wider the market for a group of technology providers; the ideal is global standards and global products, which are far more difficult in practice than in theory. It then follows that the more widely adopted a standard becomes, the greater the amount of synergy will exist between technology providers and the marketplace they service.

Standards also define by law what certain values are, such as the foot, the meter, time, a gallon or liter of gasoline, and so on. Standards define what a specific device or service is or is not, and enable both technology and service providers to legitimately claim, use, and adhere to well-defined standards.

GLOBALIZATION OF STANDARDS

One of the earliest examples of a standard was the *cubit,* which was considered to be the length of a man's arm from his elbow to the tips of his fingers, and was widely agreed to be from 17½ inches in 8 B.C. to approximately 21½ inches at 1 A.D. By this standard, we know that Noah's ark was a relatively large vessel, at 437.5 feet long by 72.9 feet wide by 43.75 feet high. Many other ancient records of measurements provide information regarding the sizes of buildings and the weights and measures used for commodities such as grain and currency.

The measurement of time is a very important standard, and the progress of technology is directly proportional to how well this resource is measured. Consider how the ancient traders and merchants of the Mediterranean engaged the owners of ships and caravans to coordinate the transfer of goods and money. It was common for either the ship's crew or the traders to wait for weeks or even months at a specified site to make their exchanges, because the standard of time used during that period was generally measured by the phases of the month or season. Predating this, navigation on the open ocean from the ancient Mediterranean world to the West could not occur until it was possible to measure time with some degree of accuracy and consistency.

Standards today have a tremendous impact on how we select and use products. In the United States, we are very familiar with the inch, the pound, the ounce, and the dollar, while most of the rest of the world is more familiar with the metric system and their native forms of currency. As the world continues to become a truly global community, many people have had to become fluent in the standards used in places far from where they work or reside.

As associated with the central theme of this book, the United Nations has been struggling with various countries for years in terms of which standard to use for mobile wireless communications, and it is only relatively recently that a world traveler could take their cellular phone from London, England and travel to Singapore through the U.S. while using the same phone. This has far less to do with the use of a single standard than it has to do with manufacturers providing cellular phones that can automatically switch between as many as three or four different modulation schemes and frequencies. Standardization of these mobile wireless modulations makes it possible to build a cell phone in Norway that will function properly on three or more continents.

The concept of defining what a standard is can be entirely nontrivial. The even larger issue of *when* it is adopted, in a world where technology is conceptualized, developed, and deployed with rapidly decreasing amounts of time, is as important as the actual standard itself.

In the world of BBFW, we are fortunate to have certain fundamental standards that have been ratified at the global level, such as frequency, power, and time. A WLAN link that operates at 2.4 GHz is equally understood and accepted in Japan, England, Canada, and the U.S. In fact, this frequency is, in most nations, set aside for the purpose of having an unlicensed WLAN. In contrast, the idea of fixed wireless standards is somewhat new, and the same number of channels acceptable for use in one country is still somewhat different from the number acceptable in other countries.

Further, since the idea of standardization for broadband, two-way BBFW usage is still somewhat new, regulatory groups, international bodies, and developing standards groups are struggling for control and authority to allocate and/or dedicate specific frequencies for particular uses in numerous countries.

Another issue not widely agreed upon on a global basis is that of *channelization,* which is how radios break up the amount of spectrum available into usable subsets for communication. Further, except for 2.4 GHz, spectrum allocations vary widely from country to country in many cases, which makes it difficult for radio manufacturers and service providers to provide truly globally usable products, because they are commonly unwilling to design, test, and submit radios for local federal approval for every country that has a substantial market.

The lack of frequency standards at the international level has forced some hardware vendors to adopt multiple products, designing specific RF units to work in allocated frequency bands for specific countries. This approach raises the cost of goods and inventories. Without this approach, countries with frequency bands different from the key revenue-generating superpowers, such as the U.S., England, Germany, France, and Japan, are left without alternatives for broadband wireless.

This is one of the major reasons some companies have adopted architectures that allow indoor equipment, such as the routers and switches at each end of the radio links, to be independent of the carrier frequency, channelization, power requirements, and other settings that reside in the actual radios that operate the outdoor portion of the network. This approach still requires multiple outdoor units, depending on the specific country.

At this point, the technology does not exist, primarily for financial reasons, that would enable *ultrawideband* microwave products that would cover multiple radio bands. Until the countries of the world and the governing bodies and regulators get together on frequency standards, there will be limited global acceptance of broadband wireless technology.

EARLY TECHNOLOGY ADOPTERS AND STANDARDS

In the open commercial market, the *early technology adopters*, commonly referred to as early adopters, are the first ones to adopt cutting-edge technology. They constitute the portion of the market that is willing to exchange the risk of using an unproven technology for a prospective gain on their competitors. The objective for them is to acquire as much market share as possible at the earliest date possible. Early adopters are generally very tolerant of the teething problems common to cutting-edge technology and are often selected as the first sites for equipment and technology trials.

The early adopter market does not typically comprise the bulk of the market and is generally made up of the smaller customers that want to take market share from the large, established companies. By accepting higher risk, new technologies can give them a technological edge in both established and emerging markets.

The early adopter market does not typically comprise the bulk of the market and is generally comprised of the smaller customers that want to give their larger or more adept competitors a run for their money by adopting higher risk. This risk is taken in the form of new technologies that can give them a technological edge as the market develops. The bulk of a market opportunity for technology providers occurs later, when the early adopters have seeded the market with the latest technology and has usually gone through significant growing pains and learning curves.

These growing pains are usually fed back to the manufacturers to revise/improve/revisit their products to evolve better solutions that are, hopefully, made available to the risk takers first, which can, once again hopefully, enhance/extend the new providers a technology lead in the emerging market. If successful, or even apparently promising, larger companies will then follow suit in using the technology, which by then has generally been suitably debugged.

The largest customers tend to require standards-based products, because this ensures several key issues for them:

- Technological maturity
- Stability of basic design
- Interoperability

Technological Maturity

The major customers, such as the telcos, are generally not highly flexible or even fast-moving entities. Once they have begun to deploy a project based on a particular technology, they are very reluctant to retrieve deployments due to fundamental flaws in the design, because doing so is costly and time consuming. Their reluctance is based in part on the fact that there are financial and time costs associated with training and equipping their deployment teams for the new technology. In short, the largest customers generally wait for proven technologies.

Stability of Basic Design

The stability of a technology is directly related to its maturity. During the earliest phases of deployment, the fundamental design is more likely to change than during the middle phases of a technology's existence, which is where most of the sales occur. During the latter phases of a technological phase, the design may again change substantially, or even be abandoned. Early adopters of technology are small enough and nimble enough to withstand the changes in the basic technology in terms of features, capabilities, scaling, migration, and interoperability.

Interoperability

One of the most important issues the largest customers face is that of interoperability with existing equipment. This is a particular issue with the telcos, because they generally have spent large sums of money deploying vast amounts of existing equipment that has very high rates of reliability. Of the many concerns large customers have, perhaps none is more important to them than maintaining very high reliability.

This reliability has become the key metric for emerging technologies and a base requirement for incumbent operators, because of the performance record of the plain old telephone. Most people in highly industrialized countries believe the telephone will work regardless of the conditions in which it is used. The legacy this has left the providers of technology is that the consumer (not the early adaptor) fully expects this level of performance with every solution deployed by their service provider.

This inherent reliability is achieved in large measure by having full interoperability between the many different types of equipment used from one end of a network to the other. Redundant networks, although not cost effective, are mandatory in communication networks, to ensure that there is always a circuit between point A and point B, regardless of where these points may be. This customer-driven perspective mandates that, from the beginning of deployment, the newest technology has to work seamlessly with existing equipment in a wide range of conditions, such as traffic loading and varying environmental conditions, while being tolerant of a reasonably wide range of expertise among the network managers and technicians.

HOW BBFW AND OTHER STANDARDS ARE DEVELOPED

As stated previously, contrary to common understanding, standards occur in the early to late stages of the deployment of a technology, rather than in the introductory phase of a technology to a marketplace. The order in which a standard occurs is generally as follows:

1. Introduction of new technology
2. Relatively high interest by developers
3. Deployment of technology to early adopters
4. Definition of the standard by one or more technology providers
5. Establishment of the standard by a standards body
6. Ratification of the standard by technology providers
7. Updates of the standard

The preceding order is that which generally reflects standards for the telco world and that of some Internet standards. Other standards are essentially derived from the Requests For Comments (RFCs). The input from these RFCs becomes de facto standards due to fairly wide adoption even though they do not always undergo the full standards process generally managed by the IEEE.

According to some individuals, most new technologies in the network industry are conceptualized and evolved by a working group, committee, or brainstorming organization. These groups are comprised of members from multiple manufacturers, who then take the in-process documents to their companies for internal development.

Introduction of New Technology

Companies are pushed to advance the state of the art in any technology. Cutting-edge technology is rarely compliant with any standard, and in many cases, standards don't exist for the newest of technologies. The introduction of a new technology may or may not occur with much advanced notice to a marketplace, because some companies do not wish to reveal their development plans and jeopardize their opportunity to take advantage of the early adopters followed by general market adoption. This is particularly true when there is the possibility for a major shift in the capabilities available to the market. As a rule, the bigger the shift, the more clandestine the effort. Other technology providers often discuss their plans openly, occasionally to discourage competition by having the marketplace delay purchases until their particular spin on the technology is made available, or when the development only enhances the existing product.

Announcement of a new technology prior to its availability is sometimes referred to as *vaporware* because, although the technology provider sometimes is willing to discuss the technology at length, the technology is either immature or, in some cases, does not even exist. If the company is a very large and well-known provider of technology, this can have detrimental effects on smaller companies, because the customer base may wait for the technology to be provided by a name-brand company.

It is common for a new technology to be provided by more than one company at a time, or to have several or more companies introduce a new technology within months of each other. The number of examples of this across many industries is fascinating to consider because it seems as though completely unrelated technology providers develop the same new idea at virtually the same time without the slightest degree of collusion, cooperation, or even communication between them.

Relatively High Interest by Developers

Before a standard has the opportunity to be developed, there has to be sufficient interest among a fairly wide group of developers. Indeed, there is little to no use for a standard if the technology is provided by only a single developer, so there must be sufficient interest among a relatively large number of technology providers, or a few providers possessing sufficient resources to reach the bulk of a market prior to a need for a standard.

While many cutting-edge technologies are developed by startups formed by employees of large companies that refused to pursue a promising technology, the size and number of the prospective technology providers must be taken into consideration. One example of this is the Bluetooth standard. At this writing, very few Bluetooth devices are on the market, despite a considerable amount of press coverage over the last five or so years. While there is not an appreciable number of Bluetooth standards-based products on the market right now, the level of interest is high for developing products that will incorporate the Bluetooth standard.

Deployment of Technology to Early Adopters

Cutting-edge technology is often deployed first to early adopters who have either had design input or have formally expressed a willingness to use unproven technology. Part of the exchange for state-of-the-art technology often includes certain nondisclosure limitations on the part of the early adopter not to reveal product weaknesses or apparent displeasure with the performance of the equipment. Equipment is commonly, but not always, released in three phases: alpha, beta, and preproduction versions. After these phases, a version sometimes called *general acceptance* or *general availability* (GA) is released.

The engagement of early adopters also enables the technology providers to determine how their equipment will perform in real-world conditions, to debug nontechnical aspects of the product, such as deployment practices and training of deployment and network management personnel, and to determine how the system will perform under real-world traffic loads and QoS mixes. Further, early adopters tend to be tolerant of technology bugs and other deficiencies, such as the high cost of the initially released technology. Finally, early adopters tend to have small markets, which provide solid baselines for statistical analysis and the building of hands-on experience and documentation.

Standards Definition by One or More Technology Providers

Following either limited or extensive field trials, one or more technology providers will work with standards bodies to define the standard. This is often accomplished by having personnel from various technology providers operate within the standards bodies in key positions. This does not necessarily lead to a standard that is preferable to one vendor over another or all others, in spite of loyalties by standards body members, because there are often checks and balances within the standards organizations to prevent this from happening.

Establishment of the Standard by a Standards Body

As it is common for multiple technology providers to develop a technology in parallel, it is also common for there to be more than one standard codified in parallel. Well-known examples of this in the world of BBFW are the wireless Ethernet, which became the 802.11 standard, and the orthogonal frequency division multiplexing spreading technique, which currently is being codified into two standards, one by the OFDM Forum and the other by the Broadband Wireless Internet Forum (BWIF).

Although forums such as BWIF is not a standards body, many are taken seriously by the market and technology providers because of their close relationship with the IEEE. In addition, forums like BWIF are generally comprised of some of the leading technology providers such as Cisco. Forums also offer the advantage of access to open architectures. If no standard exists and a new technology has been introduced, companies waiting for full standardization can completely miss a market window. Importantly, forums allow groups of companies to band together quickly to begin voluntarily complying with a particular architecture and begin making solutions that will interoperate seamlessly years before a ratified standard can be put in place. These can eventually turn into standards or de facto standards.

It is generally more common than not for a marketplace to have competing standards, and as the time required to develop and deploy new technologies decreases, the probability of competing standards increases. In a technological world where innovations are constant and occasionally revolutionary, the application of a standard is perhaps more greatly affected by how widely deployed a technology becomes than by the standards bodies.

Once in possession of a standards definition, the standards organization will commonly announce to the technology industry and market of its intent to publish a standard. A period of time from three to six months is commonly provided for technology providers and the market in general to provide input to the proposed standard. A balloting process (detailed later in this chapter) commonly occurs next, wherein the standard is voted upon.

Once the standards organization has received input from the industry, or the period for comment has passed without substantial input from the technology providers or market, the standard is published and distributed.

In some cases, a technology becomes standard even though a considerable amount of the market place believes that a different approach or standard would better serve the needs of the end users. Examples of this include the VHS versus Betamax standard for video cassettes, and Macintosh (Apple) computers versus IBM (Windows). VHS and Windows have dominated world markets because of how the company strategy was implemented, as opposed to winning the predominant market shares based purely on technological merits.

Ratification of the Standard by Technology Providers

Once the balloting process has been completed and the standard has been approved as written by one or more committees or workgroups set up to evaluate the feasibility of the standard, the real work begins. The most dynamic phase of the standards effort is when the technology providers decide whether or not to adopt the proposed standard. More importantly, the life and distribution of the standard depends on how much of the market adopts the new standard.

There are many standards available in many industries, but the ones that are the most important are those that are in the process of gaining mind share from technology providers and market share from the end users who pay for the standard-based equipment.

The development of a standard can take two years or longer and then requires acceptance by an array of manufacturers, so that they design their products in a manner that is compliant with the standard. A relatively large array of manufacturers is required in order to provide enough product to an open market, and the concept of getting a large number of independent and competing companies to agree to a standard is a time-consuming and challenging objective.

A number of successful standards initiatives relative to the Internet, LANs, and BBFW have emerged. Perhaps the best known, understood, and widely deployed example is the Ethernet standard.

NOTE: Just because a standard exists doesn't mean that every vendor must adhere to it. A company is completely at liberty to develop its own proprietary version of the standard. However, for the public, standards typically mean competition, commoditization, and, generally, lower prices.

BBFW STANDARDS BODIES AND SUPPORTING BODIES

It is important to note that while BBFW has four notable standards bodies associated with it, only two types of entities exist, with a considerable difference between them; namely, all the BBFW entities mentioned herein except for the IEEE operate to promote the products that support certain IEEE standards. HomeRF is technically a broadband entity, but is not categorized as such because it does not provide products to the service provider industry. Similarly, Bluetooth, while well known throughout the world, is based on essentially a very short-range RF protocol of approximately 1 meter with data rates approaching 1.5 Mbps.

While each of them is independent of the other, there are occasionally efforts that occur in parallel. As stated previously, some technology providers act as the nucleus for a standards effort in an attempt to have their equipment design or architectures ratified as an industry standard. These standards bodies are sometimes referred to as *clustered* standards bodies, as in the case of IBM at the center of the HomeRF standards initiative, and Intel at the center of the Bluetooth initiative, to name a couple of examples.

Institute of Electrical and Electronics Engineers

The IEEE is an enormous organization comprised of both professionals and students with over 350,000 members in approximately 25 countries. The IEEE has an unspecified number of full-time employees, volunteers, and interns. Additionally, the IEEE utilizes the support of hundreds of employees of other companies and interested parties that meet in groups, chapters, and sections. Its effect on electrical engineering in general, the Internet, BBFW, and many other electronics disciplines cannot be overestimated. Very few electrical engineering students the world over have not heard of this organization, and virtually no experienced engineer has been in the industry very long without studying at least one IEEE paper.

Another way to identify the scope and impact of this organization is its dictionary, published in four- to five-year intervals, and currently in its seventh edition. This edition contains over 35,000 technical terms relating to various electrical, computing, networking, and communications efforts from over 800 IEEE standards initiatives. Simply stated, the scope of this organization is vast.

The full process of either bringing a new standard into existence or updating an existing one is enormous and beyond the scope of a single chapter in a book on BBFW. In summary, however, it involves various processes and phases of standards sponsorship, drafting, balloting, and committee review.

Further, the process is one of careful deliberation, in-depth discussion, and political wrangling worthy of the highest levels of most governments. These are international bodies that are comprised of members from many different countries and companies with varying viewpoints, agendas, and competitive concerns, and there is often energetic disagreement between committee members and the sponsors of standards. As a result, the completed standards often include compromises that don't wholly appeal to all interested parties.

The following outline of how a standard is developed should be viewed as an ideal series of events that occur prior to the establishment of a standard. In many cases, there are iterations between the steps indicated.

NOTE: This is the process sponsored by the IEEE; the processes of other bodies may be somewhat different, but the concept and the high-level flow is mostly similar.

1. Sponsor selection and duties
2. Submit Project Authorization Request (PAR)
3. Approve PAR
4. Organize working group
5. Develop draft of standard
6. Ballot draft of standard
7. Approve draft standard
8. Publish approved standard

Sponsor Selection and Duties

The *sponsor* is the organization that assumes responsibility for the work of drafting the standards document and ensuring its timely passage through the various phases of deploying a standard. The IEEE retains ownership for the *technical* merits of the proposed standard, but it is the sponsor that shoulders the effort regarding the scope and technical content of the standard. The sponsor is generally an entity outside of the IEEE, while the IEEE *Standards Board* ensures that the proper rules and procedures for assembling and ratifying a standard are followed.

The bulk of the work takes place with an organization, not an individual. Once the concept for a standard has been conceptualized, the work of enabling a standard is accomplished by groups of individuals knowledgeable in the field in which the standard is being proposed. In many circumstances, working groups exist for new standards, but in the event no working group exists for a proposed standard, a working group is established by the IEEE through an invitation process.

The first step a sponsoring group takes is to become familiar with the various IEEE societies to determine if there is an existing committee or society that is familiar with the proposed standard. The IEEE has a large number of sponsors within various societies in the form of committees. A *society* is a group dedicated to a specific discipline, such as antennas or a modulation scheme. If the proposed standard comes within the scope of one of these committees, this is where the standards work begins, and one would contact the IEEE by e-mail or telephone to determine whether a certain society or committee would be appropriate for the proposed standard.

If no committee appears appropriate for a proposed standard, the group or person interested in the new standard would contact the IEEE's governing body for standards either by e-mail or telephone and make the request to a staff liaison to determine whether an existing committee would be interested in working with the sponsor in the new standard. In some cases, a standard would interest the members of more than one society, in which case the IEEE Standards Board has coordinating committees that exist to accommodate this type of development.

Submitting a Project Authorization Request (PAR)

The PAR is a concise document on the order of three or four pages that indicates the new standard has been approved for consideration within the IEEE. This document is approved by the IEEE *New Standards Committee* (NesCom), which is one of a number of Standards Board's committees.

The PAR is a highly detailed document that usually conforms to a template provided by the IEEE. It states the intent of the proposed standard, but is not a draft of the standard itself. While officially the IEEE doesn't recognize the work of a proposed standard until a PAR has been approved and logged in to the IEEE, it is common to have *study groups*, which are entities that guide the standards process, and are comprised of various industry experts from an array of technology providers. Once a PAR has been approved, the project must be completed within four years. If the standard requires longer than this period of time, an application for extension of time must be acquired from the NesCom.

Approving the PAR

PARs are typically reviewed once a calendar quarter by the NesCom. The PAR is generally provided to the NesCom at least 40 days prior to review, and in most cases the committee offers feedback and input for improvement to the PAR. This is important, because the PAR details the scope and intent of the standard. Standards even in their earliest phases are embodied in living documents so that, as the standard matures and changes in detail, scope, and intent, the PAR is updated in parallel.

Organizing a Working Group

Once the PAR has been approved by the NesCom, a *working group* is organized. The typical positions within a working group may commonly include the following:

- Chair
- Vice chair
- Secretary
- Treasurer
- Technical editor
- Ballot coordinator
- International standards liaison

The working group operates very much like a company board in terms of the agenda, and generally follows the *Robert's Rules of Order*, which is a widely used standard for meetings and covers the manner in which the agenda is managed in terms of review of minutes, new business, old business, and so on. One of the key purposes of the working group is to ensure consensus among the working group relative to the standard and the responsibility of the officers is to ensure the work is moving forward at an appropriate pace in addition to ensuring the interface with various committees and societies within the IEEE moves along smoothly. Like the PAR, the standard duration of a working group is four years although this period can be extended.

Developing a Draft of the Standard

The draft of the standard is completed by the working group and is generally a complex and even daunting task. It is based on the PAR, and begins with an outline. Because it is fairly common to have numerous authors of the draft, the coordination of the tone, content, and scope is coordinated by the technical editor and the officers of the working group.

From time to time, personnel from the industry affected by the standard may request a draft of the standard and desire to provide input. Working groups commonly send copies of the draft to organizations and personnel outside the working group, but generally this is done in a manner such that the working group retains tight control of the draft.

Balloting the Draft of the Standard

Following the completion of a standards draft, the work of sending it out for voting is provided through a process called *ballot work,* which takes anywhere from 30 to 90 days. The purpose of voting on a proposed standard is to ensure the greatest degree of consensus among those who will be affected by the standard, or if the balloting remains purely internal to the IEEE, to ensure that the greatest degree of consensus is achieved among the various committees. During the balloting process, the IEEE does not look for a unanimous agreement on the draft of a proposed standard, and in fact deems that a standards draft has achieved consensus when 75 percent of the voters approve the standard.

There does not appear to be a single, uniform method for balloting, but in general, the ballot group is an active committee that exists exclusively within the IEEE. However, there are many examples of when the balloting is opened to a wide array of interested parties, both from within the IEEE and from the industries that the standard may affect. In the latter case, the working group may use various forms of media to provide notice to all interested parties.

The balloting group is generally comprised of three types of groups: technology providers, end users, and other interested parties who may or may not be directly associated with the proposed standard. Ballots may be sent to the following entities:

- Corporations
- Government agencies
- Associations and societies
- Consultants
- Academic institutions
- User groups
- Other standards development organizations

The working group will commonly achieve the best balance possible among the groups, with no company or balloting group comprising more than 50 percent of the entire vote.

The ballot for proposal of a standard includes an area in which the voter may provide either technical or editorial feedback with regard to the proposed standard. If the number of negative votes is substantial, the work begins of resolving the differences between the working group developing the standard and the voters. This negotiation can be the most time consuming of the entire process. After the working group has integrated the concerns from those casting ballots, the standards draft is usually sent out once again for ballot.

Approving the Draft of the Standard

Subsequent to the successful approval of the standards draft through the balloting and comments process, the draft is then sent to a *Review Committee* (RevCom), which reviews the entire process that has occurred to date with regard to the attempt to deploy a new standard. This review board is more focused on the process than the technical content of the draft, and generally reviews the following items:

- Was the balloting group balanced?
- If the ballot was delegated to a subordinate committee, is there a record of that delegation?
- Was the ballot valid, with at least a 75 percent return and less than 30 percent abstentions?
- Did the ballot pass by at least 75 percent?
- Does the document match the title/scope/purpose of the PAR authorizing the work? (Note that title changes that are still within the PAR's scope and purpose are acceptable.)
- Was coordination with the required organizations achieved? Was coordination accomplished via the required method?
- Were all members of the balloting group given an opportunity to see all the outstanding negatives and the reasons why they could not be resolved?
- Were all members of the balloting group given an opportunity to change their vote as a result of changes made to resolve negative ballots?
- Are there any major technical or procedural oversights?

Once the RevCom is satisfied that the process has been compliant with the rules and procedures of the IEEE, it makes a recommendation for approval to the NesCom.

Publishing the Approved Standard

Prior to publication, there is still a considerable amount of work to be completed on the draft. Once the IEEE NesCom has approved the recommendation by the RevCom, the standard is then reviewed by a standards editor who ensures the standard is grammatically and syntactically correct in American English as well as ensuring it adheres to the formatting as set forth in a document entitled *IEEE Standards Style Manual*. The editor does not alter the technical content of the standard, but in the event the editor finds glaring errors in formulas or other key elements of the draft standard, the editor will offer opinions that are then reviewed by the RevCom. If the RevCom approves the change, the completed document is then sent for publication. If the proposed changes do not make it past the RevCom, the changes may be incorporated into a subsequent revision of the standard or corrected through errata appendices.

The final review of the standards draft is then performed by the working group chair or assigned delegate to ensure the edited and formatted version of the standards draft remains consistent with the PAR and technical content of the draft.

Following this final review, the draft then becomes a standard, which is then distributed through normal media channels both inside and outside the IEEE. Upon distribution, questions on the interpretation of the standard are common. Upon receipt of a request by the IEEE to interpret a standard, the request is sent to the working group and an interpretation is provided and distributed in a similar manner to the original standard. In some cases, the interpretations are cumulatively sufficient in scope that a volume of the interpretations is assembled and distributed. These interpretations do not alter the standard even upon discovery of an error, but they are important because the IEEE does not want issues with standards to remain unaddressed.

Three Key BBFW Entities That Support IEEE Standards

Three key BBFW entities are well known for supporting IEEE standards. Most of the world's leading BBFW technology providers are closely associated with one or more of these groups through financial contribution via membership dues, the "loaning" of some of their best technical staff members, and by adherence or commitment to one or more of these forums.

Wireless Ethernet Compatibility Alliance (WECA) and Wi-Fi

WECA's mission is to certify interoperability of IEEE 802.11 products and to promote Wi-Fi as the global wireless LAN standard across all market segments. WECA is one of the groups leading the charge to refer to "802.11b" as "Wi-Fi" or (*Wireless Fidelity*). But WECA is also trying to ensure that if you buy a product (particularly a PCMCIA card with the Wi-Fi badge on it) from any vendor, it will be compatible with other similar cards and, more importantly, with the access points with which it connects to the LAN or Internet at large.

Companies currently supporting this standards group include 3Com, Acere, Nokia, Apple Computer, Atmel, and Cisco Systems; currently, a total of approximately 60 companies. One of the distinguishing characteristics about WECA is that it is embracing and promoting a standard (802.11) as opposed to implementing a standard of its own. It is also focused primarily on business users, SMB, highly utilized public areas such as airports, and hospitality venues such as hotels and convention centers.

By comparison, HomeRF is focused primarily on the residential market, although it is interesting to note that WECA also works with the Bluetooth Special Interest Groups

(SIGs), as the 802.11b frequency and the Bluetooth frequencies are similar and both use spread spectrum. The alliance between WECA and Bluetooth is to promote the ability for both protocols to operate in a common physical environment. While WECA devices will primarily connect PCs and printers, Bluetooth's intent is to primarily focus on smaller electronic items such as digital cameras, PDAs, and mobile phones, although there will certainly be Bluetooth devices installed in PCs.

At this writing, WECA is focused primarily on the 802.11b standard and to ensure interoperability between products from different manufacturers. The 802.11a standard, which operates at 5 GHz as opposed to 802.11a's 2.4 GHz, will be considered a supplement to the current group focus. There is also a third working group at this writing, the 802.11g group, which is an interim standard between 802.11b and 802.11a in that it also incorporates OFDM.

WECA recognizes that while both the "b" specification and "a" specification utilize unique Layer 1 hardware, they share the same Layer 2 MAC protocol. It is the opinion of this author that the 802.11a standard will supercede the 802.11b standard and become the more widely deployed technology and protocol.

It costs approximately $20,000 for a company to join WECA, and $100,000 to become a sponsoring company. Membership in WECA ensures a voice in the promotion and deployment of the 802.11 IEEE standard.

Broadband Wireless Internet Forum (BWIF)

While BWIF currently has approximately 50 members, many of them are large companies, including Cisco Systems, Toshiba and Texas Instruments, Motorola and National Semiconductor, to name a few. The organization is comprised of a mix of companies that provide elements ranging from silicon chips, to deployment and integration services, CPE devices, service providers, and headend/ODU manufacturers.

This organization operates to support the non-line-of-site aspect of BBFW and is focused on promoting Vector Orthogonal Frequency Division Multiplexing (VOFDM), which uses a modified version of the well-known cable television standard at Layer 2, called Data Over Cable Service Interface Specification (DOCSIS), for its point-to-multipoint products. DOCSIS is a well-known standard for managing a wide number of subscribers from a single cable television headend.

BWIF membership costs from $2,500 to as much as $15,000, depending on the amount of participation requested by the joining member. Each BWIF member agrees to cross-license their technologies with each other to minimize the cost of NLOS BBFW products.

As with WECA, BWIF supports the associated developing IEEE standards that enable NLOS capabilities through VOFDM, and this group does not compete with IEEE or intend to provide standards that are independent of the IEEE. Also, as with WECA, the BWIF organization focuses on the SMB market, but also extends to the residential market with products in both the MMDS and U-NII frequencies. BWIF is closely associated with the IEEE through the latter organization's Industry Standards and Technology Organization (IEEE-ISTO).

OFDM Forum

The OFDM Forum is approximately the same size as BWIF and was founded by Philips Semiconductors and a small Canadian BBFW company called Wi-Lan. Like BWIF, this organization enjoys the participation of a number of institutional entities such as Carnegie Mellon University, Alcatel, Philips, Samsung, and Sony.

One of the differences between the OFDM Forum and BWIF is that the OFDM Forum does not restrict its membership based on the endorsement of one brand of OFDM, although in fairness it should be stated that the Wi-Lan competes vigorously with Cisco Systems with its own proprietary version of OFDM. Currently, there are at least three versions of OFDM in the market, which are W-OFDM as offered by Wi-Lan, C-OFDM as offered by Unique Broadband Systems, and VOFDM as offered by Cisco Systems.

While both the OFDM Forum and BWIF have worldwide interests, at this writing, the OFDM Forum has a more international representation with its membership and appears more focused in its attempt to have a worldwide OFDM standard.

PART VII

Government and Regulatory Issues

CHAPTER 17

The FCC

The Federal Communications Commission (FCC) is a U.S. federal agency headquartered in Washington D.C. that has numerous responsibilities, and that reports directly to the U.S. Congress. It was established by the Communications Act in 1934 and is responsible for regulating local, regional, interstate, and international communications by radio, television, wireline, satellite, and cable.

This chapter refers only to the FCC, which controls the spectrum only in the United States. Other countries are governed by their own regulatory bodies. Some spectrum management is exhibited by the ITU, but as a general rule, spectrum is managed independently by each country. These governing entities commonly look to the FCC for guidance, because the United States often encounters various issues earlier than developing nations, but countries outside the United States are not required to adhere to U.S. standards and spectrum laws and, in fact, rarely do. Currently, no single body governs spectrum internationally, although the United Nations has attempted, unsuccessfully, to ensure international unanimity on spectrum.

FCC ORGANIZATION AND BUREAUS

The FCC has five commissioners who are appointed by the president of the United States. The nominees for these positions require confirmation by the U.S. Senate, and the persons in these positions serve for five-year terms. One of the commissioners is designated by the president as the chairperson of the FCC. No more than three of the commissioners may be members of the same political party, and none of them can have financial interests in any FCC-related business.

The FCC is organized by function and has seven bureaus. The following list of bureaus is essentially directly quoted from the FCC:

- **Cable Services Bureau** Serves as the single point of contact for consumers, community officials, and the industry for cable-related issues.
- **Common Carrier Bureau** Responsible for rules and policies concerning telephone companies that provide interstate and, under certain circumstances, intrastate telecommunications services to the public through the use of wire-based transmission facilities, which include both corded and cordless telephones.
- **Consumer Information Bureau** Communicates information to the public regarding FCC policies, programs, and activities. This bureau is also charged with overseeing disability mandates.
- **Enforcement Bureau** Enforces the Communications Act, as well as the FCC's rules, orders, and authorizations.
- **International Bureau** Represents the FCC in satellite and international matters.

- **Mass Media Bureau** Regulates AM, FM radio, and television broadcast stations, as well as Multipoint Distribution (cable and satellite) and Instructional Television Fixed Services.
- **Wireless Telecommunications Bureau** Oversees cellular and PCS phones, pagers, and two-way radios. This bureau also regulates the use of radio spectrum to fulfill the communications needs of businesses, local and state governments, public safety service providers, aircraft and ship operators, and individuals.

FCC STAFF OFFICES

In addition to the seven bureaus, the FCC has ten units that work closely with the bureaus but do not participate in congressional hearings. These units are referred to as *Staff Offices*. The following Staff Office descriptions are essentially quoted directly from the FCC:

- **Office of Administrative Judges** Presides over hearings on disputed matters (but not congressional hearings) and issues Initial Decisions to the FCC and adversarial parties.
- **Office of Communications and Business Opportunities** Provides advice to the FCC on issues and policies concerning opportunities for communications businesses that are small or owned by minorities or women.
- **Office of Engineering and Technology** Allocates spectrum for non-government use and provides expert advice on technical issues before the FCC.
- **Office of the General Counsel** Serves as chief legal advisor to the FCC's various bureaus and offices.
- **Office of the Inspector General** Conducts and supervises audits and investigations relating to the operations of the FCC and is the internal "watch dog" unit of the FCC.
- **Office of Legislative and Intergovernmental Affairs** This unit is the FCC's main point of contact with Congress and other governmental entities.
- **Office of the Managing Director** Functions as a chief operating official, serving under the direction and supervision of the Chairman.
- **Office of Media Relations** Informs the news media of FCC decisions and serves as the FCC's main point of contact with the media.
- **Office of Plans and Policy** Serves as the FCC's chief economic policy advisor.
- **Office of Work Place Diversity** Advises the FCC on all issues related to workforce diversity, affirmative recruitment, and equal employment opportunity.

THE WIRELESS TELECOMMUNICATIONS BUREAU

The Wireless Telecommunications Bureau (WTB) is the agency within the FCC that handles all domestic wireless telecommunications programs and policies, except those involving satellite communications or broadcasting. The WTB scope of involvement in commercial wireless includes licensing, enforcement, and regulatory functions and is the entity most closely associated with BBFW. The agency is also is responsible for implementing the competitive bidding authority for spectrum auctions, given to the FCC by the 1993 Omnibus Budget Reconciliation Act.

The WTB is comprised of seven major divisions:

- Office of the Bureau Chief
- Management and Planning Staff
- Auctions and Industry Analysis Division
- Commercial Wireless Division
- Data Management Division
- Policy Division
- Public Safety and Private Wireless Division

There are subdivisions within some of the aforementioned divisions, and the group en masse has the following stated goals within the BBFW industry:

- Foster competition among different broadband services
- Promote universal service, public safety, and service to individuals with disabilities
- Maximize efficient use of spectrum
- Develop a framework for analyzing market conditions for wireless services
- Minimize regulation where appropriate
- Facilitate innovative service and product offerings, particularly by small businesses and new entrants
- Serve WTB customers efficiently (including improving licensing, eliminating backlogs, disseminating information, and making staff accessible)
- Enhance consumer outreach and protection; improve enforcement process

The WTB plays two key roles: prohibiting interference between different radio and other broadcast systems (cable and television), and managing fair competition practices and auctions to provide the greatest probability of multiple suppliers of information to the general public. The WTB accomplishes these goals primarily through the issuance of licenses that are granted on a regional basis in what are known as Basic Trading Areas (BTAs). Limiting the areas where RF frequencies can exist is difficult if not impossible, but is made possible by placing limits on the effective output power of the licensed system. The BTA map was established by Rand McNally, which licenses this mapping to the FCC and WTB.

As an example of fostering the maximum amount of competition in the market, when the MMDS licenses were auctioned in the mid 1980s, CLECs were prohibited from bidding on the licenses or owning companies that had purchased MMDS licenses. However, through consolidation of companies, which occurred because many companies were underfunded or had deficiencies of other resources (such as experienced management and technical personnel), most MMDS license holders were purchased by Sprint, MCI WorldCom, and BellSouth. In keeping with a policy of maximizing competition, LMDS licenses were awarded to various non-ILEC entities, such as NextLink (which owns 95 percent of the top 30 service areas in the United States), Baker Creek, Winstar, and Quest.

Unlicensed Systems

One issue that has gained a surprising relevance is that of licensed versus unlicensed systems. Unlicensed BBFW systems undoubtedly will play an important role in BBFW on a worldwide basis, because the cost of licenses is prohibitive to many prospective ISPs. However, the realm of unlicensed broadcasts does have some requirements for compliance.

A fundamental FCC rule regarding unlicensed wireless systems is that no source, licensed or unlicensed, can interfere with another radiating source. There are also regulations that ensure a system used for revenue-generation purposes does not unduly affect other locally radiating systems and, specifically, private users on unlicensed bands. A number of market experts predict that U-NII will comprise a substantial portion of the RF. In fact, a March 2001 report by the Strategis Group included the statement, "Despite billions of dollars poured into U.S. broadband wireless licenses, unlicensed wireless technologies maintain a clear lead in terms of deployments to date." You can find more information on unlicensed systems in Chapter 12.

The FCC is not a government entity historically held in high regard by end users or providers of wireless equipment, because the agency is typically underfunded and inefficient as a source of information. The agency has also gained considerable criticism for the manner in which it auctions and reclaims licenses. Not surprisingly, the FCC is not well known for a high level of talent given the level of competition in the open market for wireless talent.

However, at the time of this writing, the agency has considerably improved its efficiency and general state of operation. In March 2000, the WTB reported that it had reduced its backlog of items by 99 percent. This is a remarkable accomplishment given that the WTB handles all of the U.S. domestic wireless telecommunications programs, which extend well beyond BBFW to other wireless areas, including cellular, ship-to-shore, and paging.

In December 1998, the WTB had more than 64,000 items pending; at the time of this writing, the backlog number is approximately 500 items. As the world continues a very rapid transition to various wireless technologies, the WTB now finds itself managing over two million licenses while processing 400,000 applications for licenses, in addition to other matters, on an annual basis. The typical length of time for processing a WTB issue is now on the order of 90 days or less from the date of filing.

Universal Licensing System (ULS)

A commonly held misconception is that the FCC issues and regulates wireless licenses. While this is essentially a true statement, the FCC is an organization much broader in scope than one that merely doles out and monitors broadcast licenses. In fact, the FCC has assigned wireless matters to the WTB.

Arguably, no issue is more central to the WTB than the issuance and management of BBFW and other licenses. As stated previously, this organization processes hundreds of thousands of licenses on an annual basis, and the database on license holders, along with the licensing process system, has become enormously complex because it has no less than 11 different licensing management and monitoring systems in place. In response to this unwieldy assortment of processes, databases, processes, and polices, the WTB is streamlining this disjointed array into a single system called the Universal Licensing System (ULS). The intent of the WTB is to dramatically improve the way it engages those seeking a license.

The ULS is a browser-based system, for ease of use, and will include security features to enable businesses to apply for, bid, and receive a BBFW license. Additionally, the ULS can be used for antenna structure registration. The information requested by the WTB will be specific to the type of license being requested, and the owners of licenses will be able to update their records on a forward-going basis. Checking on the status of a license application or other matter will be as simple as contacting the WTB through a standard browser interface.

A key feature of the ULS will be the ability to review the markets already covered by a license, and the ability to quickly identify those markets that have not yet been issued a license, which is perhaps one of the most sought-after types of information by BBFW-interested parties. This information will be displayed in a map format, providing an easy user interface that is quickly understood by the inquirer.

The regulated aspects of a BBFW system, whether licensed or unlicensed, include:

- Frequency
- Bandwidth
- One- or two-way data path
- Maximum power output
- Spurious radiations (unintentional or unlicensed radio emissions)
- Safety concerns relative to the system design
- Antenna height from ground
- Angle of transmission relative to the ground or rooftop
- Antennas and towers

While the issues of frequency, bandwidth, and other items previously listed have been covered elsewhere in this book, one of the WTB's little-known areas of responsibility to the public is that of antennas and issues associated with antennas, such as the antenna tower/mast, the angle at which an antenna can be tilted downward (downtilt), and the maximum allowable height that an antenna can be placed from the bottom of the base upon which it sits. Thus, those issues are discussed next.

Antennas and Towers

The FCC requires that owners of towers register all towers greater than 200 feet in height, and requires the painting and lighting of antenna structures that may pose a hazard to air navigation. All towers or structures that have antennas attached to them within approximately one-half mile of an airport are required to be registered with the FCC in order to promote air navigation safety. The FCC accommodates this registration process with its *Antenna Structure Registration* (ASR) registration page, which is online. The structure or tower owner must first receive approval from the Federal Aviation Administration (FAA) before they can register the structure with the FCC.

It is important to note that while approximately 85 percent of all antennas are mounted on buildings, as opposed to stand-alone towers, a great preponderance of antenna structures (towers) are owned by an entity other than the BBFW provider. The FCC defines antenna structures as "The radiating or receiving system, its supporting structures, and any appurtenances mounted thereon." This refers not only to stand-alone structures but also, as a practical matter, to any structure built specifically to support the antenna or act as an antenna or a manmade device, such as a building, bridge, water tower, or any other such structure capable of supporting BBFW outdoor equipment. The FCC does not require the owner to register the structure unless it supports BBFW outdoor equipment.

Angle of Transmission Relative to the Ground or Rooftop

The signals from a BBFW tower are essentially directed toward the horizon with approximately five degrees of downtilt to accommodate CPE units that are within a mile of the head end. As stated in Chapter 2, the power density drops precipitously over even short distances (by the square of the distance within a lobe), and this is even more true with regard to the area at the base of the tower or antenna structure.

The measured power density at the base of a tower or antenna structure is typically thousands of times less than the FCC limits for safe exposure. A person would have to stand virtually in front of the antenna for an extended period of time to incur even a modest amount of radiation in a properly deployed residential system (though, as a rule, a person should not stand in front of any antenna they are not professionally licensed or trained to handle).

With residential installations, however, the antennas are commonly mounted on either the rooftop or the side of the building near the roofline. This type of antenna installation can provide opportunities for increased exposure that approach the safety limits as set forth by the FCC. It is also noteworthy that the approved maximum power transmissions of the unlicensed band transmitters are far below that of licensed band transmitters, which makes U-NII and ISM transmissions somewhat safer than MMDS and LMDS from a radiation-exposure point of view. Having stated that, exposure to RF radiation should always be minimized. The FCC sets clear limits on the physical approachability and the radiation patterns around all antennas.

Auctions

In spite of its profile in the BBFW industry, perhaps one of the least-understood events regarding the issuance of spectrum is the auction process. The relative obscurity of the licensing process continues in spite of the continual process by the FCC and WTB to bring the matter under the full and bright light of the general public.

The key elements in the auction of a spectrum are the following:

- Single-round vs. multiple-round bidding
- Sequential vs. simultaneous bidding
- Auction automation
- Auction tracking
- Auction event countdown

Single-Round versus Multiple-Round Bidding

One of the most important elements of a government auction is for the auctioning body to provide as much fairness and openness as possible so that all bidders may have a relatively equal opportunity to purchase a license. This, of course, does not take into consideration that many licenses are enormously expensive, commonly running in the tens of millions of dollars, and that there is a considerable amount of disparity between the financial resources of the bidders. That notwithstanding, certain conditions such as substantial differences in financial resources among bidders is not generally controlled by the auctioning body and is known and understood by all participating parties.

In single-round bidding, the bids are commonly *sealed*, meaning that the bids are offered in a nondisclosed manner among the bidders, and the highest bidder is awarded the spectrum. In this type of bidding, the bidders place a bid on a single block of spectrum. The disadvantage to this process is that the bidders do not know the value others are placing upon the spectrum being auctioned. However, it is reasonable to expect that the spectrums that most closely match those that have been previously auctioned will be awarded at premium pricing, primarily because the providers of discrete components, such as RF chipsets, will already have made the products available that will greatly shorten the time to market for the spectrum bidders.

In multiple-round bidding, the process is similar to that observed at most auctions; in other words, each of the bidders can observe the value others place on a certain spectrum. The advantage to this is that the bidders who place the greatest value on a spectrum and, importantly, have the resources to outbid all other contestants will usually win the auction. This is not necessarily the case with a single round of bidding, because all parties have to make an educated evaluation as to the value all bidders are placing on a certain block of spectrum. This should, of course, be based on in-depth financial modeling of the proposed trading area, the cost of acquiring and deploying capitol equipment, the probability of success in a given market, the services to be rendered, and the price set on those services.

However, it is more commonly just a guess. In sealed, or single-round, bidding, the party that most desires the spectrum may guess incorrectly as to the amount to bid. As collusion is strictly illegal in sealed bidding, multiple rounds of bidding will provide the greatest likelihood that the party who places the highest value on a block of spectrum will be awarded that spectrum.

Multiple-round bidding, therefore, virtually always is deemed as the most fair type of auction by the participants. Multiple-round bidding is also greatly preferred by the government, because it almost always generates far more revenue, which has historically run into the billions for licenses such as LMDS in the United States and mobile licenses in Europe. Auction prices have been so high for some mobile licenses in Europe that the market has voiced great concern about having enough capital to build out the networks following the purchase of the rights to broadcast in the acquired RF spectrum.

Sequential versus Simultaneous Bidding

Whether sequential bidding has advantages over simultaneous bidding depends on the *interdependence* among licenses, which is to say that the value of the licenses is affected in part by whether they are geographically adjacent to each other or whether they are in prime BTAs, such as the BTA that includes Oakland, San Jose, and San Francisco, California. The value in having a contiguous footprint is obvious in terms of marketing, deploying, and managing the BBFW links.

Licenses therefore are referred to as either compliments or substitutes. A certain license may be a compliment to another for the reasons stated in the previous paragraph. Conversely, it may also be deemed a substitute if the bidder does not particularly care in which city or location the license may be offered. In other words, a license for downtown Las Vegas may be about as valuable as a license for a metropolitan area of a similar size anywhere in the United States.

A license that was sold in San Jose may therefore have a value to a bidder who already has the license for either Oakland or San Francisco, California, because the two latter cities are relatively adjacent to San Jose. By offering the San Jose license subsequent to the award for the San Francisco or Oakland licenses, the government is almost assured of receiving a higher price for the San Jose license, as long as the bidder for the Oakland and San Francisco licenses is participating in the San Jose license bid. Because BBFW operators place a high value on contiguous footprints, the probability is high that the same bidder who won either the San Francisco or Oakland license would bid on the San Jose license. For this reason, it is common for the government to offer sequential bids in a manner such that a BBFW provider ends up building a contiguous footprint over time—and at a premium for each additional area that would increase the footprint.

The exception to this, to some degree, is when license holders purchase licenses from entities who either outbid them at the original auction or purchased the license at a time before the party interested in building a contiguous footprint entered into the BBFW market. It is certainly also true that many entities that won licenses at auctions are later forced to sell them for unrelated financial reasons, in which case the party looking to build a contiguous footprint may be able to increase their footprint at a discounted price. The strategy of building a contiguous BBFW footprint is also referred to as the *aggregation of licenses*.

Auction Automation

The U.S. government is no stranger to auctions. It has disposed of billions of dollars of assets through auctions, from used office equipment and furniture to items seized from illegal activities and foreclosed properties. These auctions are carried out in both the sealed bid and oral auction manner. The large number of licenses, combined with the number of prospective purchasers, required the institution of a new system, called *Automated Auction System* (AAS).

One other key element called for the AAS, which is a requirement unique to the sale of this federal asset that makes it different from the process of disposing of government items: the requirement to track the purchasers of the asset. When the government sells property or furniture, it generally has no need for further information as to what happens to the asset or the entity that purchased it. RF spectrum, on the other hand, is an enormously valuable asset that increases in value over time and must be used in compliance with a wide array of regulations.

The AAS is a browser-driven tool that tracks all bidder activity from initial contact through the award of the license. It offers auction-tracking capabilities, summary maps, and bulletin board information with two tools that reside within the AAS, one of which is the *Automated Tracking Tool* (ATT), shown in Figure 17-1.

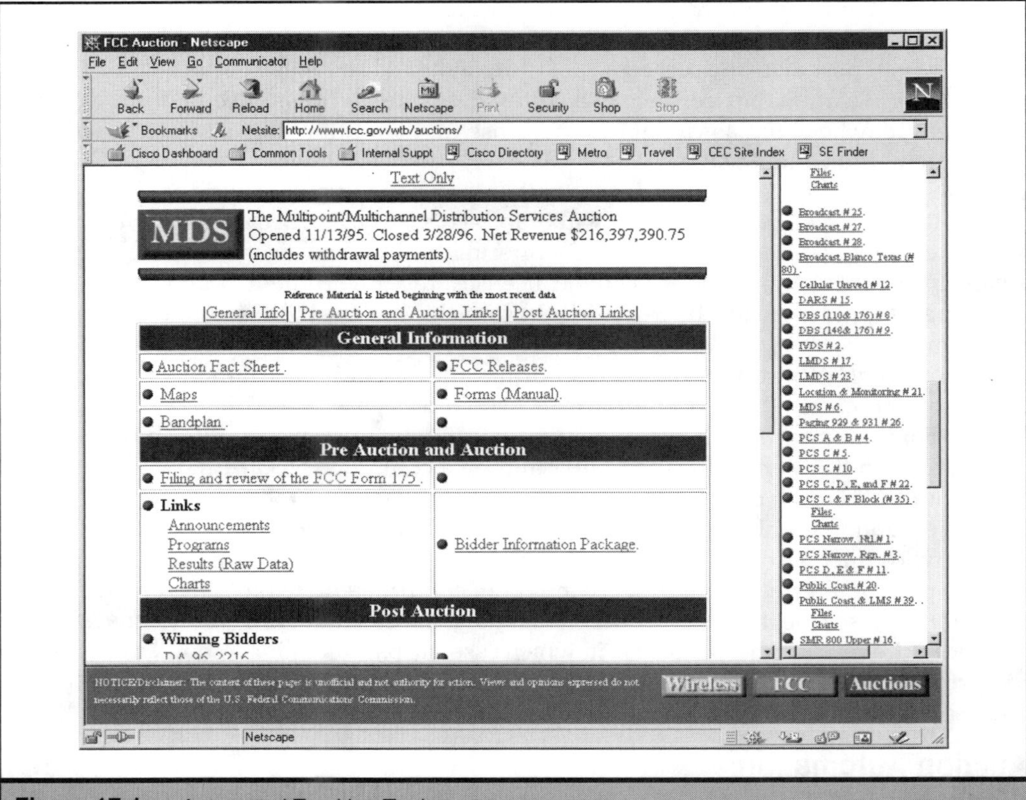

Figure 17-1. Automated Tracking Tool

The ATT provides key information to all interested parties, including the Auction ID number, the sequential bid round number, the geographic location of the licenses, the identification of the bidders, and the amounts the bidders are offering for a specific spectrum.

The other tool that can be accessed as part of the AAS is the Maps of Auction Results, shown in Figure 17-2. This tool is also known as the Geographical Information System (GIS), which includes three key types of information: a Market Analysis by Number of Bids, Round Results Summary, and Bidder Activity.

The Market Analysis by Number of Bids enables interested parties to determine which licenses received bids in a given round and how many bids each market received. The Round Results summary provides a high-level view of the activity that occurred for a specific sequential round of bidding, which bids were increased, and which bids were withdrawn. Markets in which no activity occurred are also shown, which can provide prospective bidders with valuable information in terms of the general value placed on a specific BTA, which in turn can lead to limiting the probability of paying a premium for a BTA. This can provide highly valuable information in particular to BBFW providers, which have a strategy that focuses on substitute licenses as opposed to a strategy that relies on the more expensive contiguous footprints. Having stated that, many BBFW

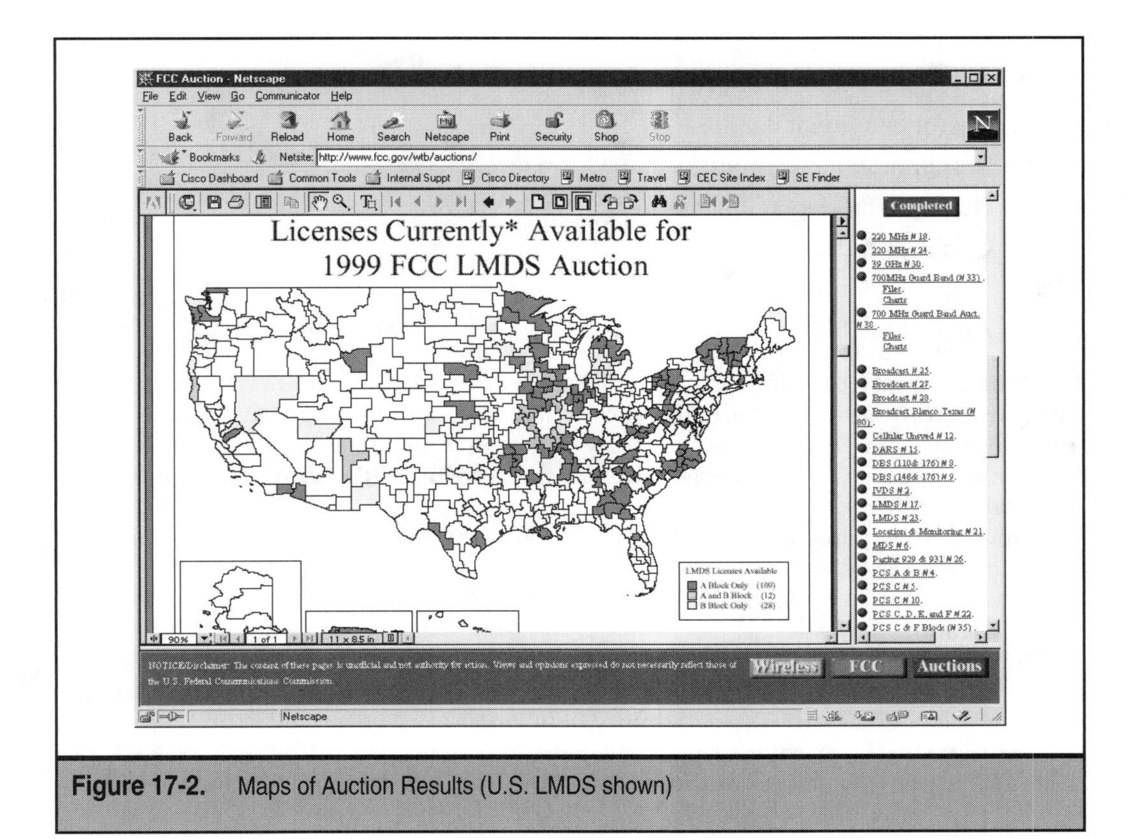

Figure 17-2. Maps of Auction Results (U.S. LMDS shown)

providers have migrated from a substitute license strategy to a complimentary license strategy, depending on opportunistic situations that arise out of a free marketing of licenses. The Bidder Activity indicates the high bid for a selected geography or BTA. This activity is based on the information received on the latest sequential bid round.

Auction Event Countdown

The sequential bid process has an array of steps that occur over a period of approximately six months prior to the beginning of an auction. The auctions can range in duration from a single day to three or four months, but most auctions last approximately one month. The following steps and chronology represent an overview of what takes place prior to an auction:

1. First announcement
2. Auction team established
3. Ask for comment
4. Announce terms and conditions
5. Seminar
6. Short form filing deadline
7. Filing status public notice released
8. Upfront payment deadline
9. Resubmission deadlines for short forms
10. Qualified bidders public notice released
11. Qualified bidders registration
12. Mock auction
13. Auction begins

First Announcement (180—120 days) The government provides an announcement anywhere from four to six months before an auction, to provide enough time for the industry to prepare for the prospect of acquiring spectrum or, as in the case of many companies, to plan for the augmentation of their existing BBFW initiatives by adding additional spectrum.

Time is also required to raise sufficient capital, acquire or reassign personnel, and corroborate the prospect of acquiring new spectrum with market studies, board meetings, and other studies and internal initiatives that enable the prospective buyers to prepare for the auction.

One other key item to verify is the availability of equipment. In certain cases where equipment does not exist that matches the spectrum, lead times and costs of purchasing or developing equipment must be determined or alliances must be formed to develop or purchase equipment. Ecopartner systems must also be put in place with the associated contracts and agreements. While six months may seem like a suitably lengthy time to prepare for the acquisition of new spectrum, the prospective buyers must move quickly and decisively to be prepared for the prospect of new spectrum.

Auction Team Established The government must also prepare for the upcoming auction by assembling a team of specialists, which includes attorneys, economists, project managers, and support staff. Milestones unrelated to staffing include assessing the automation initiatives for the browser-based tools (which includes loading specific spectrum-based information about BTAs and spectrum) and installing a wide array of templates for bidding, tracking, and reporting purposes.

The government must also determine the amount of any required upfront payments, as well as the minimum bid amounts. A complete study and report must also be made of the incumbent spectrum holders, if the auction is for spectrum that has already been allocated in other BTAs.

Ask for Comment The government then by law is required to solicit comments from the public. The solicitation extends to comments on the rules for the impending auction along with comments on the upfront payments, opening bids, and any other material aspect of the upcoming auction, to ensure fairness among the bidders. This is accomplished to alleviate or address concerns that may arise regarding the ability of all interested parties to adequately compete.

Announce Terms and Conditions Following the receipt and addressing of any comments or concerns from the public, the WTB releases the procedures, terms, and conditions of the upcoming auction to the public approximately 90 days prior to the auction. This is done on the FCC/WTB Web sites as well as through print media to the industry.

Seminar Approximately 45 days prior to the beginning of the auction, the government holds a seminar for all bidders, to ensure they understand the processes and procedures involved in the upcoming bidding process for the spectrum. Attendance is not mandatory and is provided at no charge to the attendees and interested parties.

Short Form Filing Deadline The prospective bidders then face their first deadline at 30 days prior to the inception of the auction to file FCC Form 175, which is a form that provides basic information to the government about the bidders and the markets and licenses in which they are interested. This form is filed electronically with the government.

Filing Status Public Notice Following the receipt of the short form, the government staff reviews the forms and makes a determination as to whether the form is accepted, incomplete, or rejected. A public notice is then filed 21 days in advance of the auction on the Internet and in print to the industry as to the status of the short forms. A package is then sent out to each applicant by overnight delivery that also informs them of the status of their forms.

For those applicants whose form where either rejected or deemed incomplete, the government includes a letter specifying the deficiency. The letter also includes contact information for further assistance if necessary to complete the form.

Upfront Payment Deadline The second deadline must then be met by the applicants, which is the payment of upfront monies to the government on or before 14 days in advance of the beginning of the auction. This refundable deposit then enables the applicants to participate in the subsequent auction.

Resubmission Deadlines for Short Forms At the same time as the deposit of upfront monies, short-form deficiencies must be amended and filed with the government. Missing this deadline forfeits the right to participate in the auction, although the government has been known to exercise some leniency on this point if required by extenuating circumstances.

Qualified Bidders Public Notice Released With the auction approximately 10 to 12 days away, the government publishes a list of the bidders with their FCC Account Number and the target markets for which the bidders have filed an intent to bid. The government also publishes notice of the mock auction and the start date of the auction.

Importantly, at this time the government also publishes the number of bidding rounds if the auction is to be a sequential bidding affair, the number of allowed rounds, and any other information germane to the impending auction.

Qualified Bidders Registration Immediately following the public notice of the qualified bidders, an overnight package is sent to the bidders that contains the confidential login ID, password, and Bidder ID Number (BIN) necessary to access the Automated Auction System. Client software is also provided to the bidders along with a manual for use and dedicated phone numbers, in case the bidders decide to participate by telephone as opposed to the AAS, which is browser-based.

Mock Auction The purpose of the mock auction is to allow the bidders time to work with the AAS without risk of making a mistake that could impact their participation in the bid process. The mock auction usually takes place approximately five days prior to the beginning of the auction. The bidders are free to use the AAS up until the actual auction begins.

Auction Begins The auction begins on the scheduled and notified day. Each of the bidders initiates their bids in the opening round or rounds either by entering their bid information to the AAS or by using a dedicated telephone number. A report at the end of each round is issued by the government, and a round ends either based on a deadline or when no more bids have been received over a certain period of time.

The bidders do not actually bid money; rather, they bid *credits* or *bidding units*, which are purchased by the bidders in advance of the auction with their deposits and are declared by the bidder in their FCC Form 175. Typically, one dollar equals one bidding unit, and the government assigns a specific number of units to a license prior to it being auctioned. A bidder's eligibility therefore is based on the size of their deposit. They are then qualified to bid on certain licenses, which helps to ensure that the most valuable licenses are purchased by the most interested and capable parties.

Bidders are required to be active in each round by bidding a specified percentage of their bidding units. This is part of what the government has established as *activity rules*, which ensure fairness, in part, by preventing a bidder from waiting until the final rounds before bidding. The required amount of activity by each bidder increases in each round if they wish to remain viable in the auction of the spectrum they are pursuing.

To place a new high standing bid on a license during a round, or at the opening of a new round, a bidder must have a sufficient number of unused bidding units left over from the previous round of bidding. The number of units offered by the bidder must be equal to or greater than the units assigned to the license.

To ensure the bidding does not increase by too small of an amount for each round, the government has established what is called an *exponential smoothing formula*. This formula establishes that the incremental bid must be a certain percentage of the standing high bid. Incremental bids are generally in the range of 10 to 20 percent in excess of the standing high bid. The formula also accounts for how much activity there is for a specific license. The greater the amount of bidding for a license, the greater the incremental bid set by the government at the end of each round.

A bidder may withdraw their bid, even if their bid is the standing high bid. If the license is sold for an amount that is less than the high bid they established prior to withdrawal, the withdrawing bidder then forfeits a portion of their deposit. This discourages insincere bidding.

Auctions are carefully measured to determine the competitive nature of the bidders. This assists the government in refining the bidding process for future auctions. The metrics for the competitive aspect of an auction are determined by the ratio of bidding units purchased against the bidding units assigned to all the spectrum being sold in that particular auction. In a typical auction, bidding units held by bidders outnumber the bidding units assigned to the spectrum anywhere from 7:1 to as low as 1.5:1.

The government also has a bidding rule called the *simultaneous stopping rule*, which is an activity rule stating that an auction will end if there are no new bids or withdrawals in a subsequent round. While unusual, the government retains the right to keep the auction open for a specified period of time even if there is no new activity.

DEPLOYMENT REQUIREMENTS

Once the license has been purchased, the real work begins in terms of deploying the asset and moving toward the profitability phase. Once the license has been purchased, the license holder has three years to begin deploying the license. The FCC has shown leniency in terms of what "deployment" actually means, but it is clear the government does not want licenses to languish unused.

A considerable number of companies purchase licenses only to subsequently go out of business. If they hold a license, it is deemed an asset, such as in the high-profile case of NextWave Telecom, which won 63 LMDS licenses for approximately $5 billion in 1996. When NextWave went bankrupt, the FCC rescinded its licenses and announced its intent to reauction them in 2000.

The reauction generated approximately $17 billion in revenues, mostly by Verizon Wireless from the resale of the spectrum. This occurred despite NextWave's attempts to block the sale and delay the auction. The matter became very unsettled when the Federal

Appeals Court determined that the licenses were an asset of a bankrupt company and, therefore, exempt from being taken from the company. NextWave was then able to successfully raise approximately $200 million from a German company, and then won a $100 million contract from Lucent.

This left the FCC in a poor light, because Verizon had requested several times during and after the solicitation for comments phase of the auction that the FCC hold off on the reauction until the NextWave matter had been completely settled.

A relatively similar incident occurred with General Wireless, which purchased 14 LMDS licenses for approximately $1 billion in the same 1996 auctions, but also subsequently filed for bankruptcy and was able to retain the licenses for approximately $166 million based on a ruling of the bankruptcy court.

The point of these two incidents is that the purchase, deployment, and occasional bankruptcy make for a very challenging environment at the FCC and WTB. All in all, they have done a very good job in ensuring competition, reducing backlog, and retaining order for one of the most valuable and limited assets in the technology industry today, that of spectrum.

CHAPTER 18

Licensing

The penultimate chapter in this work will center on spectrum licenses and frequency allocation. It also reviews a few of the regulations to which FCC license holders must adhere to remain in compliance. This chapter represents only a small sampling of the full scope of issues associated with BBFW, and in no way should be considered comprehensive. While the author is much in favor of unlicensed BBFW systems and their advantages, they too have a substantial number of compliance issues. In short, no BBFW effort should take place without due diligence and education relative to federal, state, and local regulatory agencies and laws. This chapter should serve to initiate you to a few of the regulations with which you must comply if you are to successfully operate a network that includes BBFW components.

U.S. BBFW LICENSES

There are 39 licensed frequencies in the United States, ranging from 928 MHz to 40 GHz. These frequencies are used for a wide array of purposes, from cellular telephone to satellite communications, depending on where they are in the frequency bands and, in many cases, according to the specific allocations of spectrum, which may be subsets of the licensed spectrum. There are other frequencies in use by the military, police, and other government entities that are not listed but are included in the overall spectrum. As a side note, significant debate exists today regarding whether some of the military (restricted) bands should be opened and/or shared for commercial purposes.

Table 18-1 illustrates the array of frequencies available for use with a license in the United States. Those that include BBFW are indicated as such.

928–929	MAS, MAS, PRS	Comments
932.0–932.5	MAS, PRS	
932.5–935.0	CC, OFS	(1)
941.0–941.5	MAS, MAS, PRS	
941.5–944.0	CC, OFS, Aural BAS	(1)
952–958	OFS/MAS, PRS	
958–960	MAS, OFS	
1850–1990	OFS, PCS	
2110–2130	CC, PET	

Table 18-1. U.S. Licensed Frequencies (in MHz)

2130–2150	OFS, PET	
2150–2160	OFS, MDS	BBFW
2160–2180	CC, ET	
2180–2200	OFS, PET	
2450–2500	LTTS, OFS, TV BAS, ISM	F/M/TF, BBFW
2650–2690	OFS, MDS/ITFS	BBFW
3700–4200	CC LTTS, OFS, SAT	
5150–5250	MDS	BBFW
5250–5350	MDS	BBFW
5725–5825	MDS	BBFW
5925–6425	CC, LTTS, OFS, SAT	
6425–6525	LTTS, OFS, TV BAS, CARS	M
6525–6875	CC, OFS	
10,550–10,680	CC, OFS, DEMS	
10,700–11,700	CC, OFS, SAT	
11,700–12,200	LTTS, SAT	
12,200–12,700	OFS, DBS	
12,700–13,250	CC LTTS, OFS, TV BAS, CARS	F/M/TF
14,200–14,400	LTTS, SAT	
17,700–18,580	CC, OFS, TV BAS, SAT, CARS	
17,700–18,300	CC, OFS, TV BAS, CARS	
18,300–18,580	CC, OFS, TV, BAS, CARS, SAT	
18,580–18,820	CC, OFS, Aural BAS, SAT	
18,820–18,920	DEMS, OFS, DEMS, SAT	
18,920–19,160	CC, OFS, Aural BAS, SAT	
19,160–19,260	DEMS, OFS, DEMS, SAT	
19,260–19,700	CC, OFS, TV, BAS, CARS, SAT	
21,200–23,600	CC LTTS, OFS	TF
24,250–25,250	DEMS	
27,500–28,350	LMDS	BBFW
29,100–29,250	LMDS, SAT	BBFW
31,000–31,300	CC, LMDS, LTTS, OFS, LMDS	F/M/TF, BBFW
38,600–40,000	CC, OFS, TV BAS	F/M/TF, BBFW

Table 18-1. U.S. Licensed Frequencies (in MHz) *(continued)*

The following are the acronyms and abbreviations that appear in Table 18-1:

- **BAS** Broadcast Auxiliary Service
- **BBFW** Broadband Fixed Wireless (not an FCC designation)
- **CARS** Cable Television Relay Service
- **CC** Common Carrier Fixed Point-to-Point Microwave Service
- **DBS** Direct Broadcast Satellite
- **DEMS** Digital Electronic Message Service
- **F** Fixed
- **ISM** Industrial, Scientific & Medical
- **ITFS** Instructional Television Fixed Service
- **LTTS** Local Television Transmission Service
- **M** Mobile
- **MAS** Multiple Address System
- **MDS** Multipoint Distribution Service
- **OFS** Private Operational Fixed Point-to-Point Microwave Service
- **PCS** Personal Communications Service
- **PET** Emerging Technologies
- **PRS** Paging and Radiotelephone Service
- **SAT** Fixed Satellite Service
- **TF** Temporary Fixed

Prior to 1998, the FCC required that information transmitted below 19 GHz and in the 24 to 25 GHz bands be modulated at a minimum prescribed rate. This bit rate, measured in bits per second, was to be equal to or greater than the amount of allocated bandwidth; for example, if a licensee was granted 20 MHz of spectrum, users had to operate equipment that was capable of transmitting at rates not less than 20 Mbps. This was done to ensure the licensee was operating at a minimum level of efficiency so as to minimize the total amount of spectrum used. As of December 1998, this rule was discarded by the FCC.

The issue of minimizing the total amount of spectrum used is a key one, because one of the fundamental tenets of wireless use is akin to the availability of real estate: the total amount available doesn't increase over time. Having stated that, in addition to continually evaluating and reallocating existing frequency bands, there is a continual effort to increase spectral density through more sophisticated modulation schemes, and manufacturers are in an endless pursuit to provide equipment with steadily improving bit recovery mechanisms and architectures that are cost effective.

The FCC has retained regulations that specify equipment placed in service after June 1, 1997 in the 3700–4200 MHz (4 GHz), 5925–6425 and 6525–6875 MHz (6 GHz), 10,550–10,680 MHz (10 GHz), and 10,700–11,700 MHz (11 GHz) bands must meet the minimal payload capacities set forth in Table 18-2.

Nominal Channel Bandwidth (MHz)	Min. Payload Capacity (Mbps)	Comparable Voice Circuit Utilization
0.4	1.54	1 DS-1
0.8	3.08	2 DS-1
1.25	3.08	2 DS-1
1.6	6.17	4 DS-1
2.5	6.17	4 DS-1
3.75	12.3	8 DS-1
5	18.5	12 DS-1
10	44	1 DS-3/STS-1
20	89.4	2 DS-3/STS-1
30	89.4	2 DS-3/STS-1
30	134.1	3 DS-3/STS-1
40	134.1	3 DS-3/STS-1

Table 18-2. Minimum Voice Circuits for a Given Bandwidth

In Table 18-2, DS and STS refer to the number of voice circuits a channel can accommodate:

- 1 DS-1 = 24 voice circuits
- 2 DS-1 = 48 voice circuits
- 4 DS-1 = 96 voice circuits
- 8 DS-1 = 192 voice circuits
- 12 DS-1 = 288 voice circuits
- 1 DS-3/STS-1 = 672 voice circuits
- 2 DS-3/STS-1 = 1344 voice circuits
- 3 DS-3/STS-1 = 2688 voice circuits

Maximum Power Output

The FCC is clear in its regulations about the maximum amount of power a BBFW system can emit. This limitation is placed on users in order to minimize the effects of interfering with other license holders as well as to minimize the potential for health hazards to humans. In the event the FCC receives a complaint that is verified through a hearing, it may require license holders to reduce their power outputs.

While there is an array of exceptions to Table 18-3, it may be used as a general guide for maximum allowable outputs for BBFW-licensed systems in the frequencies listed.

Frequency Band	DbW
928.0–929.0	+17
932.0–932.5	+17
932.5–935.0	+40
941.0–941.5	+30
941.5–944.0	+40
952.0–960.0	+40
1850–1990	+45
2110–2150	+45
2150–2180	+45
2180–2200	+45
2450–2500	+45
2500–2686	+45
2686–2690	+45
3700–4200	+55
5925–6425	+55
6425-6525	+55
6525–6875	+55
10550–10680	+55
10700–11700	+55
12200–12700	+50
12700–13250	+50
14200–14400	+45
17700–18600	+55
18600–18800	+35
18800–19700	+55
21200–23600	+55
24250–25250	+55
27500–28350	+55
29100–29250	+55
31000–31075	+30
31075–31225	+30
31225–31300	+30
38600–40000	+55

Table 18-3. Maximum Allowable Power Output

The Most Common U.S. BBFW Licenses

The most widely deployed licenses of interest to the BBFW community are the MMDS and LMDS licenses, because the majority of BBFW technology providers that have been underwritten by venture capitalists have elected to provide radios in one of those frequencies. The concept of which frequencies are supported by the VC community should not be minimalized because virtually every BBFW service provider requires outside capital to execute its business plan.

In general, you don't see many equipment designers or manufacturers provide both MMDS and LMDS equipment, because the technical, marketing, and business expertise required to build them are quite different. However, occasionally, as the result of an acquisition, an equipment designer or manufacturer will provide both MMDS and LMDS equipment.

Geographic References Used by the FCC

Licenses are assigned on a geographic basis for a limited period of time of approximately ten years. However, the length of time a license is deemed valid can vary from license to license and can be based on items such as when the license was originally issued. The FCC uses a dizzying array of geographic references to the geographies in which licenses are allocated. These references include Basic Trading Area (BTA), Major Trading Area (MTA), Economic Area (EA), Regional Economic Area (REA), and Major Economic Area (MEA). Figures 18-1 and 18-2 indicate BTAs and MEAs, respectively, in the United States.

In addition to the preceding geographic references, the FCC also allocates licenses based on Public Coast Station Areas (PCSAs), Location and Monitoring Service Areas (LMSAs), Metropolitan Statistical Areas (MSAs), and Rural Service Areas (RSAs), the latter two of which are shown in Figure 18-3.

One of the interesting aspects of the MMDS licenses is that most of them have ended up in the possession of either Sprint or MCI WorldCom. These two corporations currently own approximately 83 percent of all MMDS licenses in the United States. While this represents an asset of considerable magnitude, costing a collective $2 billion, these companies have had to negotiate the rights to the entire spectrum in a contiguous manner, because 60 MHz of that spectrum was grandfathered to schools.

MMDS Licenses

MMDS channels were originally referred to as H channels, and the licenses for these channels were handled by the WTB until 1992, at which time the processing functions were transferred to the FCC's Common Carrier Bureau. Subsequent to that, the processing functions for the licenses were transferred to the Mass Media Bureau in 1994, where these processes continue to be handled today.

The Mass Media Bureau is better known for its regulation of commercial broadcast AM and FM radio stations. It ensures that the licensed broadcasts will be received without co-channel interference within a specified Basic Trading Area.

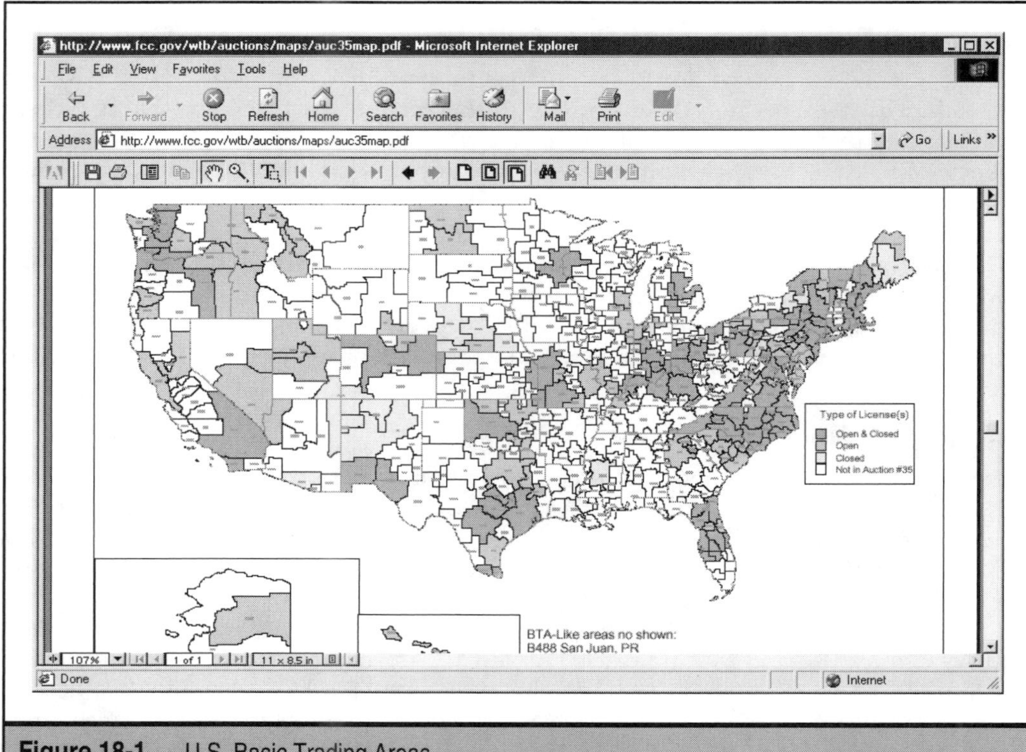

Figure 18-1. U.S. Basic Trading Areas

The MMDS spectrum is comprised of 200 MHz of frequency at 2.5 GHz to 2.7 GHz, and was originally comprised of 33 analog video channels that were 6 MHz wide. There is also a small amount of spectrum at 2.15 GHz to 2.162 GHz that is referred to as MDS by the FCC, but is sometimes referred to as part of the MMDS spectrum. Current equipment providers have been able to leverage this channelization into as many as 99 digital data channels. With power settings as high as 100 watts, an MMDS can be broadcast on a point-to-point basis at ranges of up to 35 miles on a line-of-sight basis.

MMDS license holders in the U.S. have to consider the presence of schools (or other users) within the BTA for which they have been granted the license, because these groups have the right to transmit in the Instructional Television Fixed Service (ITFS) band. This spectrum is a band of 20 television channels at 6 MHz apiece which range from 2.5 GHz to 2.686 GHz, which makes this a shared spectrum with MMDS. This frequency overlap has brought on hundreds of lease agreements between educational entities and wireless service providers. These arrangements run the gamut from the school selling a portion of its spectrum, to leases that enable wireless service providers to use the spectrum.

ITFS is typically used for distance learning, with the spectrum being used on a one-way basis only. The channels can also be used for entertainment purposes, to compete

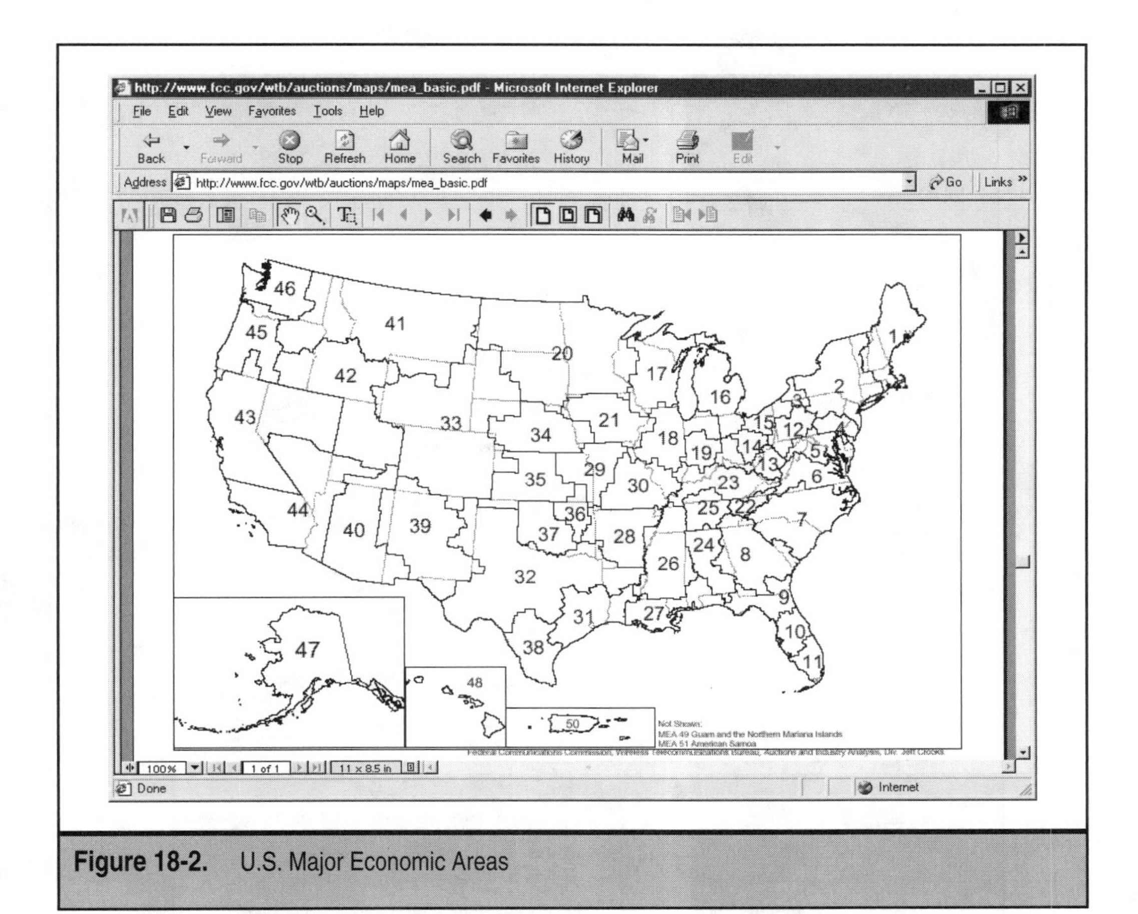

Figure 18-2. U.S. Major Economic Areas

with local cable companies. Educational institutions have historically not harnessed the financial opportunities afforded them by either selling distance learning or by exchanging their ownership in the spectrum for cash or equipment.

If the school is using that spectrum and does not wish to partner with a service provider (which is rare, given that schools rarely have an excess of financial assets), the MMDS license holder must then adopt its frequency reuse plan to accommodate the spectrum owned by the schools. In some instances, the schools have partnered with local business entities to exchange the frequency for telecommunications equipment, services, or both that are of value to the school.

The service providers are required by the FCC to preserve, or make available in the future, at least eight ITFS channels for future use by non-service providers such as schools. Further, wireless service providers may not operate on more than eight channels. They must also own at least four MMDS channels in that BTA and demonstrate to the FCC that no MMDS or MDS channels are available in that BTA in lieu of using the ITFS spectrum.

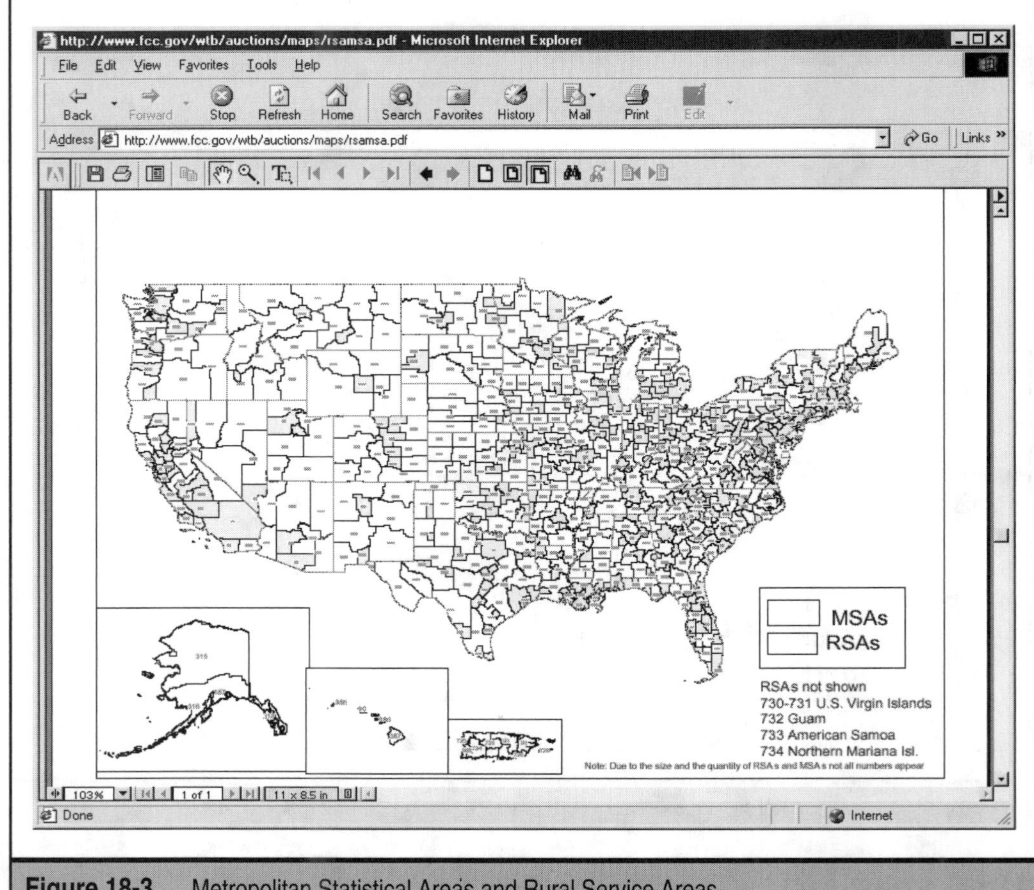

Figure 18-3. Metropolitan Statistical Areas and Rural Service Areas

LMDS Licenses

LMDS licenses are considered part of the millimeter portion of the electromagnetic spectrum, and generally are used for point-to-point communications of up to two or three miles. Most of this technology is derived from work done during World War II, and most of the millimeter wave engineers today have military radio backgrounds.

Over the years, these systems have evolved from those used for voice to those used today for data and video, and these radios have been deployed as key elements in the public switched telephone network since the 1960s. They now more commonly see duty for backhaul or high-bandwidth campus type use, although some equipment manufacturers have offered P2MP systems in the LMDS frequency.

Historically, the FCC has regulated both millimeter and microwave services under what was known as Part 21, which was designated for service providers, and Part 94, which was designated for networks that carry traffic only for private service. A private service is a service where entities utilize frequencies purely for internal business purposes or public safety communications, and not as a service provider. These services have been consolidated under a new set of rules since August of 1996 in what is now known as Part 101. Along with this streamlining, the rules have been amended to allow BBFW users to commence operation immediately upon filing of their applications instead of awaiting formal approval, provided certain compliance conditions are met.

LMDS is deployed in 493 BTAs in the United States, which are defined by Rand McNally and cover the 50 continental states, as well as BTA-like areas, which include United States Virgin Islands, American Samoa, Guam, Mayaguez/Aguadilla-Ponce, Puerto Rico, San Juan, Puerto Rico, and the Commonwealth of Northern Marinas. Figure 18-4 illustrates the BTAs for LMDS.

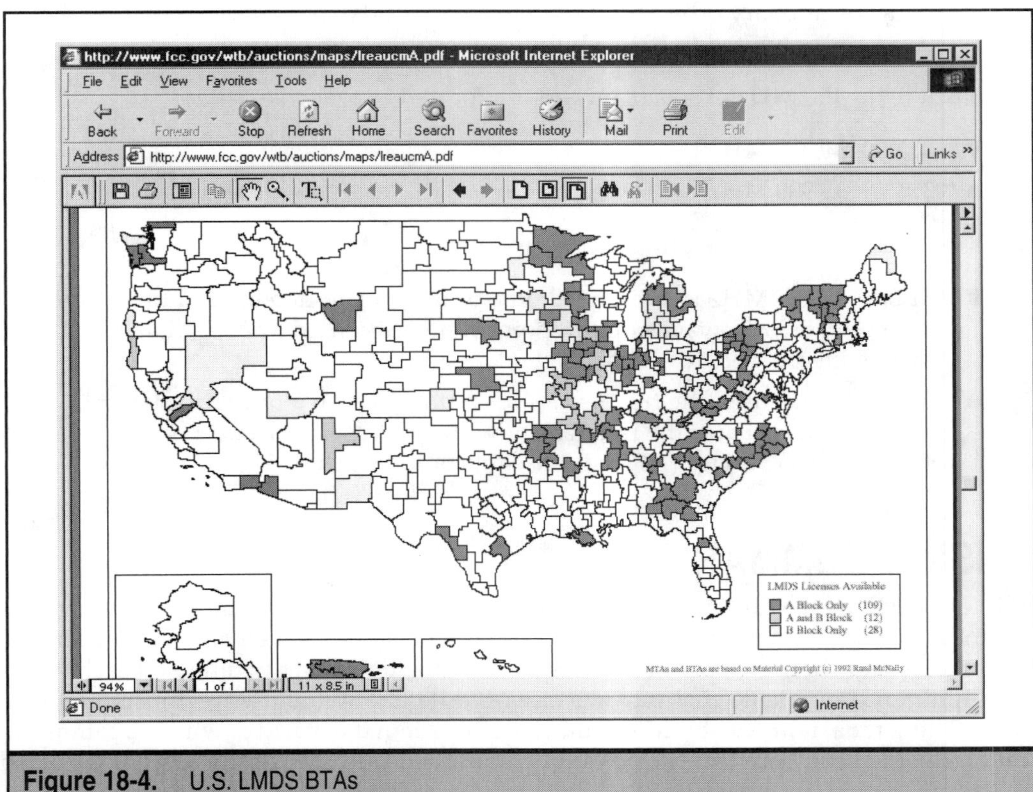

Figure 18-4. U.S. LMDS BTAs

A number of countries around the world offer "LMDS" frequencies, and this spectrum differs from country to country. In the United States, this spectrum is available in two license blocks. The first and much larger of the two is Block A, which has 1150 MHz of spectrum in the following allocations:

- 27,500–28,350 MHz
- 29,100–29,250 MHz
- 31,075–31,225 MHz

The Block A licenses are authorized as follows:

- 27,500–28,350 MHz is authorized on a primary protected basis and is shared with Fixed Satellite Service (FSS) systems.
- 29,100–29,250 MHz is shared on a co-primary basis with backhaul links for nongeostationary orbit Mobile Satellite Service (NGSO/MSS) systems in the band and is limited to LMDS hub-to-subscriber transmissions.

Block B has 150 MHz of spectrum and is allocated as follows:

- 31,000–31,075 MHz
- 31,225–31,300 MHz

The Block B licenses are authorized as follows:

- 31,075–31,225 MHz is authorized on a primary protected basis, and is shared with private microwave point-to-point systems licensed prior to March 11, 1997, as provided in Sec. 101.103(b).
- In Block B licenses, the frequencies are authorized on a primary protected basis if LMDS shares the frequencies with systems licensed as Local Television Transmission Service (LTTS) licensed prior to March, 1997.

LICENSING SUMMARY

While the FCC has certainly attracted criticism by the general public as well as numerous business and license holders, in general, it has performed very well given the enormous complexity and sheer number of licenses that are auctioned and maintained each year.

Think for a moment what our world would be like without wireless communications. Think for a moment about how much less efficient the world of wireless communications would be if there were no governing body, and the providers of technology could develop and sell equipment without regard for an organized approach. While some would be tempted to say that standards bodies would emerge without the intervention, monitoring, and regulation of the federal government in this day and age, one must remember that the world has enjoyed the fruits of wireless communications since the days of Tesla and Marconi in the late 1800s. In general, the world has enjoyed these advantages (which have indeed become life-saving necessities) without much confusion or delay.

While attempting to reduce the number of regulations, the FCC has, at the same time, had to manage licenses in the most efficient and fair manner possible in order to preserve one of our most important national assets: that of wireless spectrum.

PART VIII

airBand Networks: An Interview

CHAPTER 19

Deploying a Network: An Interview with airBand Networks of Dallas, TX

This final chapter focuses on a number of issues faced by a real-world service provider: airBand Communications in Dallas, TX. airBand is a company that provides broadband services through several mediums, including BBFW, DSL, and ISDN. While it resells DSL and ISDN to SMBs, its broadband crown jewel is airBandits U-NII offering to SMBs.

airBand is one of the largest service providers in the United States that has a primary focus on the U-NII band architecture and technology. This technology enables airBand to address the tremendous demand for broadband from the SMB market. Currently, less than 10 percent of all businesses in the United States, and less than 5 percent of businesses in other industrialized countries, have broadband access of any kind. Yet, the demand for this business tool greatly exceeds the supply worldwide.

airBand's 100 employees and its CEO Andy Lombard operate out of a modern facility just off the LBJ Freeway in north Dallas. airBand has been evaluating U-NII technologies and vendors since 1999. What follows is based on an interview with Mr. Lombard, who provided some interesting and relevant points regarding U-NII and the role of a wireless service provider.

Mr. Lombard states airBand's executive summary to its business plan as follows, "airBand's objective is to provide high-speed Internet services to businesses of all sizes, and we're focused on urban environments throughout the country. We've developed an advanced core broadband network that includes BBFW for the last mile, and we utilize the U-NII band so we don't have to purchase a billion dollars worth of licenses to cover the whole United States. As a result, our business model is considerably more cost effective from an infrastructure perspective and cost effective from a customer perspective than if we used licensed spectrum. As a result, we're able to offer high-bandwidth services at a fraction of the price and a fraction of the time to deploy than if we utilized licensed spectrum."

In response to being asked about his vision for providing services such as storage networks, CDN (content data networking), and remote host management, Mr. Lombard states, "It's appropriate to take a step back from that question to address the real issue, which is the final-mile bottleneck for broadband. We don't think many people have gotten that part right yet. While providing services may be more interesting to some, broadband access remains a huge hurdle for telecommunications companies in the United States.

"Our vision in terms of services is relatively rudimentary and we want to keep it simple. We think that once you put the IP layers in, we can add services such as storage devices, Web hosting, managed extranet, managed security systems, and VPNs. These can all be part of the capabilities of a wireless infrastructure that extends the metro environments. In addition, we've gone out and have been fairly pragmatic with regard to the market and where the technologies are today. For example, voice over wireless infrastructure is close to being ready. It's certainly been tested and many of us have talked over it. I've done it over ATM and I've done it over IP, but it's not yet standardized and it's not yet at the point where the economics for the support can be driven from the vendor perspective. We're excited about that for the future, but it's not something we'd want to roll out before mid-2002 or so. Perhaps at that time we can start to add voice as a complimentary service, but not as a leading service for our capability.

"In a vision perspective, we think voice plays an important role to a bundling strategy and a value-added strategy, but the prime services that will become the meat and potatoes of our service strategy by generating the most revenue at the earliest date will be managed extranets and security, storage, VPNs."

The author agrees with Mr. Lombard's perspective. While the state-of-the-art BBFW systems operate at Layer 3 over the air, a number of studies have determined what customers value the most from broadband. At least one notable Virginia ILEC (which has been in existence for more than 100 years) invested in a detailed survey that found an overwhelming response in favor of simply having broadband access as opposed to immediate access to various services that only broadband can carry. The question asked most often by the respondents of the survey was, "When can I get broadband?" The author has also polled most of the accounts with which he has been associated in the two years prior to writing this book, and a similar response was received from not only service providers but also SMBs and even prospective residential customers.

As for the BBFW equipment providers that are providing links at Layer 3, it may be too late in the market and product lifecycles to return to the drawing board and release a Layer 2 version of their products. An appropriate strategy may be to have their customers focus on the concept that the time until services will be demanded is less than originally thought; possibly on the order of the next 18 to 24 months. Therefore, they may wish to take the stance that their customers will not have to migrate from a Layer 2 architecture to a Layer 3 architecture. If they have Layer 2 equipment in place, they can migrate with routers at each end of the BBFW link until they install Layer 3 BBFW links. The upside for the customers is that those who have already purchased Layer 3 BBFW equipment will not have to migrate from Layer 2 or deploy extra routers to convert Layer 2 links to Layer 3 links. Virtually all participants in the service provider market space believe that services will be important in the upcoming years and that links will need to have the bandwidth, intelligence, speed, and delay sensitivity that will enable services to operate efficiently and reliably.

"So when we look at providing access versus providing services, there are two areas we focus a lot of our time on. The first is the core network and the metro area networks. If you start to look at the SONET rings (fiber optics) or gigabit Ethernet rings in urban environments, there's a cost associated with developing the infrastructure beyond a certain geographical point. Those economics become prohibitively expensive when you start to extend the laterals off the rings, which may not in fact make a lot of sense for a fiber provider to light up a specific building. Those expenses can be upwards of $500,000 to $1 million to light up that building. So we think there's a natural partnership that exists between the metro area core networks and our extension of access networks by BBFW. Once you get that part done, we think that access starts to drive itself towards nomadicity or at least some level of roaming on a person or on a laptop, PDA, or some device that drives bandwidth through those access links. This strategy that drives from the last mile to the last yards and the last feet with regards to technologies available such as 802.11 WLANs.

"I think I'm going to be at a fairly mature age before fiber to the SMB market has even a 10 percent penetration rate, as the number of buildings being lit even compared to the rate of construction for new buildings isn't that great over the last four or five years in spite of the incredible investment made to install laterals and light up buildings. Further, recent studies have indicated that only about 30 percent of SMBs can access fiber, and there's certainly not 30 percent of SMBs that would access that fiber because of cost constraints, because the in-building networks are even more expensive than the laterals, which are far more expensive than metro rings in terms of deployment.

"The way we see it, the more metro fiber rings that are deployed, the better it is for us, because all we have to do is connect to a point on that ring. With BBFW, you can radiate a sector that is comprised of SMBs, whereas a fiber lateral by nature can only target one building at a time. Where a fiber lateral build requires an attack on a building-by-building basis, we can blanket a downtown area at a fraction of the cost and in a fraction of the time. Now the reality of deploying BBFW to a building is that it requires a very specific strategy, but the strategy is tied more to the general geographic area than on a building-by-building basis, which is the requirement for running a lateral from a ring into a building. One has to remember that there's no such thing as a point-to-multipoint fiber on a metro basis. On the other hand, it's a fundamental architectural and strategic approach for BBFW.

"Fiber-based companies come to us because they like the notion of us seeding their future by linking customers to their rings and increasing their traffic loads. Further, I'm okay with playing the seeing-eye dog with my counterparts in the fiber provider companies. We like playing this role for them because they can help determine, through the use of our BBFW access capabilities, whether or not a certain geography is worth making the enormous investment for installing laterals."

Regarding competing with ubiquitous fiber, Mr. Lombard states, "If fiber became ubiquitous in every single neighborhood and every single building, then we'd better get on the fiber bandwagon somehow and convert our customers over to that, but again, that's going to take a very long time and untold billions of capital investment to create that environment, and I don't think that's going to happen in the next five to ten years in any significant way. If you look at the money that's fueled that space in the last three years, we should be reaping those benefits today on a broad basis, but the fact is, that's not the reality, because fiber has perhaps picked up one additional percentage point of penetration into the SMB space. If the general technology market takes a downturn, that penetration rate will only get worse, not better. The market will require more and more ways to get an economic broadband access."

The author would add that even if the various LECs or utilities were to decide to invest the billions of dollars to deploy fiber laterals, there would be at least one other significant bottleneck—the lack of deployment resources necessary to support such a buildout. The only entities with the resources to consider such a buildout, purely from a manpower and equipment resources perspective, are the utilities, which, at best, currently are showing signs of a wavering commitment to challenging the phone company incumbents for bringing fiber to the buildings. To that observation, Mr. Lombard sagely added with a chuckle, "Right, and the real winner in fiber deployment has probably been Caterpillar and John Deere. Corning was a winner for a while but found out you can only buy so much glass."

"So the combination of wireless LAN technologies and wireless WAN technologies that extend the last mile and connect that into the metro environment is airBand's key sweet spot. As far as additional services go, once you get the access part, once you put an IP network together that is highly reliable, is efficient, and will carry the bits we need to carry across the network, then we can talk services. But for now, we need to nail the last mile."

When asked for his vision on an appropriate strategic response for a BBFW service provider who goes into an area where there is no broadband and is subsequently followed by a copper broadband competitor, Mr. Lombard responds, "Most copper is going to offer speeds on the order of 256 Kbps, which isn't very interesting to us. BBFW offerings will start at around 1 Mbps and up. We think as advancements to DSL occur, the associated speeds will increase, but they'll also increase with BBFW. If you're getting around a megabyte of speed from a cable modem, you're either very close to the plant or your neighborhood is relatively barren in terms of others who are also using cable modems for data." Both of those scenarios are relatively rare, and most people who use cable modems for high-speed data transmission find an erosion of speed and reliability over time as increasing numbers of users are added. Mr. Lombard adds the following thought, "I think it's going to be about scaling, and it's easier, faster, and less expensive to scale a U-NII-band BBFW network than a cable plant."

When asked for his vision of broadband copper continued market penetration in both residential and SMB markets, Mr. Lombard states, "BBFW will remain fairly far behind the copper broadband market for some period of time to come, based on the current capital markets. I do think that DSL will start to work some of its kinks out but will remain limited in terms of deployment. Cable modems—where the plant is in place—they'll upgrade, but they don't have to go well beyond their current 60 million homes to be successful.

"I would say if there is going to be a winner over the next four or five years, cable will probably edge out DSL, because a lot of the investment on the DSL side is going to have to be driven by the ILECs, so it's a tough call to bet between the ILECs and the cable companies. However, I do think that after about a four-year period, you'll start to see a pretty good penetration on the fixed wireless side once you see some larger-volume CPE purchases made."

He also pointed out an interesting perspective on how broadband access endures during both strong and weak periods of capitalization, "In spite of the current capital market, which as we speak in September of 2001 is probably the worst I've seen, the demand for broadband access remains very strong. This speaks volumes on the true value proposition a company like airBand can provide. Service providers that offer the U-NII band can provide a superior value proposition to the market, because their offering is more cost effective. And once the word from one U-NII end user spreads to others, the market will perceive U-NII as good because they'll not only listen to the U-NII service providers, they'll listen to the U-NII end users. Most end users have heard offers for broadband services from three or four providers, and most end users have used two or three broadband providers because the service providers have gone out of business. These end users need to hear from someone offering something other than DSL or cable, because they've heard these pitches before and it hasn't worked out as well as it could have in a number of SMB cases. U-NII band end users tend to speak positively about this type of access medium."

Mr. Lombard more than believes in U-NII and the unlicensed BBFW technology. His commitment to this is demonstrated in his statement, "Something fundamental to us at airBand is that you have to be able to use your own service and know it very well. Every one of us at airBand is reliant on a U-NII BBFW link every single day, and we have no T1 into our facility, and we have no data connection other than our U-NII wireless link. That's the way we're connected to the Internet, and it flies, it works really well. I want to know first-hand what can go wrong so we're prepared when a customer calls with a problem."

On the issue of the profitability for a BBFW service provider versus a well-run CLEC, Mr. Lombard highlights a very favorable scenario in favor of the U-NII BBFW service provider, "If you look at our model, which is a Cisco Powered Network MAN with a fixed wireless U-NII last mile, compared to a well-built CLEC model, the difference in the financial return is startling. On the top line, the U-NII provider will see about a $400 to $500 per customer investment for a fixed wireless model that is focused on a data and value-added services model, no voice. A well-run CLEC will run about $700 to $800 per customer, because they need the extra investment to ensure a voice-capable network. One other difference between the two comes out loud and clear, which is that the return on capital for our model versus the well-run CLEC is substantially better; in fact, sometimes 20 to 30 percentage points better.

"If you look at the incremental costs of adding each new SMB customer for each of the models, and considering all the shared networking costs, the CLEC will spend more money because of their required investments in co-location expenses, leasing lines for the POP, and other costs for building out their core network. Where a CLEC can easily spend $500,000 to light up a building, the most we've spent on lighting up a building is $15,000. This means we don't have to make a large capital investment each time we add a new SMB customer. In summary, we are seeing margins that are off the scale compared to telecommunications companies of the past."

The author would add that there is also a time-value aspect, as a BBFW service provider can install a link to an SMB building in one to three days (assuming the right of way is in place) in comparison to lead times of weeks or months for a CLEC to provide copper. The lead time to install fiber to that same building can easily exceed a year, depending on the building location. However, even if the SMB site is only one or two thousand yards from a fiber ring node, the lead time can be months, if not a year or more if the fiber trenching crew has to cross a stream, hits a bedrock vein, or uncovers an archeologically significant underground site.

Mr. Lombard expands further on the economic advantage of a U-NII service provider with, "There is also something very strategically favorable, which is in avoiding the ILEC. They're difficult to deal with because they're competitors to the CLEC. They're being smart competitors by slow-rolling the CLECs in terms of access to the ILEC equipment. The ILECs want to ensure their competitors have the least opportunity to succeed and they will ensure this to the extent the law allows. Secondly, the ILEC can easily delay providing the full range of services to the CLEC that the ILEC itself can provide. In general, there's no difference between what a CLEC can offer and what the ILEC will offer in terms of services."

The author would add that the CLECs entered into a pricing war with the ILECs and generally did so without having the very deep pockets of the ILECs. In short, the CLECs showed up to do battle with far less long-term resources than the ILECs had available.

Mr. Lombard continues along this thread by stating, "I think that bypassing the ILEC obviously gives you the provisioning time frame advantage, gives you a clear right of way to the Internet, but it also allows you the ability to do something end to end that they can't do in a co-location scenario with the ILECs."

His vision for the future of BBFW service providers is rosy: "The market studies indicate a huge opportunity. Forrester Research indicates the SMB market for fixed wireless will grow from $1 billion today to $7.4 billion in 2003. Again, even in the very difficult capital market we're presently experiencing, the SMB demand for broadband is just rocking. The bad news with regard to the capital markets is that they've seen a lot of fixed wireless service providers fail. The good news, however, is that there is a big void, especially when you consider all the data LECs like Covad, Northpointe, and Rhythm that have ceased to exist; you know who your competition is, which is the two or three remaining CLECs and the ILECs. While they're bigger companies, there are fewer of them."

Mr. Lombard's view on the issues of security relative to U-NII links is seasoned and rational. He states, "First, from a security perspective, over-the-air DES encryption is part of the capability and secondly, from a network perspective, you ensure that you do the right things to provide security. But there is no inherent weakness to an FCC Part 15 (unlicensed) device in comparing U-NII to licensed bands."

The author would add that some markets, such as the financial industry, rely on very comprehensive security techniques, such as DES encryption. For this type of customer, the equipment providers can provide additional security enhancements through the use of a VPN tunnel over the air.

The reality of the situation is that even if a hacker were able to crack the spreading technique along with the modulation technique, it would be an unusually sophisticated attack. In many cases, it would be easier to infiltrate the target site in other ways. What is common to U-NII and licensed alike is that no BBFW link is secure from jamming. While jamming is not the same as sniffing for information, it has a disruptive effect on the link. It should be noted that there are very few reported incidents of jamming. Mr. Lombard adds, "Jamming can happen over any wireless device. There is no advantage to an MMDS link over a U-NII link in this regard. While companies like Sprint can go to the FCC with this type of complaint, it still won't help a financial institution in the short run. There are also options such as path diversity to ensure high reliability under all circumstances."

When asked what he would say to a room full of prospective U-NII users if he had ten minutes of their time, Mr. Lombard responded, "I think the most important thing is not to focus on whether it should be U-NII or by another medium that the customer should focus on a very highly scalable service that is cost effective. U-NII can meet this almost 90 percent of the time. The idea of using U-NII, versus fiber, versus copper, versus ISM, versus 38 GHz systems, quite frankly, I believe is a moot point. The customer wants three 9s to five 9s reliability. This means the end customer must have not only a technology in place that will provide high reliability, but equally important, they must have a service provider partnership that believes it's in the customer service business. Whether that means providing new services or responding to issues or concerns they have, the service provider must be world-class at this.

"In the future, airBand will look at evolving into fiber plays, 802.11 buildout into MTUs, and adding voice and managed services and other services such as storage networking. Right now, as stated earlier, we're focusing on providing broadband access, and 5.8 GHz is very well positioned to provide that in the United States. It's a phenomenally clean spectrum, there's an abundance of it, it has great channelization and QoS to provide great services, and etcetera. What more could you ask for?"

It's on this general note that this work concludes. While U-NII, and indeed BBFW in general, is a great tool, it has its place and its attendant advantages and disadvantages. On the whole, however, it's an exceptional access medium that can provide more than enough bandwidth, speed, reliability, and differentiated services for all but a few customers.

Above all, U-NII should not be considered as an ancillary technology to a network; rather, the more fully it is integrated into the WAN and LAN, the faster it will become an indispensable part of the wonder we call the Internet experience.

APPENDIX

BBFW-Related Web Sites

The following is a list of some good Web sites that pertain to BBFW. As Web sites tend to move, change, disappear, or merge into other sites, this list is a good place to start.

http://lucent.com/search97cgi/vtopc?collection=ALL&queryString=wireless
A good site for Lucent BBFW products

http://search.calpoly.edu/search97cgi/s97_cgi.exe?Action=Search&QueryMode=Internet&ResultTemplate=calpoly%2Fcalpolygenlrsl.hts&Collection=Cal+Poly+General&QueryText=wireless
An interesting look at what the California Institute of Technology is doing with BBFW

http://www.cisco.com/warp/public/44/jump/wireless.shtml
A good site for Cisco BBFW and other wireless products

http://www.cisco.com/warp/public/cc/pd/witc/wt2700/2750/bbfw_wp.htm
A good white paper on unlicensed BBFW

http://www.cisco.com/warp/public/cc/pd/witc/wt2700/nwwbb_sd.htm
A good white paper on general BBFW issues

http://www.isp-planet.com/fixed_wireless/index.html
A good general site for BBFW and it has entertainment value

http://www.radio.gov.uk/topics/bfwa/bfwa-index.htm
An interesting site for 28GHz and 40GHz—mostly for European interests

http://www.shorecliffcommunications.com/magazine/index.asp
A very good online BBFW industry magazine

http://www.fcc.gov/
The FCC site (U.S.). Good information but you have to dig somewhat.

http://www.bbwexchange.com/
A very good BBFW online magazine

http://standards.ieee.org/announcements/wirelesshuman.html
IEEE standards issues on BBFW MAN

http://www.dslreports.com/forum/dslalt
An interesting BBFW bulletin board

http://www.itu.int/was/was.html
ITU Website

http://www.nortelnetworks.com/products/01/fwa/
Nortel Networks BBFW with a focus on 3.5GHz

http://www.totaltele.com/links/list.asp?CategoryID=843
An international listing of some BBFW service providers

http://www.strategisgroup.com/publications/publication.asp?ItemId=391
Strategis Group publications; a site for industry reports

http://isp-lists.isp-planet.com/isp-wireless/
ISP BBFW bulletin board

http://www.business.com/directory/telecommunications/wireless/fixed_wireless/
Business oriented search engine with BBFW ISP and equipment provider list

http://www.delloro.com/SERVICES/wire_service.shtml
Dell'Oro Group industry BBFW industry reports

http://www.bwif.org/in_the_news2.html
BWIF Forum; Focuses on OFDM issues

http://www.ofdm-forum.com/index.asp?ID=92
OFDM Forum; competing effort to BWIF

http://www.convergedigest.com/StartupDirectoryX/Wireless.asp
Listing of BBFW Startups; includes some 802.11 companies

http://www.portal.com/markets/fixed_wireless/top_reasons.htm
An OSS site for BBFW; well known provider of OSS

http://www.gbmarks.com/wireless.htm
Goodman's bookmarks on BBFW

http://www.igigroup.com/st/pages/wireaccess00.html
IGI Group industry reports

http://80211b.weblogger.com/
802.11 Networking news—good site.

http://www.wi-fi.net/
Wireless Ethernet Compatibility Alliance (WECA) site—802.11

http://www.WLANA.org/
Wireless LAN site—802.11

http://www.ieee802.org/
IEEE 802.11 protocol page

http://www.personaltelco.net/index.cgi/WirelessCommunities
Informal site with listings of 802.11 final mile networks

http://www.mobilian.com/whitepaper_frame.htm
Some of the best 802.11 white papers and presentations to be found

GLOSSARY

Access Method Generally, the way in which network devices access the network at large; in other words, the medium that connects LANs. Examples include broadband fixed wireless, DSL, and cable modems.

Adjacent Channel A channel or frequency that is directly above or below a specific channel or frequency.

Algorithm Well-defined rule or process for arriving at a solution to a problem.

Amplitude The magnitude or strength of a varying waveform.

Analog Signal The representation of information with a continuously variable physical quantity such as voltage. Because of this constant changing of the wave shape with regard to its passing a given point in time or space, an analog signal may have an infinite number of states or values. This contrasts with a digital signal, which has a very limited number of discrete states.

ANSI American National Standards Institute. Voluntary organization composed of corporate, government, and other members that coordinates standards-related activities, approves U.S. national standards, and develops positions for the United States in international standards organizations. ANSI helps develop international and U.S. standards relating to, among other things, communications and networking. ANSI is a member of IEC and ISO.

Antenna A device for transmitting or receiving a radio frequency (RF). Antennas are usually designed for specific and relatively tightly defined frequencies and are quite varied in design. As an example, an antenna for a 2.5 GHz (MMDS) system will generally not work for a 28 GHz (LMDS) design.

Antenna Gain The measure of an antenna *assembly* performance relative to a theoretically perfect antenna called an isotropic radiator (*radiator* is another term for antenna). Certain antenna designs feature higher performance relative to radiating a specific area or with regard to frequencies.

Application Layer Layer 7 of the OSI reference model. This layer provides services to application processes (such as e-mail, file transfer, and terminal emulation) that are outside of the OSI model. The application layer identifies and establishes the availability of intended communication partners (and the resources required to connect with them), synchronizes cooperating applications, and establishes agreement on procedures for error recovery and control of data integrity. Corresponds roughly with the transaction services layer in the SNA model. *See also* Data Link Layer, Network Layer, Physical Layer, Session Layer, and Transport Layer.

ARQ Automatic Repeat reQuest. Communication technique in which the receiving device detects errors and requests retransmissions.

ASCII American Standard Code for Information Interchange. Specifies 8-bit code for character representation (7 bits plus parity).

ATM Asynchronous Transfer Mode. International standard for cell relay in which multiple service types (such as voice, video, or data) are conveyed in fixed-length (53-byte) cells. Fixed-length cells allow cell processing to occur in hardware, thereby reducing transit delays. ATM is designed to take advantage of high-speed transmission media such as E3, SONET, and T3.

Attenuation Loss of communication signal energy, whether by equipment design, operator manipulation, or transmission through a medium such as the atmosphere, copper, or fiber.

Authentication In security, the verification of the identity of a person or process.

Backbone Part of a network that acts as the primary path for traffic that is most often sourced from, and destined for, other networks.

Backplane Physical connection between an interface processor or card and the data buses and the power distribution buses inside a chassis.

Bandwidth The frequency range necessary to convey a signal, measured in units of hertz (Hz). For example, voice signals typically require approximately 7 kHz of bandwidth, and data traffic typically requires approximately 50 kHz of bandwidth, but this depends greatly on modulation scheme, data rates, and how many channels of a radio spectrum are used.

Baseband Characteristic of a network technology where only one carrier frequency is used. Ethernet is an example of a baseband network. Also called *narrowband*.

Baud Unit of signaling speed equal to the number of discrete signal elements transmitted per second. Baud is synonymous with bits per second (bps) if each signal element represents exactly 1 bit.

BBFW Broadband Fixed Wireless. One of the most commonly used terms in the fixed wireless industry. In general, it implies data transfers in excess of 1.5 Mbps.

BER Bit Error Rate. Ratio of received bits that contain errors compared to bits received without error.

Best Effort The type of traffic that has the lowest priority between two or more devices. BE traffic is commonly data that is not sensitive to delay. E-mail is generally the best example of this.

Bridge Device that connects and passes packets between two network segments that use the same communications protocol. Bridges operate at the data link layer (Layer 2) of the OSI reference model. In general, a bridge will filter, forward, or flood an incoming frame based on the MAC address of that frame.

Broadband In general, a data system is deemed "broadband" if it has a constant data rate at or in excess of 1.5 Mbps. Its corresponding opposite is *narrowband*. Historically, it refers to a transmission system that multiplexes multiple independent signals onto one cable, or a telecommunications terminology that refers to any channel having a bandwidth greater than a voice-grade channel (4 kHz). In LAN terms, it can refer to a coaxial cable on which analog signaling is used. Also called wideband (by LAN definition).

Broadcast Not to be confused with the term "multicast," in which a single originating point connects with multiple sites; in the networking world, it is the same term as broadcast.
In general, it is the opposite of "narrowcast" and infers that a signal is sent to many points at the same time and/or is transmitted in an omnidirectional pattern. In the radio world, "broadcast" is a term of art, which means it has a special meaning relative to a specific technology. A "broadcast" signal is one that's intended for reception by the general public.

BTA Basic Trading Area. The geographical area frequently used by the FCC for assigning licensed frequencies. BTAs are typically contiguous counties or trading areas and were first described by the Rand McNally mapping company. Rand McNally eventually licensed these area descriptions to the FCC.

Buffer Storage area used for handling data in transit. Buffers are used in internetworking to compensate for differences in processing speed between network devices. Bursts of data can be stored in buffers until they can be handled by slower-processing devices. Sometimes referred to as a *packet buffer*.

Byte A series of consecutive binary digits that are operated upon as a unit (for example, an 8-bit byte).

Caching Form of replication in which information learned during a previous transaction is used to process later transactions.

Carrier Frequency The frequency of a transmitted signal that would be transmitted if it were not modulated. Some BBFW systems also have intermediate frequencies, which reside between the indoor equipment and the outdoor equipment. Carrier "frequency" can be either a single frequency or a range of frequencies carried at one time between the transmitter and receiver.

Cat 5 Cabling Category 5 Cabling. One of five grades of UTP cabling described in the EIA/TIA-586 standard. Category 5 cabling can transmit data at speeds up to 100 Mbps.

CBR Committed Bit Rate. A prioritization of information that is higher than BE-type traffic, but lower than Unsolicited Grant Service (UGS). In ATM networks, CBR refers to Constant Bit Rate and is used for connections that depend on precise clocking to ensure undistorted delivery.

CDMA Code Division Multiple Access. A transmission scheme that allows multiple users to share the same RF range of frequencies. In effect, the system divides a small range of frequencies out of a larger set and divides the data transmission among them. The transmitting device divides the data among a preselected set of nonsequential frequencies. The receiver then collates the various data "pieces" from the disparate frequencies and into a coherent data stream. As part of the RF system setup, the receiver components are "advised" of the scrambled order of the incoming frequencies.An important aspect of this scheme is that the receiver system filters out any signal other than the ones specified for a given transmission.

Channel A communications path. Multiple channels can be multiplexed over a single contiguous amount of spectrum. It also refers to a specific frequency allocation and bandwidth. As an example, downstream channels used for television in the United States are 6 MHz wide.

Checksum Method for checking the integrity of transmitted data. A checksum is an integer value computed from a sequence of octets taken through a series of arithmetic operations. The value is recomputed at the receiving end and compared for verification.

Coaxial Cable The type of cable used to connect BBFW equipment to antennas and indoor/outdoor gear. Coaxial cable, or *coax*, usually consists of a center wire surrounded by a metal shield with an insulator separating the two. The "axis" of the cable is located down the center of the cable. "Coaxial" means that there is more than one conductor oriented around a common axis for the length of the cable. Coaxial cable is one of the primary means for transporting cable TV and radio signals.

Collision Domain In Ethernet, the network area within which frames that have collided are propagated. Repeaters and hubs propagate collisions; LAN switches, bridges, and routers do not.

Convergence Speed and ability of a group of internetworking devices running a specific routing protocol to agree on the topology of an internetwork after a change in that topology.

Converter Also referred to as "up/down converter" or "transverter." Some RF systems have two fundamental frequencies: one that is sent over the air (carrier frequency) and another that is sent back and forth between the indoor equipment and the outdoor equipment (intermediate frequency). The intermediate frequencies are converted to and from the carrier frequency by a converter.

Cookie Piece of information sent by a Web server to a Web browser that the browser is expected to save and send back to the Web server whenever the browser makes additional requests of the Web server.

CRC Cyclic Redundancy Check. Error-checking technique in which the frame recipient calculates a remainder by dividing frame contents by a prime binary divisor and compares the calculated remainder to a value stored in the frame by the sending node.

Data Link Layer Layer 2 of the OSI reference model. Provides reliable transit of data across a physical link. The data link layer is concerned with physical addressing, network topology, line discipline, error notification, ordered delivery of frames, and flow control. IEEE divides this layer into two sublayers: the MAC sublayer and the LLC sublayer. Sometimes this is simply called the link layer. Roughly corresponds to the data link control layer of the SNA model.

Data Encryption Key Used for the encryption of message text and for the computation of message integrity checks (signatures).

dB Decibel. A unit for expressing a ratio of power or voltage in terms of gain or loss. Units are expressed logarithmically and typically expressed in watts. dB is not an absolute value, but rather is the measure of power loss or gain between two devices. For example, –3dB indicates a 50 percent loss in power, and +3dB indicates a doubling of power. The rule of thumb to remember is that 10 dB indicates an increase (or loss) by a factor of 10; 20 dB indicates an increase (or loss) of a factor of 100; 30 dB indicates an increase (or loss) by a factor of 1,000. Gain or loss is expressed with a "+" or "-" sign before the number. Because antennas and other RF devices/systems commonly have power gains or losses on the orders of magnitude or even orders of four orders of magnitude, dB is a more easily used expression.

dBi Decibels of antenna gain referenced to the gain of an isotropic antenna (hence the *i*). An isotropic antenna is a theoretical antenna that radiates with perfect symmetry in all three dimensions. Real-world antennas have radiation patterns that are far from symmetric, but this effect is generally used to the advantage by the system designer to optimize coverage over a specific geographic area.

dBm Decibels of power referenced to a milliwatt; 0 dBm is 1 mW.

dBW Decibels of power referenced to 1 watt.

Demodulator The part of a receiver for assembling signals from the radio into a format usable by the network or device attached to the radio. The corresponding device on the transmission side of a system is a *modulator*.

DES Data Encryption Standard. Standard cryptographic algorithm developed by the U.S. National Bureau of Standards. In networking terms, also stands for Destination End Station.

DHCP Dynamic Host Configuration Protocol. Provides a mechanism for allocating IP addresses dynamically so that addresses can be reused when hosts no longer need them.

DOCSIS Data Over Cable Service Interface Specification. Defines technical specifications for equipment at both subscriber locations and cable operators' headends. Adoption of DOCSIS will accelerate deployment of data-over-cable services and ensure interoperability of equipment throughout system operators' infrastructures.

Domain A general grouping of LANs based on organization type or geography.

DS-0 Digital Signal level 0. Framing specification used in transmitting digital signals over a single channel at 64 Kbps. Compare with

DS-1 at 1.544 Mbps (commonly referred to as 1.5 Mbps), and DS-3 at 44.736 Mbps (commonly referred to as 45 Mbps).

E-1 Wide-area digital transmission scheme used predominantly in Europe that carries data at a rate of 2.048 Mbps. E1 lines can be leased for private use from common carriers.

EIRP Effective Isotropic Radiated Power. Expresses the performance of a transmitting system in a given direction. EIRP is the power that a system using an isotropic antenna would use to send the same amount of power in a given direction that a system with a directional antenna uses. EIRP is usually expressed in watts or dBW. EIRP is the sum of the power at the antenna input plus antenna gain in dBi.

Electromagnetic Spectrum The full range of electromagnetic (same as magnetic) frequencies, a subset of which is used in commercial RF systems.

Encapsulation Wrapping of data in a particular protocol header. For example, Ethernet data is wrapped in a specific Ethernet header before network transit. Also, when bridging dissimilar networks, the entire frame from one network is simply placed in the header used by the data link layer protocol of the other network.

Encryption Application of a specific algorithm to data so as to alter the appearance of the data, making it incomprehensible to those who are not authorized to see the information.

Equalization Technique used to compensate for communications channel distortions.

Ethernet Baseband LAN specification invented by Xerox Corporation and developed jointly by Xerox, Intel, and Digital Equipment Corporation. Ethernet networks use CSMA/CD and run over a variety of cable types at 10 Mbps. Ethernet is similar to the IEEE 802.3 series of standards.

ETSI European Telecommunication Standards Institute. Organization created by the European PTTs and the EC to propose telecommunications standards for Europe.

Fast Ethernet Any of a number of 100-Mbps Ethernet specifications. Fast Ethernet offers a speed increase ten times that of the 10BaseT Ethernet specification, while preserving such qualities as frame format, MAC mechanisms, and MTU. Such similarities allow the use of existing 10BaseT applications and network management tools on Fast Ethernet networks. Based on an extension to the IEEE 802.3 specification.

FCC Federal Communications Commission. U.S. government agency that supervises, licenses, and controls electronic and electromagnetic transmission standards.

FDM Frequency Division Multiplexing. The modulation scheme that divides the total available spectrum into subsets, which are commonly used in parallel across one more links.

Firewall Router or access server, or several routers or access servers, designated as a buffer between any connected public networks and a private network. A firewall router uses access lists and other methods to ensure the security of the private network.

Fixed Wireless The type of wireless in which both the transmitter and receiver are nonmobile. BBFW wireless is always broadband wireless; in other words, it is capable of data rates in excess of 1.5 Mbps, though the links can be throttled to data rates below that, but typically not less than 256 Kbps downstream and 128 Kbps upstream.

Flow Control Technique for ensuring that a transmitting entity, such as a modem, does not overwhelm a receiving entity with data. When the buffers on the receiving device are full, a message is sent to the sending device to suspend the transmission until the data in the buffers has been processed.

Footprint The geographical area covered by a transmitter.

Fourier Transform Technique used to evaluate the importance of various frequency cycles in a time series pattern.

Fragmentation Process of breaking a packet into smaller units when transmitting over a network medium that cannot support the original size of the packet.

Frame Logical grouping of information sent as a data link layer unit over a transmission medium. Often refers to the header and trailer, used for synchronization and error control, that surround the user data contained in the unit. The terms cell, datagram, message, packet, and segment are also used to describe logical information groupings at various layers of the OSI reference model and in various technology circles.

Frequency Number of cycles, measured in hertz (once per second), of an alternating current signal per unit of time.

Frequency Reuse One of the fundamental concepts on which commercial wireless systems are based, which involves the partitioning of an RF radiating area (cell) into segments of a cell, which for BBFW purposes means the cell can be broken into up to 13 or more equal segments. Notably, most RF cells are segmented into either three or four segments. One segment of the cell uses a frequency that is far enough away from the frequency in the bordering segment that it does not provide interference problems. Frequency reuse in mobile cellular systems means that each cell has a frequency that is far enough away from the frequency in the bordering cell that it does not provide interference problems. Identical frequencies are used at least two cells apart from each other. This practice enables cellular providers to have many more customers for a given site license.

Fresnel Zones Theoretically, ellipsoid-shaped volumes that reside in the space between a transmitting and receiving antenna. The industry rule of thumb for line-of-sight links is to leave 60 percent of the centermost part of the first Fresnel zone free from obstruction. There are many Fresnel zones within an RF link, and they are often referred to as the First Fresnel Zone, Second Fresnel Zone, and so on as the area referred to is farther from the center of the beam path.

FTP File Transfer Protocol. Application protocol, part of the TCP/IP protocol stack, used for transferring files between network nodes.

Full Duplex Capability for simultaneous data transmission between a sending station and a receiving station. Half duplex is where only one side of a link can transmit at a time; simplex is where there is only one transmitter and one receiver in a link.

FWA Fixed Wireless Access. Also referred to as BBFWA (Broadband Fixed Wireless Access).

Gain For an amplifier, "gain" is the ratio of the output amplitude of a signal to the input amplitude of a signal. This ratio is typically expressed in decibels. For an antenna, "gain" is the ratio of its directivity in a given direction compared to some reference antenna. The higher the gain, the more directional the antenna pattern.

Gb Gigabit. Approximately one billion bits.

Gbps Gigabits per second.

GHz One billion cycles per second.

Goodput Sometimes referred to as *throughput*. The net amount of data transmitted minus the overhead traffic to manage the link or connection.

Headend Main point of a BBFW network. All CPE units transmit toward the headend; the headend then transmits toward a number of CPE devices.

Header Control information placed before data when encapsulating that data for network transmission.

Interference Unwanted communication noise that decreases the performance of a link or prevents a link from occurring.

Isochronous Transmission Asynchronous transmission over a synchronous data link. Isochronous signals require a constant bit rate for reliable transport. Compare with asynchronous transmission.

ITU International Telecommunication Union. International body that develops worldwide standards for telecommunications technologies.

IXC Inter-eXchange Carrier. Common carrier providing long-distance connectivity between LATAs. The three major IXCs are AT&T, MCI, and Sprint, but several hundred IXCs offer long-distance service in the United States.

Jitter Analog communication line distortion caused by a signal that is sent in random time occurrences or excessive variances in signal timing. Jitter can cause data loss, particularly at high speeds.

Kb Kilobit. Approximately 1,000 bits.

Kbps Kilobits per second.

LAN Local Area Network. High-speed, low-error data network covering a relatively small geographic area (typically up to a few thousand meters). LANs connect workstations, peripherals, terminals, and other devices in a single building or other geographically limited area. LAN standards specify cabling and signaling at the physical and data link layers of the OSI model. Ethernet, FDDI, and Token Ring are widely used LAN technologies. Compare with MAN and WAN.

Latency

- Delay between the time a device requests access to a network and the time it is granted permission to transmit.

- Delay between the time a device receives a frame and the time that frame is forwarded out the destination port.

Excessive latency is not generally a problem with e-mail, but can readily become a problem with latency-sensitive applications such as voice and video streaming.

License The purchased right to transmit RF waves over a given BTA on certain frequencies for a certain period of time. The license tightly governs the design parameters of an RF system and its use. Licenses are usually granted in a way that ensures a greatly reduced probability of interference from other users of the same spectrum. Depending on the licensed service and the country in which the license is issued, the license may be issued as the result of an auction, or as the result of a "beauty contest" in which the regulator evaluates the merits of proposals to use the spectrum.
The theory behind auctions is that they use free market forces so that spectrum is put to its best use, and it uses market forces to set requirements for spectral efficiency, which is another way of stating efficient use of the RF spectrum.

Line of Sight Characteristic of certain transmission systems such as fixed wireless, laser, microwave, and infrared systems in which no obstructions in a direct path between transmitter and receiver can exist.

Load Balancing In routing, the ability of a router to distribute traffic over all of its network ports that are the same distance from the destination address. Good load-balancing algorithms use both line speed and reliability information. Load balancing increases the use of network segments, thus increasing effective network bandwidth.

LMDS Local Multipoint Distribution Service. A relatively low power license for transmitting voice, video, and data. In the United States, there are typically two licenses granted in three frequencies each to separate entities within a BTA. These licenses are known as Block A or Block B licenses. In the United States, Block A licenses are from 27.5 to 28.35 GHz, 29.10 to 29.25 GHz, and 31.075 to 31.225 GHz for a total of 1.159 GHz of bandwidth. Block B licenses operate from 31.00 to 31.075 GHz and 31.225 to 31.300 GHz for a total of 150 MHz of bandwidth. LMDS systems have a typical maximum transmission range of approximately 3 miles, as opposed to the transmission range of an MMDS system, which is typically 25 miles. This difference in range is primarily a function of absorption due to precipitation and other physical phenomena, as well as FCC-allocated output power limits.

LOS Line of Sight. Refers to the fact that there must be a clear, unobstructed path between the transmitters and receivers. This is essential for millimeter wave products like LMDS and most microwave products lacking modulation and other schemes specifically designed to overcome the effects of a partially occluded (blocked) beam path. Having an LOS path enhances general performance in every RF deployment, as opposed to partially obstructed data paths. The opposite to LOS is NLOS, or Non Line of Sight (also referred to as Near Line of Sight).

MAC Media Access Control. Lower of the two sublayers of the data link layer defined by the IEEE. The MAC sublayer handles access to shared media, such as whether token passing or contention will be used.

MAC Address Standardized data link layer address that is required for every port or device that connects to a LAN. Other devices in the network use these addresses to locate specific ports in the network and to create and update routing tables and data structures. MAC addresses are 6 bytes long and are controlled by the IEEE. Also known as a hardware address, MAC layer address, and physical address.

Mb Megabit. Approximately 1 million bits.

Mbps Megabits per second.

MDU Multiple Dwelling Unit. Condominium or apartment building.

MMDS Multichannel Multipoint Distribution Service. A licensed frequency in the United States. The FCC has allocated two bands of frequencies to this service, which are 2.15 to 2.161 GHz and 2.5 to 2.686 GHz. Licenses have been assigned by BTA.

Mobile Wireless The type of wireless utilized in mobile phones, PDAs, pagers, and other small, portable, battery-powered devices that can transmit and/or receive information by radio.

Modem MODulator-DEModulator. Device that converts digital and analog signals. At the source, a modem converts digital signals to a form suitable for transmission over analog communication facilities. At the destination, the analog signals are returned to their digital form. Modems allow data to be transmitted over voice-grade telephone lines.

Modulation Process by which the characteristics of electrical signals are transformed to represent information.

MTU Multiple Tenant Unit. Building with multiple business tenants.

MxU Multiple Tenant Unit or Multiple Dwelling Unit.

NAT Network Address Translation. Mechanism for reducing the need for globally unique IP addresses. NAT allows an organization with addresses that are not globally unique to connect to the Internet by translating those addresses into globally routable address space. Also known as Network Address Translator.

NEBS Network Equipment Building Systems. Covers spatial, hardware, craftsperson interface, thermal, fire resistance, handling and transportation, earthquake and vibration, airborne contaminants, grounding, acoustical noise, illumination, EMC, and ESD requirements.

Network Layer Layer 3 of the OSI reference model. This layer provides connectivity and path selection between two end systems. The network layer is the layer at which routing occurs. *See also* Application Layer, Data Link Layer, Physical Layer, Session Layer, and Transport Layer.

Network Management Generic term used to describe systems or actions that help maintain, characterize, or troubleshoot a network.

NLOS Non Line of Sight. Also known as an obstructed RF path or pathway. Often also referred to as Near Line of Sight. *See* Fresnel Zone and LOS.

NOC Network Operations Center. Organization responsible for maintaining a network.

OSI Open System Interconnection reference model. Sometimes referred to as the OSI reference stack. Network architectural model developed by ISO and ITU-T. The model consists of seven layers, each of which specifies particular network functions such as addressing, flow control, error control, encapsulation, and reliable message transfer. The lowest layer (the physical layer) is closest to the media technology. The lower two layers are implemented in hardware and software, while the upper five layers are implemented only in software. The highest layer (the application layer) is closest to the user. The OSI reference model is used universally as a method for teaching and understanding network functionality. Similar in some respects to SNA. *See also* Application Layer,

Data Link Layer, Network Layer, Physical Layer, Session Layer, and Transport Layer.

OSS Operations Support System. Network management system supporting a specific management function, such as alarm surveillance and provisioning, in a carrier network. Many OSSs are large centralized systems running on mainframes or minicomputers. Common OSSs used within an RBOC include NMA, OPS/INE, and TIRKS.

Out-of-Band Signaling Transmission using frequencies or channels outside the frequencies or channels normally used for information transfer. Out-of-band signaling is often used for error reporting in situations in which in-band signaling can be affected by whatever problems the network might be experiencing. Contrast with in-band signaling.

Oversubscription The method of having more users on a network than the network can accommodate if all the users were to use the network simultaneously. What makes this work is the premise that rarely, if ever, do all users actually use the network at the same time. Oversubscription is mission-critical to the financial models used by Internet Service Providers (ISPs) and other entities, and in many cases, oversubscription is what keeps an entity solvent. Oversubscription rates can be anywhere from a factor of 6 to a factor of 50 or more depending on the class of service the subscriber has agreed to and other factors, including how much bandwidth the subscribers use.

Packet Logical grouping of information that includes a header containing control information and (usually) user data. Packets are most often used to refer to network layer units of data. The terms datagram, frame, message, and segment are also used to describe logical information groupings at various layers of the OSI reference model and in various technology circles.

Parabolic Antenna A dish-like antenna that sends and receives radio waves in a highly focused manner. Such antennas provide very large antenna gains and are highly efficient. This antenna is typical to most point-to-point RF systems, but is not the only design available or appropriate for a given RF link. The primary task of an antenna is to provide gain (signal boost) and to radiate in particular directions in accordance with the network's intended use; for example, point-to-multipoint or point-to-point, or to cover a prescribed geographic area.

Passband The frequencies that a radio allows to pass from its input to its output. If a receiver or transmitter uses filters with narrow passbands, then only the desired frequency and nearby frequencies are of concern to the system designer. If a receiver or transmitter uses filters with wide passbands, then many more frequencies in the vicinity of desired frequency are of concern to the system designer. In a frequency division multiplexing (FDM) system, the transmit and receive passbands will be different. In a time division multiplexing (TDM) system, the transmit and receive passbands are the same.

Path Loss The power loss that occurs when RF waves are transmitted through the air. This loss occurs because RF waves expand as they travel through the air and the receiver antenna only captures a small portion of the total radiated energy. In addition, a significant amount of energy may be absorbed by molecules in the atmosphere or by precipitation when the carrier frequencies are above 10 GHz. The amount of absorption due to precipitation depends on the amount of precipitation and is usually a factor only for systems that operate at frequencies above 10 GHz.
The amount of atmospheric absorption depends greatly on the particular frequency used. At 12 GHz, water vapor absorbs a great deal of energy, and at 60 GHz, oxygen molecules absorb even more energy. Systems that operate at those frequencies have very limited ranges.

PDU Protocol Data Unit. OSI term for packet.

Physical Layer Layer 1 of the OSI reference model. The physical layer defines the electrical, mechanical, procedural, and functional specifications for activating, maintaining, and deactivating the physical link between end systems. Corresponds with the physical control layer in the SNA model. *See also* Application Layer, Data Link Layer, Network Layer, Session Layer, and Transport Layer.

POP Point of Presence. A term commonly used to describe a centralized facility that subscribers use to access the Internet.

POTS Plain Old Telephone Service.

Presentation Layer Layer 6 of the OSI reference model. This layer ensures that information sent by the application layer of one system will be readable by the application layer of another. The presentation layer is also concerned with the data structures used by programs, and therefore negotiates data transfer syntax for the application layer. Corresponds roughly with the presentation services layer of the SNA model. *See also* Application Layer, Data Link Layer, Network Layer, Physical Layer, Session Layer, and Transport Layer.

Propagation Delay Time required for data to travel over a network, from its source to its ultimate destination.

Protocol Formal description of a set of rules and conventions that governs how devices on a network exchange information.

Protocol Stack Set of related communications protocols that operate together and, as a group, address communication at some or all of the seven layers of the OSI reference model. Not every protocol stack covers each layer of the model, and often a single protocol in the stack will address a number of layers at once. TCP/IP is a typical protocol stack.

PSTN Public Switched Telephone Network. General term referring to the variety of telephone networks and services in place worldwide.

PTM Point To Multipoint. A common acronym that has another version "P2MPT." However, all versions denote the same concept which is a communication between a group of sites that interfaces a single hub site. PTM is commonly set up in three or four segments to enable frequency reuse, but can be designed for as many as a dozen or more segments within a single cell.

PTP Point To Point. A common acronym that has another version which is "P2P." However, all versions denote the same concept which is to provide communication between two end points. In the United States, PTP systems are typically found in the ISM, U-NII, and LMDS bands.

PTT Post, Telephone, and Telegraph. Government agency that provides telephone services. PTTs exist in most areas outside North America and provide both local and long-distance telephone services.

QAM Quadrature Amplitude Modulation. Method of modulating digital signals onto a radio-frequency carrier signal involving both amplitude and phase coding. QAM is a modulation scheme mostly used in the downstream direction (QAM-64, QAM-256). QAM-16 is generally more prominent in the upstream direction. Numbers indicate number of code points per symbol. The QAM rate or the number of points in the QAM constellation can be computed by 2 raised to the power of *<number of bits/symbol>*.

QoS Quality of Service. A feature of certain networking protocols that treats different types of network traffic differently to ensure required levels of reliability and latency according to the type of traffic. Certain kinds of traffic, such as voice and video, are more sensitive to transmission delays and are therefore given priority over data that is less sensitive to delay.
As an example, Cisco Systems pt-mpt BBFW systems traditionally have 4 levels of QoS, but some systems have as many as 13 levels of QoS depending on how many bits are used to prioritize the traffic. Most systems use either three or four levels of QoS and are commonly referred to as Unsolicited Grant Service (USG), CBR (Committed Bit Rate; sometimes referred to as CIR or Committed Information Rate), and BER (Best Effort Rate). USG has priority over CIR/CBR, which has priority over BER. QoS levels are set in Layer 2 (data link) of the OSI reference stack.

QPSK Quadrature Phase-Shift Keying. A method of modulating digital signals onto a radio-frequency carrier signal using four phase states to code two digital bits.

Repeater Device that regenerates and propagates radio or electrical signals between two network segments.

RF Radio Frequency. Generally refers to wireless communications with frequencies below 300 GHz. The term RF is commonly used too broadly to cover all types of wireless.

RFC Request For Comments. Document series used as the primary means for communicating information about the Internet. Probably the best known versions are from the IEEE. Some RFCs are designated as Internet standards. Most RFCs document protocol specifications such as Telnet and FTP, but some are humorous or historical. RFCs are available online from numerous sources.

RJ Connector Registered Jack connector. Standard connectors originally used to connect telephone lines. RJ connectors are now used for telephone connections and for 10BaseT and other types of network connections. RJ-11, RJ-12, and RJ-45 are popular types of RJ connectors.

RSA Rivest, Shamir, and Adelman, the inventors of the RSA security technique. RSA is a public-key cryptographic system that may be used for encryption and authentication.

RTS Request To Send. EIA/TIA-232 control signal that requests a data transmission on a communications line. As opposed to CTS, Clear To Send.

Session Layer Layer 5 of the OSI reference model. This layer establishes, manages, and terminates sessions between applications and manages data exchange between presentation layer entities. Corresponds to the data flow control layer of the SNA model. *See also* Application Layer, Data Link Layer, Network Layer, Physical Layer, and Transport Layer.

SID Service ID. A term used in cable standards such as DOCSIS and is a number that defines (at the MAC sublayer) a particular mapping between two network devices. The SID is used for the purpose of upstream bandwidth allocation and class-of-service management.

SNA Systems Network Architecture. Large, complex, feature-rich network architecture developed in the 1970s by IBM. Similar in some respects to the OSI reference model, but with a number of differences. SNA is essentially composed of seven layers. *See also* Data Flow Control Layer, Data Link Control Layer, Path Control Layer, Physical Control Layer, Presentation Services Layer, Transaction Services Layer, and Transmission Control Layer.

SNMP Simple Network Management Protocol. Network management protocol used almost exclusively in TCP/IP networks. SNMP provides a means to monitor and control network devices, and to manage configurations, statistics collection, performance, and security.

Spoofing 1. Scheme used by routers to cause a host to treat an interface as if it were up and supporting a session. The router spoofs replies to keepalive messages from the host in order to convince that host that the session still exists. Spoofing is useful in routing environments such as DDR, in which a circuit-switched link is taken down when there is no traffic to be sent across it in order to save toll charges. *See also* DDR.
2. The input of a hacker illegitimately claiming to be from an address from which it was not actually sent. Spoofing is designed to foil network security mechanisms such as filters and access lists.

T1 Transmits DS-1-formatted data at 1.544 Mbps through the telephone-switching network.

T3 Transmits DS-3-formatted data at 44.736 Mbps through the telephone switching network.

TCP Transmission Control Protocol. Connection-oriented transport layer protocol that provides reliable full-duplex data transmission. TCP is part of the TCP/IP protocol stack.

TCP/IP Transmission Control Protocol/Internet Protocol. Common name for the suite of protocols developed by the U.S. Department of Defense (DoD) in the 1970s to support the construction of worldwide internetworks. TCP and IP are the two best-known protocols in the suite.

TDMA Time Division Multiple Access. A technique for splitting transmissions on a common frequency into time slots, which enables a greater number of users to use a given frequency. A technique commonly used as opposed to CDMA and frequency division multiplexing (FDMA).

Telnet Standard terminal emulation protocol in the TCP/IP protocol stack. Telnet is used for remote terminal connection, enabling users to log in to remote systems and use resources as if they were connected to a local system.

TFTP Trivial File Transfer Protocol. Simplified version of FTP that allows files to be transferred from one computer to another over a network.

Throughput Sometimes referred to as goodput. The net amount of data transmitted minus the overhead traffic to manage the link or connection. The more common specification for BBFW links is one that includes overhead traffic, and does not therefore clearly indicate link performance. As a general rule of thumb, overhead represents an additional thirty percent of bandwidth over throughput.

Traffic Shaping Use of queues to limit surges that can congest a network. Data is buffered and then sent into the network in regulated amounts to ensure that the traffic will fit within the promised traffic envelope for the particular connection. Traffic shaping is used in ATM, Frame Relay, and other types of networks. Also known as metering, shaping, and smoothing.

Transport *Layer* Layer 4 of the OSI reference model. This layer is responsible for reliable network communication between end nodes. The transport layer provides mechanisms for the establishment, maintenance, and termination of virtual circuits, transport fault detection and recovery, and information flow control. Corresponds to the transmission control layer of the SNA model. *See also* Application Layer, Data Link Layer, Network Layer, Physical Layer, and Session Layer.

Truck Roll The concept of "rolling" trucks to the installation site to install, repair, or upgrade equipment.

U-NII Unlicensed National Information Infrastructure. Primarily a U.S. frequency band. The wireless products for this are in the 5.725 GHz to 5.825 GHz frequency for outdoor use. There are two other U-NII bands, which are used indoors: 5.25 GHz to 5.35 GHz and 5.5 GHz to 5.6 GHz. The indoor U-NII frequencies are transmitted at lower power levels than the 5.725 to 5.825 GHz. These frequencies do not require the use or purchase of a site license, but the gear does require certification by the FCC and strict compliance to its regulations. U-NII was a term coined by federal regulators to describe access to an information network by citizens and businesses. Equivalent to the term "information superhighway," it does not describe system architecture, protocol, or topology.

Unix Operating system developed in 1969 at Bell Laboratories. Unix has gone through several iterations since its inception.

VDSL Very-high-Data-rate digital Subscriber Line. One of four DSL technologies. VDSL delivers 13 to 52 Mbps downstream and 1.5 to 2.3 Mbps upstream over a single twisted copper pair. The operating range of VDSL is limited to 1,000 to 4,500 feet (304.8 to 1,372 meters).

VoIP Voice over IP. Enables a router to carry voice traffic (for example, telephone calls and faxes) over an IP network. In VoIP, the DSP segments the voice signal into frames, which are then coupled in groups of two and stored in voice packets. These voice packets are transported using IP in compliance with ITU-T specification H.323.

WAN Wide Area Network. Data communications network that serves users across a broad geographic area and often uses transmission devices provided by common carriers.

Wireless Access Protocol A language used for writing Web pages that uses far less overhead compared to HTML and XML, which makes it more preferable for low-bandwidth wireless access to the Internet from devices such as PDAs and cellular phones that also have small viewing screens. WAP's corresponding OS is the OS created by 3Com in its Palm Pilot. Nokia has recently adopted the Palm OS for its Web-capable cellular phone. WAP is based on the Extensible Markup Language (XML). XML dictates *how* data is shown, whereas HTML dictates *where* data is located within a browser page.

Wireline The use of copper phone, cable lines, or fiber. Wireline advantages include high reliability, high tolerance to interference, and generally easier troubleshooting. In the case of fiber, wireline also has exceptionally high bandwidth. Wireline is the technological opposite of wireless.

XDSL Group term used to refer to ADSL, HDSL, SDSL, and VDSL. All are emerging digital technologies using the existing copper infrastructure provided by the telephone companies. XDSL is a high-speed alternative to ISDN.

INDEX

 Numbers

5.725 to 5.825 band, 232
64 QAM, 159, 162
802.11 architecture for homeowners association, 264–267
802.11 DSSS and chipping, 267–268
802.11 FHSS, 268–269
802.11 IEEE standard, 252, 255–256
802.11 indoor antennas example, 278
802.11 MAC (media access control) layer, 306–308
802.11 network installations
requirements for, 273–276
sample deployments, 280–283
site survey steps for, 276–279
802.11 radios
components of, 256
versus HomeRF radios, 253
modulation schemes for, 267–272
role of autorate in, 272
as service provider medium, 265–267
802.11 site surveys, 272–273
802.11 WLAN standard, 304–308
802.11 WLAN summary, 284
802.11b IEEE standard for WLANs, 103
802.11b IEEE standard
versus 802.11a, 308
access points and client adapters in, 213
autorate negotiation, 146, 148
deployment issues, 144–151
deployment tools, 146–148
impact of AP utilization increase with number of associated clients, 145
impact of higher speeds from smaller cells, 145
in-building design considerations, 144–148
indoor deployments versus outdoor deployments, 149
number of channels in, 148
sharing with Ethernet LAN, 145

 A

AAS (Automated Auction System), 340–342
ABR (available bit rate) ATM class of service, 22
Access devices, BBFW systems as, 88
Access points in 802.11b networks, 213

Access registrar in CPEs, 201
Access, role in total network solutions, 32
Access techniques for duplexing technologies, 169
ActiveX vulnerability, 222
Ad hoc architecture for indoor WLANs, 262
Addressing, role in TCP/IP, 17–18
Adjacent channel, definition of, 4
Administrative Judges staff office of FCC, 333
AGC (automatic gain control), 68, 162
Aggregation, role in total network solutions, 32
airBand interview, 364–370
Aironet Client Utility, 146–148
Aironet Site Survey client utility, 273
Aironet wireless security solution, 213
AM (amplitude) RF modulation scheme, 63–64
Amplitude, definition of, 4
Amplitude modulation scheme, 156–157
AMPS (Advanced Mobile Phone Service), 66
Analog signal, definition of, 4
Analog systems and multipath signals, 162
Antenna gain, definition of, 5
Antenna selection, role in transmission range, 270
Antennas, 73, 195–196
 definition of, 4
 and electromagnetic fields, 57
 FCC requirements for, 337
 monthly charges for, 139
 in outdoor WLAN radios, 254
 selecting for BBFW deployment, 136–137
 types of, 136
AOL (America Online) ISP, 111–112
AP (access point) in 802.11 radios, 256, 259
Apple Data-PCS initiative, 233
Application layer
 of OSI reference model, 14
 of TCP/IP stack, 15
Application platform providers, 102
Area coverage for 802.11 network installation, 278
ASICs (application-specific integrated circuits) and sine waves, 58
ASP (Application Service Providers), 101–102
ASR (Antenna Structure Registration), 337
Association between 802.11 radios, 257
ATM (asynchronous transfer mode), 21–26
 circuits and switches, 22
 classes of service, 22–24
 comparing to IP, 19–21
 versus IP, 25–26
 performance metrics, 23
 traffic management, 24–26
Atmospheric refractive conditions, effect on microwave and millimeter wave links, 180
ATT (Automated Tracking Tool) and AAS, 340–341
Auctions of spectrums, 338–345
Authentication process, establishing in CPE registration, 202
Autorate, role in 802.11 radios, 272

 B

Back off time, 307–308
Backhaul BBFW systems, 88, 106, 192–193
Ballot work, role in developing standards, 323
Bands
 below 10 GHz, 59
 measuring, 57
Bandwidth
 definition of, 5
 increasing, 48
 minimum voice circuits for, 351
 optimizing in IP BBW links, 20
 role in link performance, 186–189
BAS (Broadcast Auxiliary Service) item in U.S. licensed frequencies, 350
BBFW (broadband fixed wireless) networks, 84, 350
 advantages of, 184
 advantages over copper, 42
 benefits to service providers, 125
 Big 3 versus pure RF companies, 35
 block diagram of, 45–47
 and colocation, 125
 common terms and acronyms, 4–11
 comparable speeds for frequencies of, 86
 as complement to DSL, 126
 components and services required for installation of, 33, 131
 configurations for MDUs, 43
 cons, 86–90
 considerations, 114
 cost factors of, 88
 definition of, 5
 deployment, 130–139
 deployment in dense urban environments, 167
 elements of, 45
 as emerging broadband medium, 246
 equipment providers for, 34
 versus fiber and copper, 125
 including value-added services to, 89–90

increasing profit contribution of employees with, 291
Layer 3 versus Layer 2 solutions in, 26–28
major providers of, 33, 131
maximum power output allowed on, 351–352
as means of reducing costs of doing business, 291
pros, 85
PT-PT RF system block diagram, 71–73
PTP data rates, 85
relative element values for, 34
residential versus commercial, 40–44
sample of, 73–74
site evaluation prior to installation of, 132
and SMB, 126
types of, 88
value proposition for, 125–126
vendor engagement template for, 118–122

BBFW customer types, 106
BBFW deployment
customer financing considerations, 142–143
customizing lengths and connectors for, 134
exceptional logistics and project management considerations, 141–142
final assembly, testing, and documentation of, 137
forward-going risk profile assessment, 143–144
importance of depth of resources in, 141
importance of established relationships with subcontractors, 142
importance of local environment experience in, 141
installation map for, 148
lightening protection systems for, 134–135
local permit requirements, 142
negotiating right-of-way access to, 139
obtaining FCC certification for, 140
performing preliminary site evaluation for, 132
performing site survey for, 132–134
providing emergency spares for, 138
role of antennas in, 136–137
role of cable routing in, 134
role of construction in, 136
role of mechanical assembly in, 135
role of pointing in, 136–137
role of weatherproofing in, 136–137
selection for minimal signal loss in, 134
selection, integration, configuration, and documentation of routers and switches for, 137–138
systems integration, 143
verifying financial solvency of subcontractors involved in, 140–141

BBFW deployment partners, selecting, 139–144
BBFW entities supporting IEEE standards, 325–327
BBFW link access as hacking medium, 208–216
BBFW link security, 212–216
BBFW links
challenges to penetration of, 210–211
denial of service attacks on, 208–209
effects of interference on, 240
functionality of, 20
hacking, 211
role in reducing time to market for products and services, 292

BBFW market opportunities, 108
BBFW-MMDS wireless technology, 39
BBFW modulation schemes. *See* Modulation schemes, 156
BBFW product comparisons, 89
BBFW security conclusions, 227
BBFW solutions, use frequencies for, 46
BBFW standards bodies, 319–327
BBFW standards, development of, 316–319
BBFW system performance metrics, purpose of, 176
BBFW targets, prioritization by service providers, 123–124
BBFW-UNII wireless technology, 39
BE (Best Effort), role in differentiated services for Layer 3 solutions, 27
BER (bit error rate), 87, 170
role in link performance, 186–187
role in modulation schemes, 159

BFSK (binary frequency shift keying) modulation scheme, 157
Bidding in spectrum auctions, 338–342
Bits per symbol, determining spectral density with, 160
Bits, role in DES, 223
BLECs (building local exchange carriers), 112–113
Block A and Block B LMDS frequencies, 358
Block error rate, role in link metrics, 178, 179
Block interleaving, 172
Blocks of information role in DES, 223–224
Bluetooth WLAN standard, 304–305, 309–310
BMAC (type B multiplexed analog components) RF modulation scheme, 64

BPSK (binary phase-shift keying) phase modulation, 63–65, 157, 162
Bridge mode of 802.11 radios, 266
Bridges
  in outdoor WLAN radios, 254
  role in BBFW networks, 45
Broadband commercial wireless portion of electromagnetic spectrum, 58
Broadband comparisons, 82
Broadband, definition of, 5
Broadband mobility provided by WLANs, 103, 149, 250–251
Broadband wireless systems. *See* BBFW (broadband fixed wireless) networks
Broadcast, definition of, 5
Browser vulnerabilities, 222–225
BT (burst tolerance) ATM performance metric, 23–24
BTAs (basic trading areas)
  definition of, 5
  role in BBFW deployment, 131
  role in licensed and unlicensed spectrums, 230
  and WTB, 334
Building drawings, role in 802.11 network installations, 274
Bursts of data, role in OFDM, 69
Bursty ILEC data traffic, 108
Business
  increasing profit contribution of employees, 291
  increasing revenue and profits of, 291
  reducing cost of, 290–291
  reducing time to market for products and services, 292
Business environment sample BBFW deployment, 151
Business issues, 289–292
Business versus technology, 288–289
BWIF (Broadband Wireless Internet Forum), 326

C

Cable broadband access, 84, 244
Cable routing, role in BBFW deployment, 134
Cable Services Bureau of FCC, 332
Cabling in outdoor WLAN radios, 254
CALEA (Communications Assistance for Law Enforcement Act), role in BBFW security, 216
Capitalization, role in selecting frequencies, 295
Carrier frequency, 63
  definition of, 5
  and ODUs, 72–73
  role in modulation schemes, 156
CARS (Cable Television Relay Service) item in U.S. licensed frequencies, 350
CBR (constant bit rate) ATM class of service, 22, 100–101
CC (Common Carrier) item in U.S. licensed frequencies, 350
CDMA (Code Division Multiple Access), 169
  advantages of, 66–67
  definition of, 5
  use on advanced IP RF systems, 163–164
CDNs (content data networks), 94–96, 98
CDV (cell delay variation) ATM performance metric, 23–24
Cell models, 51–52
Cells, role in PTP architecture, 49–50
Certificates, role in Internet security, 225–226
CGI (Computer Generated Image) vulnerability, 222
Channel, definition of, 5
Charges, decreasing, 48
Chipping, 267–268, 308
Ciphers, role in DES, 224
Ciphertext, role in DES, 224
CIR (Committed Information Rate) service, role in differentiated services for Layer 3 solutions, 27
Citizens Band wireless technology, 39
Class 5 switches, role in ILECs, 106
Class A-C IP addresses, 19
CLECs (competitive local exchange carriers), 109–111
CLI (command line interface), role in BBFW security, 215–216
Client adapters in 802.11b networks, 213
Client cards in 802.11 radios, 256
Clients in outdoor WLAN radios, 255
Cloning modems, 212
Cloud versus string comparisons, 19
CLR (cell loss ratio) ATM performance metric, 23–24
Clustered WLAN standards, 304
CO (central office), installation of DSL in relation to, 83
Co-channel interference, resolution by U-NII systems, 48–49, 241–242, 247
Coaxial cable
  definition of, 6
  in outdoor WLAN radios, 254–255
  typical length when used with ODUs, 72
Colocation of CLECs, 110

Commercial AM and FM radio wireless technologies, 39
Commercial BBFW systems versus residential, 40–44
Common Carrier Bureau of FCC, 332
Communications and Business Opportunities staff office of FCC, 333
Communications links, synchronization of, 12–13
Competition between service providers, 115–116
Condominium fiber, 83
Construction, role in BBFW deployment, 136
Consumer Information Bureau of FCC, 332
Content distribution and management in CDNs, 98
Content routing in CDNs, 99
Content switching in CDNs, 99
Converter, definition of, 6
Cookies, 221–222
Core networks, role in total network solutions, 32
CPE (customer premises equipment)
    and PT-MPT links, 73
    role in total network solutions, 31–32, 193
    using parabolic antennas in, 197
CPE radios
    versus HE radios, 195–196
    role in link metrics, 179
CPE registration process, 201–203
CPE routers, TFTP download service configuration for, 203
CPEs
    costs of, 199–200
    data flow in, 198–199
    slave status in, 198
    usage of, 200–201
    using routers and switches in, 197–198
CRC (Cyclic Redundancy Check) error detection scheme, 170–171
Cryptanalysis, role in DES, 225
CSES (Consecutive Severely Errored Seconds)
    error rates, 182
    role in link metrics, 179–180
CSMA/CD protocol, role in WLANs, 307
CTD (cell transfer delay) ATM performance metric, 23–24
Customer's target markets, service providers' consideration of, 115
CW (continuous wave) RF modulation scheme, 63–64
Cycles expressed as frequencies, 57
Cycles per bit, 160
Cycles per second, comparing, 160

D

Dark fiber, 83
Data encapsulation, role in OSI and TCP/IP, 16–17
Data expansion ciphering, 224
Data expansion enciphering, 224
Data flow in HEs and CPEs, 198–199
Data Link layer of OSI reference model, 14
Data loads, splitting and routing, 18
Data-PCS initiative by Apple, 233
Data rate versus throughput, role in 802.11 links, 270–271
Data rates, role in selecting frequencies, 295
Data transmissions, processing of, 46
Data versus distances in 802.11 specifications, 255–256
dB (decibels)
    definition of, 6
    expressing path loss numbers as, 60–61
dBi, definition of, 6
dBm, definition of, 6
DBS (Direct Broadcast Satellite) item in U.S. licensed frequencies, 350
dBw, definition of, 6
DECT (Digital Enhanced Cordless Telephone) standard, 308
Delay spread between signals, 68
Demodulator, definition of, 6
DEMS (Digital Electronic Message Service) item in U.S. licensed frequencies, 350
Demultiplexing headers, 17
Denial of service attacks on BBFW links, 208–210
Deployment phase, 130–139
DES (data encryption standard), 223–225
Design stability, role in early adoption, 315
DFE (decision feedback equalization), role in digital modulation schemes, 163
DHCP servers, connecting in CPE registration process, 202
Differentiated services in Layer 3 solutions, 26–27, 100–101
Digital modulation schemes, optimizing, 163
Disassociation between 802.11 radios, 257
Discrete electrical components, 58
Distance, role in selecting frequencies, 294
Distribution between 802.11 radios, 258

Diversity and multipath signals, 69
DM (downtime minutes)
error rates, 182
role in link metrics, 178
DOCSIS (Data Over Cable Interface Specification), role in IP differentiated services, 101
Doppler effect, role in RF transmissions, 60–61
DSL (digital subscriber line) broadband access, 41–42, 83–84, 244
DSSS (direct sequence spread spectrum), 65–67, 70–71, 168, 267–268, 270–271
versus FHSS, 267–269
role in digital modulation schemes, 163
Duplexing technologies, 168–170
Dynamic keys versus static keys, 213
Dynamic WEP keys, 214

E

Early adopters of standards, 314–315
Ecopartners, deployment considerations for, 131
Ecosystems, entities of, 44
Edge devices, BBFW systems as, 88
Educational facility design of 802.11 network, 282
EFS error rates, 182
EHF (extremely high frequency) band, 62
EIRP (effective isotropic radiated power), 6
Electromagnetic spectrum, definition of, 7
Electromagnetism, role in RF, 56–57
Element managers, providing BBFW security with, 215
Elevation, improving range with, 186
Emergency spares, providing for BBFW deployment, 138
End-user applications running on services, service providers' consideration of, 115
Enforcement Bureau of FCC, 332
Engagements between service providers and customers, 114–122
Engineering and Technology staff office of FCC, 333
Equipment analysis, role in selecting frequencies, 296–297
Equipment settings, role in link metrics, 183–184
Error control schemes, 170–173
Error correction, 18
versus error detection, 170, 172–173
and IP networks, 20
Error detection, 18, 170, 172–173
Error statistic table example, 182
Errors, role in link metrics, 178
ES (errored seconds)
error rates, 182
role in link metrics, 178
Ethernet LANs versus WLANs, 307
Ethernet protocol, 16
Expertise, role in selecting frequencies, 296
Extranet VPNs, 94

F

F (Fixed) item in U.S. licensed frequencies, 350
Factories, use of WLANs in, 261
Fast fading and multipath signals, 68
FCC (Federal Communications Commission), 332
allocations for U-NII bands, 232–235
certification, 140
deployment requirements, 345–346
license assignments, 353
organization and bureaus, 332–333
Proposed Rule 99-231, 269
staff offices, 333
time limit for packet acknowledgment, 255
FDD (frequency division duplex), 169
FDM (frequency division multiplexing), 7, 66
versus OFDM, 68
role in digital modulation schemes, 163–164
role in link metrics, 179
FDMA (frequency division multiple access), 66, 164, 169
FEC (forward error correction), 87, 163, 172
FHSS (Frequency Hopping Spread Spectrum), 66, 70–71, 168, 267–271
Fiber broadband access, 81–83
benefits of, 86
installations, 82
U-NII as final mile for, 242–243
File viruses, 220
Final mile management by CLECs, 109
Fixed wireless, definition of, 7
Fixed wireless systems. *See* BBFW (broadband fixed wireless) networks
Flow control, role in ATM traffic management, 24–25
FM (frequency) RF modulation scheme, 63–64
Footprint, definition of, 7
FR (frame relay networks), advisory about error correction in, 18
Frames, role in IP networks, 16

Free space path loss of RF transmissions, 60
Frequencies, 58–59, 348–351
  adjustments in CPE registration process, 202
  scanning in CPE registration process, 201
  scoring, 299
  selecting, 292–298
  role in electromagnetism, 57
Frequency bands in RF, 62
Frequency (FM) RF modulation scheme, 63–64, 156–157
Frequency reuse
  definition of, 7
  example of, 51–52
Frequency selection versus management execution, 247–248
Fresnel zones
  definition of, 7
  role in earth curvature for LOS systems, 74–75
FSK (frequency shift keying), role in HomeRF WLAN standard, 309
Full-duplex BBFW links, 170

G

G.826 specification, maximum error limits in, 183
Gain and power, adjusting in CPE registration process, 202
Gain, definition of, 7
General SMB area, role in RF site surveys, 133
Geography
  improving range with, 186
  role in selecting frequencies, 295
Gigahertz symbol and cycles per second, 57
Goodput, role in link performance, 188–189, 270–271
Grant and request policies in differentiated services for Layer 3 solutions, 27
GRE (Generic Router Encapsulation) VPNs, 93

H

Hacker software for Unix, 217–218
Hacker tools, 218
Hacking
  BBFW links, 211
  process of entering LANs for purpose of, 209
  via BBFW radios, 211
  via wireline access, 216–220
Hacking software proliferation, 206–207
Ham wireless technology, 39
Hardware store sample BBFW deployment, 148–151
Harvest rate, 166
HE (head end) and PT-MPT links, 73
HE radios versus CPE radios, 195–196
HE security, 215
Headend of P2MP links, 194
Headers in protocols
  adding and removing, 17
  contents of, 17
  functionality of, 16
Hertz symbol and cycles per second, 57–58
HEs (head ends)
  costs of, 199–200
  data flow in, 198–199
  slave status in, 198
  using routers and switches in, 197–198
  using Yagi antennas in, 196–197
HF (high frequency) band, 62
Histograms, role in forward-going risk profile assessment of BBFW deployment, 144
Homeowners association 802.11 architecture, 264–267
HomeRF radios versus 802.11b radios, 253
HomeRF WLAN standard, 304–305, 308–309
Homes, use of WLANs in, 260
Hopping patterns in 802.11 FHSS, 268
Hospitals, use of WLANs in, 261
Hosting companies, 102
HRFWG (Home RF Working Group), 269
Hybrid systems, role in selecting frequencies, 297
HyperLAN band, 233

I

IE (Internet Explorer) vulnerability, 222
IEEE (Institute of Electrical and Electronics Engineers)
  802.11 standard, 103
  BBFW standards body, 320–325
IKE (Internet Key Exchange), role in VPNs, 94
ILECs (incumbent local exchange carriers), 19, 106–110
Incident signals and multipath signals, 68
Indoor WLAN radios, 256–258, 261–267. *See also* Outdoor WLAN radios
Induction fields, role in RF transmissions, 60
Information rates, 199

Information theft, 209
Infrared 802.11, 258
Infrastructure architecture for indoor WLANs, 262
Initialization vector changes WEP key enhancement, 215
Inspector General's staff office of FCC, 333
Installation map for sample BBFW deployment, 148
Integration of 802.11 radios, 258
Intelligent network services in CDNs, 99–100
Interference, 240
    occurrence of, 178
    role in selecting frequencies, 295
    in U-NII networks, 236–242
International Bureau of FCC, 332
Internet elements and market segment ownership, 107
Internet industry, businesses and growth, 290
Internet layer of TCP/IP stack, 15
Interoperability, role in early adoption, 315
Interpolation OFDM error correction technique, 171
IP address registration, role in TCP/IP, 19
IP BBFW services, 48–49
IP BBW links, functionality of, 19–20
IP connectivity, establishing in CPE registration process, 202
IP differentiated services architecture, 100
IP (Internet protocol), comparing to ATM, 19–21
IP layers, packets in, 17
IP networks
    advantages of, 20–21
    versus ATM, 25
    cost effectiveness of, 21
    disadvantages of, 20
    role of frames in, 16
    role of packets in, 16
IP wireless security features, 212
Ipgrab hacker tool, 218
IPSec VPNs, 93
IS-IS (intermediate system-to-intermediate system)
    definition of, 7
    role in OSI reference model, 12
ISAKMP (Internet Security Association and Key Management Protocol) SAs, use with IKE, 94
ISI (inter-signal interference), 68
    and OFDM, 69
    role in multipath signals and OFDM, 162
ISM band, 232
ISM (Industrial, Scientific & Medical) item in U.S. licensed frequencies, 350
ISM statistics, 47, 86, 89
ISM unlicensed radios, 124
ISM wireless technology, data rates and pricing, 42
ISPs (Internet service providers), 19, 111–112
ITFS (Instructional Television Fixed Service) 46, 293, 350
ITU (International Telecommunications Union)
    definition of SES measurement by, 178
    Specification G.826 maximum error limits, 183
IXCs (inter-exchange carriers), 111

J

Jamming, 209
Java vulnerability, 222

K

Kerberos security protocol, 218
Keys, role in DES, 224
Kilohertz symbol and cycles per second, 57

L

L2TP VPNs, 93
Laptop thefts, 206
Last mile management by CLECs, 109
Layer 2
    encapsulation, 16
    protocols and error correction, 18
Layer 3
    BBFW solution versus Layer 2, 26–28
    options for licensed and unlicensed transmitting equipment, 241
    RF links, 26–27
Layer 4 protocols and error correction, 18
Layers
    of OSI reference model, 14
    of TCP/IP, 15
Legislative and Intergovernmental Affairs staff office of FCC, 333
Lengths and connectors, customizing in BBFW deployment, 134
LF (low frequency) band, 62
License, definition of, 8
Licensed BBFW versus unlicensed BBFW, comparison of, 238–239
Licensed frequencies in U.S., 348–351

Licensed links
- and co-channel interference, 241
- deployment plans for, 230
- footprint restrictions on, 240–241
- and in-band market competition, 242

Licensed spectrum, venture capital aspects of, 244–248
Licensed systems versus unlicensed systems, 335
Licenses
- advisory about, 295
- assignment of, 353
- role in selecting frequencies, 293–294
- role in spectrum-auction bidding, 339

Lightening protection systems for BBFWs, 134–135
Link availability, criteria for, 176–177, 180–182
Link bandwidth versus speed, 271
Link degradation, 240
Link layer of TCP/IP stack, 15–16
Link margin analysis sample, 180–182
Link margin management, 177–178
Link margins, characteristics of, 183
Link metrics, 178–185
- determining, 183
- role of site stability in, 184–185

Link performance
- factors, 176
- monitoring, 183
- references to, 185–189
- role of bandwidth in, 186–189
- role of QoS in, 189
- role of range in, 185–187

Link performance monitoring, advisory about, 144
Link security, establishing in CPE registration process, 202
Links
- characterizing, 185
- evaluating, 185–189
- measuring for performance and reliability, 180

LMDS (local multipoint distribution service)
- data rates and pricing, 42
- definition of, 8
- frequency and speeds, 86
- licenses, 299, 353, 356–358
- statistics, 47, 89

Logic bomb viruses, 219
Lombard, Andy airBand interview, 364–370
Loopback tests, monitoring link performance with, 183
LOS (line of sight) environments, 8
- earth curvature calculation for, 74–75
- multipath signals in, 68
- requirements for 802.11 radios, 266–267
- role in 802.11 radios, 272
- role in multipath signals and OFDM, 162

LTTS (Local Television Transmission Service) item in U.S. licensed frequencies, 350

M

M (Mobile) item in U.S. licensed frequencies, 350
MAC (media access control) layer in 802.11 standard, 306–308
Macro viruses, 220
Management execution versus frequency selection, 247–248
Managing Director's staff office of FCC, 333
Maps of Auction Results, role in spectrum auctions, 341–342
MAS (Multipoint Distribution Service) item in U.S. licensed frequencies, 350
Mass Media Bureau of FCC, 333
Mast construction of BBFWs, 134–135
Maxwell, James Clerk and electromagnetism, 56
MCI WorldCom IXC, 111
MDS (Multipoint Distribution Service) item in U.S. licensed frequencies, 350
MDUs (Multi-Dwelling Units)
- BBFW configurations for, 44
- example of, 19

Mechanical assembly, role in deploying BBFWs, 135
Media Relations staff office of FCC, 333
Medical facility designs of 802.11 networks, 283
Megahertz symbol and cycles per second, 57
MF (medium frequency) band, 62
Microcell networks versus supercell networks, 52–53
Microwave links, effect of atmospheric refractive conditions on, 180
Microwave oven frequencies, 61–62
Microwave spectrum, 57, 59
Microwave systems versus millimeter wave systems, 87
Migration to next generation, role in selecting frequencies, 298
Millimeter wave links, effect of atmospheric refractive conditions on, 180
Millimeter wave spectrum, 57, 59, 87
Millimeter wave systems, 88, 160

MIMO (multiple-in, multiple-out) technology, 166
Minimal signal loss, selecting for BBFW deployment, 134
MMDS licenses, 293, 299, 353–356
MMDS (multichannel multipoint distribution service)
  data rates and pricing, 42
  definition of, 8
  frequency and speeds, 86
  speed of, 46–47
  statistics, 89
Mobile IP, 252
Mobile wireless technology, 8, 39
Modem registrar in CPEs, 200
Modems, cloning, 212
Modulation of energy, 56
Modulation of sine waves, 58
Modulation schemes
  choosing, 159–160
  objective of, 156
  and RF, 63–70
  types of, 156–157
Monthly charges, decreasing, 48
MTUs (Multi-Tenant Units), 31
  and CLECs, 109
  example of, 19
Multi-Dwelling Units (MDUs)
  BBFW configurations for, 44
  example of, 19
Multifloor sample BBFW deployment, 150
Multipath environments, using spacial diversity in, 69–70
Multipath signals
  and OFDM, 160–168
  and RF, 59, 67–68
  role in microwave and millimeter wave systems, 87
Multiple-round bidding in spectrum auctions, 338–339
Multiplexing headers, 17
Mutual authentication WEP key enhancement, 214
MxU. *See* MTUs (Multi-Tenant Units)

N

Native encryption, 212
Net throughput, role in link performance, 188–189
Netscape vulnerability, 222
Network layer of OSI reference model, 14
Network-level controls in ATM traffic management, 24
Network performance, improving with CDNs, 94–96
Network registrar in CPEs, 200
Network solutions, elements of, 30–35
Networks
  function of, 12
  using as stooges, 209
Nines in link availability, 176–177
NLOS (non-line of sight)
  capability of RF equipment, 176
  definition of, 8
nmap hacker tool, 218
NMS (network management system), 31–32
Node-level controls in ATM traffic management, 24
Number of nines percentages for link availability, 176–177

O

OC-12, role in PTP architecture, 192
OC-3 aggregates, role in PTP architecture, 192
OC-3 broadband data rate, 87
ODU (outdoor units) and carrier frequencies, 72–73
OFDM Forum, 327
OFDM modulation scheme, 66–67, 68–70, 87, 164
  interpolation error correction technique in, 171
  and multipath signals, 160–168
  role in error correction, 18
  settings for, 188
  use of QPSK modulation scheme with, 165
  using VOFDM with, 166–167
  versus VOFDM, 237
OFDM products, use in obstructed environments, 167
OFS (Private Operational Fixed Service) item in U.S. licensed frequencies, 350
Omnidirectional antennas, role in BBFW deployment, 136
OSI (Open Systems Interconnection) reference model, 12–14
  seven layers of, 13–14
  versus TCP/IP stack, 15
OSS (operational system support), 28–32

Outdoor WLAN radios, 253–256. *See also* Indoor WLAN radios
Overhead in data encapsulation, 16
Oversubscription
  definition of, 8–9
  on residential services, 51

P

P2MP BBFW architecture, 192–193
P2P BBFW architecture, 192–193
Packet acknowledgment, time limit for, 255
Packets, role in protocol layers, 16
PAR (project authorization request), role in standards development, 322
Parabolic antenna, definition of, 9
Parabolic antennas
  in outdoor WLAN radios, 254
  using in CPEs, 197
  using in SMB deployments, 197
Passband, definition of, 9
Patch antennas, role in BBFW deployment, 136–137
Path diversity, mitigating catastrophic effects with, 239
Path loss
  definition of, 9
  of RF transmissions, 60
Payload of protocols, 16
  adding control data to, 17
  fragmentation of, 18
PCBs (printed circuit boards) and discrete electrical components, 58
PCM (pulse code modulation) RF modulation scheme, 64
PCR (peak cell rate) ATM performance metric, 23–24
PCS (Personal Communications Service) item in U.S. licensed frequencies, 350
Performance metrics, purpose of, 176
Pervasive computing provided by WLAN radios, 103, 149, 251–252
PET (Emerging Technologies) item in U.S. licensed frequencies, 350
PGP (Pretty Good Privacy) security protocol, 218
Phase modulation scheme, 156–157
Phase (PM) RF modulation scheme, 63–64
Physical layer of OSI reference model, 14
Plaintext, role in DES, 224
Planning sessions, role in 802.11 network installations, 275–276
Plans and Policy staff office of FCC, 333
PM (phase) RF modulation scheme, 63–64
Pointing, role in BBFW deployment, 136–137
Power and gain, adjusting in CPE registration process, 202
Power output maximum on BBFW systems, 351–352
Presentation layer of OSI reference model, 14
Processing gain, role in transmission range, 270
Profit maximization, role in selecting frequencies, 298
Proprietary OSS, 28–31
Protocol headers, importance of, 17
Protocol layers, role of packets in, 16
Protocols, functionality in OSI and TCP/IP stacks, 16
PRS (Paging and Radiotelephone Service) item in U.S. licensed frequencies, 350
PSK (phase-shift keying), role in RF modulation schemes, 63
PT-MPT links and antennas, 73
PT-PT RF links, symmetry of, 73
PT-PT RF system block diagram, 71–72
PTM (point-to-multipoint)
  definition of, 9
  location of cells in, 51
  role in link metrics, 179
  role of sectors in, 50
PTM (point-to-multipoint) millimeter wave systems, ranges of, 88
PTP (point-to-point)
  architecture, 49–52, 192–193
  data rates provided by BBFW systems, 85
  definition of, 9
  RF systems, 71–73
PTP (point-to-point) links, role in ILECs, 108
Pull through of RF components, 34
Push technology vulnerability, 222
PVCs (permanent virtual circuits)
  example of, 19
  role in ATM, 21

Q

QAM (quadrature amplitude modulation) RF modulation scheme, 64–66, 69–70, 157–159, 162
QoS (quality of service)
  and antennas, 73

considering migration of, 124
definition of, 10
role in IP systems, 101
role in link performance, 189
QPSK (quadrature phase-shift keying) RF modulation scheme, 64–65, 157, 162, 188
use with OFDM systems, 165
using for modulation schemes, 69
Quick licensed bands, 232, 294

R

Radio functions in HEs and CPEs, 201–203
Radio spectrum sales, 234
Radio waves, functionality of, 61
Radios, 195. *See also* RF entries
loss of lock between, 178
pinging in link metrics, 179
synchronization of, 178
Range, role in link performance, 185–187
Rate-shifting technique of 802.11 standard, 306–307
RC4 algorithm, 225, 304
Rdns hacker tool, 218
Re-association between 802.11 radios, 258
Re-authentication policies WEP key enhancement, 215
Receiver sensitivity, role in transmission range, 270
Reflective panels, role in BBFW deployment, 137
Refractive conditions, effect on microwave and millimeter wave links, 180
Registration events in state-of-the-art systems, 200
Repeater architecture for indoor WLANs, 263
Repeaters, using with multipath signals, 160–161
Reports of site surveys, writing, 277, 279
Residential BBFW systems versus commercial, 40–44
Retailer design of 802.11 network, 280–281
RF components, pull through of, 34
RF data links and routers, 27–28
RF equipment selection, 176–177
RF links in Layer 3, advantages of, 26–27
RF (radio frequency), 10. *See also* Radios
frequency bands, 62
modulation schemes, 63–70
and multipath signals, 67–68
power transmission, 59–62
reflected signals associated with, 59
role of electromagnetic energy in, 56
RF gains, achieving, 64
RF site survey prior to BBFW deployment, 132–134
RF spectrum, 57
RF systems
categories of, 59
purpose of, 63
types of, 160
RF transmissions, path loss and free space path loss of, 60
RF waves, changing characteristics of, 58
Right-of-way access, negotiating for BBFW deployment, 139
Rooftop BBW installations versus tower BBFW installations, 135
Rounds, role in DES, 224
Routers
in HEs and CPEs, 197–198
in PTP RF systems, 71–72
role in BBFW networks, 45
RSA (Rivest, Shamir, Adleman) public encryption keying system, role in DES, 223

S

S-HTTP (Secure HTTP) security protocol, 218
S/MIME security protocol, 218
SA (security associations), role in VPNs, 94
SAT (Fixed Satellite Service) item in U.S. licensed frequencies, 350
SATAN (Security Administrator Tool for Analyzing Networks), 217
Satellite wireless technology, 39, 42
Scaling, role in selecting frequencies, 297–298
School sample BBFW deployment, 150
Schools, use of WLANs in, 260
SCR (sustained cell rate) ATM performance metric, 23–24
Sector viruses, 220
Sectors, role in PTP architecture, 49–50
Secure key derivation WEP key enhancement, 214–215
Security breaches
common types of, 216–217
costs of, 206
Security conclusions, 226–227
Security protocols, 217–218
Security, resources for, 227
Security threats, 206–208
Segment management, role in TCP/IP addressing, 18

Semiparabolic antennas, role in BBFW deployment, 136
Sequential bidding in spectrum auctions, 339
Service architectures, service providers' consideration of, 116–118
Service delivery architecture, 101–102
Service deployment, development of customer timetable for, 118
Service provider strategies, 122–125
Service providers, 102
802.11 radios as, 265–267
target market and user benefits, 114–122
Services provided by OSI reference model layers, 13
SES (severely errored seconds)
error rates, 182
role in link metrics, 178
Session layer of OSI reference model, 14
SGC (Server Gated Cryptography) security protocol, 218
SHF (super-high frequency) band, 62
Signal spreading, 70
Simplex communications, 169
Simultaneous bidding in spectrum auctions, 339
Sine waves, role in frequency, 58
Single-round bidding in spectrum auctions, 338–339
[SINR] (signal to interference plus noise), 65
Site evaluation prior to deployment of BBFW, 132–134
Site stability, role in link metrics, 184–185
Site survey steps for 802.11 network installations, 276–279
Site tours, role in 802.11 network installations, 275
Slave status in HEs and CPEs, 198
SMB access and BBFW, 126, 133
SMB deployments, using parabolic antennas in, 197
SMB offices, use of WLANs in, 260
Sniffit hacker tool, 218
SNR, achieving significant gains in, 166
Solutions for networks, 31, 35, 46
Spacial diversity
in microwave industry, 69–70, 87–88
in multipath environments, 166–167
Specification G.826 maximum error limits, 183
Spectral density, role in BBFW modulation schemes, 156, 158–159
Spectral efficiency, role in RF modulation schemes, 63
Spectrum allocations
for common bands in U.S., 159
comparison of, 238
Spectrum auctions, 338–345
automation of, 339–342
counting down, 342–345
mapping results of, 341–342
Spectrum divisions, 57
Spectrum sharing, 234–235
Spectrums, 57
intermittent use of, 235
minimizing usage of, 350
Speed versus link bandwidth, 271
Sponsor selection, role in standards development, 321
Spoofing, 209
Spreading techniques, 66–67, 70–71
Sprint IXC, 111
SSB (single side band) RF modulation scheme, 64
SSL (Secure Sockets Layer) security protocol, 218
Standards
approving drafts of, 324
developing drafts of, 316–319, 323
globalization of, 312–314
publishing, 324–325
Static keys versus dynamic keys, 213
Stooges, networks used as, 209
Strategis Group projections on broadband use, 80
Streaming ciphers, RC4 as, 225
Subcontractors, selection criteria for BBFW networks, 33–34
Subnets, adding to registered IP addresses, 19
Subscriber limitations of PTM, 50
Supercell versus microcell networks, 52–53
SVCs (switched virtual circuits), example of, 19
Switches
explanation of, 106
in HEs and CPEs, 197–198
in PTP RF systems, 71–72
role in BBFW networks, 45
System sector viruses, 220

T

Tcdump hacker tool, 218
TCP/IP (transmission control protocol/Internet protocol), 14–21
tcp_scan hacker tool, 218
TDD (time division duplex), 169
TDMA (time division multiple access), 66, 169
benefits of, 164

definition of, 10
role in link metrics, 179
Technological maturity, role in early adoption, 315
Technology versus business, 288–289
Telecom venture entities, 245
Telemetry wireless technology, 39
Telocity DSL provider, 83
Terahertz symbol and cycles per second, 57
Test equipment, role in 802.11 network installations, 275
TF (Temporary Fixed) item in U.S. licensed frequencies, 350
TFTP download service configuration for CPE routers, 203
Theft
of information, 209
of service, 209, 212
Third-party OSS, 30–31
Throughput, role in link performance, 188–189
Throughput versus data rate, role in 802.11 links, 270–271
Tier 1-Tier 3 customers, 131
Time synchronization in CPE registration process, 202
Time variation and multipath signals, 68
TIP (Transaction Internet Protocol), 218
TLS (Transport Layer Security) protocol, 218
TOD, establishing in CPE registration process, 202
Token ring protocol, 16
Tower construction of BBFWs, 134–135
Towers, FCC requirements for, 337
Training tones, 171
Transmission angle, FCC limitations of, 337
Transmission range, components of, 269–271
Transmit power, role in transmission range, 270
Transport layer
of OSI reference model, 14
of TCP/IP stack, 15
Trenching, advisory about, 42–43
Trojan horse viruses, 219–220
Truck roll, definition of, 10
Tunnels, role in VPNs, 93–94

U

U-NII bands, 230–235
U-NII regional service providers, competition with VCs, 247
U-NII (unlicensed national information infrastructure)
data rates and pricing, 42
definition of, 10
drawbacks of, 235
for extension of DSL and cable, 244
as final mile for fiber, 242–243
frequency and speeds, 86
interference and coexistence with other U-NII networks, 236–242
role in ILECs, 108
spectrum allocation, 241–242
speed of, 46–47
statistics, 89
U-NII unlicensed radios, 124
UBR (unspecified bit rate) ATM class of service, 22, 24
UGS (Unsolicited Grant Service) policies, role in differentiated services for Layer 3 solutions, 27
UHF (ultra-high frequency) band, 62
ULS (Universal Licensing System), 336–345
Ultrawideband microwave products, 314
Unix-based hacker software, 217–218
Unlicensed bands in U.S., 232
Unlicensed BBFW versus licensed BBFW, comparison of, 238–239
Unlicensed links
and co-channel interference, 241
and in-band market competition, 242
Unlicensed Personal Communications Services, 233
Unlicensed radios, management by service providers, 124–125
Unlicensed spectrum
advantages of, 230–234
venture capital aspects of, 244–248
Unlicensed systems
advantages of, 240
versus licensed systems, 335
Upstream and downstream transmission rates, predetermining for CPE devices, 201
U.S. licensed frequencies, 348–351
User benefits for target markets, service providers' consideration of, 115
User registrar in CPEs, 200
USG policy, role in IP differentiated services, 100

Vaporware, role in standards development, 316
VBR-NRT (variable bit rate-non-real time) ATM class of service, 22
VBR-RT (variable bit rate-real time) ATM class of service, 22
VC (Venture capital), relationship to unlicensed and licensed spectrum, 244–248
Vendor engagement template for BBFWs, 118–122
Vendor offerings and analysis, role in selecting frequencies, 296
VHF (very high frequency) band, 62
Viruses, 219–222
VLF (very low frequency) band, 62
VOFDM used in OFDM modulation scheme, 69–70, 87, 166–167, 237
Voice circuits, 351
VoIP (Voice over IP) technology, 111
VPCs (virtual path connections), role in ATM, 21
VPNs (virtual private networks), 92–96
    advantages of, 94–96
    functionality of, 94
VSB (vestigial side band) RF modulation scheme, 64

WAP (wireless access protocol), definition of, 10
Warehouse design of 802.11 network, 281–282
WBFH (Wide Band Frequency Hopping) proposal, 269
Weatherproofing, role in BBFW deployment, 136–137
Web browser vulnerabilities, 222–225
WECA (Wireless Ethernet Compatibility Alliance), 325–326
WEP keys, 214
WEP (Wireless Equivalent Protocol), role in link security, 212–215
Wi-Fi (Wireless Fidelity), 325
Wireless applications, mobility of, 235
Wireless links, definition of, 45
Wireless technologies, introduction to, 38–40
Wireless Telecommunications Bureau of FCC, 333
Wireline access as hacking medium, 216–220
Wireline data transfer, advantages of, 87
Wireline, definition of, 11
WLAN bridges
    advantages of, 253
    configurations, 255–256
WLAN communication, elements of, 256–258
WLAN radios, capabilities of, 250–252
WLAN sales, 304
WLAN spread spectrums, 70–71, 168
WLAN statistics, 47
WLAN (Wireless Local Area Network) BBFW systems, 39, 88, 89, 102–103, 250–252
    autorate negotiation, 146, 148
    common uses for, 258–261
    definition of, 252
    deployment issues, 144–151
    deployment tools, 146–148
    versus Ethernet LANs, 307
    in factories, 261
    in homes, 260
    in hospitals, 260
    impact of AP utilization increase with number of associated clients, 145
    impact of higher speeds from smaller cells, 145
    in-building design considerations, 144–148
    in schools, 260
    sharing with Ethernet LAN, 145
    in SMB offices, 260
WLL (Wireless Local Loop) BBFW systems, 88, 109–110
Work Place Diversity staff office of FCC, 333
Working groups, organizing for standards development, 322–323
Worm viruses, 220
WTB (Wireless Telecommunications Bureau), 334–346

Yagi antennas
    in outdoor WLAN radios, 254
    role in BBFW deployment, 136
    using in HEs, 196–197

INTERNATIONAL CONTACT INFORMATION

McGraw-Hill Book Company Australia Pty. Ltd.
TEL +61-2-9417-9899
FAX +61-2-9417-5687
http://www.mcgraw-hill.com.au
books-it_sydney@mcgraw-hill.com

CANADA
McGraw-Hill Ryerson Ltd.
TEL +905-430-5000
FAX +905-430-5020
http://www.mcgrawhill.ca

GREECE, MIDDLE EAST, NORTHERN AFRICA
McGraw-Hill Hellas
TEL +30-1-656-0990-3-4
FAX +30-1-654-5525

MEXICO (Also serving Latin America)
McGraw-Hill Interamericana Editores S.A. de C.V.
TEL +525-117-1583
FAX +525-117-1589
http://www.mcgraw-hill.com.mx
fernando_castellanos@mcgraw-hill.com

SINGAPORE (Serving Asia)
McGraw-Hill Book Company
TEL +65-863-1580
FAX +65-862-3354
http://www.mcgraw-hill.com.sg
mghasia@mcgraw-hill.com

SOUTH AFRICA
McGraw-Hill South Africa
TEL +27-11-622-7512
FAX +27-11-622-9045
robyn_swanepoel@mcgraw-hill.com

UNITED KINGDOM & EUROPE (Excluding Southern Europe)
McGraw-Hill Education Europe
TEL +44-1-628-502500
FAX +44-1-628-770224
http://www.mcgraw-hill.co.uk
computing_neurope@mcgraw-hill.com

ALL OTHER INQUIRIES Contact:
Osborne/McGraw-Hill
TEL +1-510-549-6600
FAX +1-510-883-7600
http://www.osborne.com
omg_international@mcgraw-hill.com